Disease and the Hippo Pathway

Disease and the Hippo Pathway: Cellular and Molecular Mechanisms

Special Issue Editor

Carsten Gram Hansen

MDPI • Basel • Beijing • Wuhan • Barcelona • Belgrade

MDPI

Special Issue Editor
Carsten Gram Hansen
University of Edinburgh
Centre for Inflammation Research
UK

Editorial Office
MDPI
St. Alban-Anlage 66
4052 Basel, Switzerland

This is a reprint of articles from the Special Issue published online in the open access journal *Cells* (ISSN 2073-4409) in 2019 (available at: https://www.mdpi.com/journal/cells/special_issues/disease_hippo_pathway).

For citation purposes, cite each article independently as indicated on the article page online and as indicated below:

LastName, A.A.; LastName, B.B.; LastName, C.C. Article Title. *Journal Name* **Year**, *Article Number*, Page Range.

ISBN 978-3-03921-776-2 (Pbk)
ISBN 978-3-03921-777-9 (PDF)

Contents

About the Special Issue Editor

Carsten Gram Hansen conducted his PhD studies at the MRC-Laboratory of Molecular Biology in Cambridge under the supervision of Dr Ben Nichols. His studies focused on the biogenesis and functions of caveolae, small specialized plasma membrane domains. Dr Hansen was involved in the identification and initial characterization of several new caveolar protein families. In 2012, he was awarded a Post-Doctoral Fellowship from the Danish Science Ministry to work with Professor Kun-Liang Guan in California (UCSD). While in Kun-Liang's laboratory, he generated the first CRISPR genome-edited YAP/TAZ knockout cells and identified a strong link between the Hippo pathway, amino acid signaling, crosstalk with mTOR, and the fundamental process of cellular volume regulation. Dr Hansen was involved in several additional collaborative projects detailing the context-specific kinase regulation of the Hippo pathway. In November 2015, he was awarded a Chancellor's Fellowship at the University of Edinburgh to set up his own laboratory in the Centre for Inflammation Research. Here, he has recently discovered how the Hippo pathway and caveolae functionally interact. His work is highly interdisciplinary, and his team's work spans different biological models to uncover disease-relevant mechanistic insights into fundamental cellular processes driven by the Hippo pathway. In 2020, the Gram Hansen laboratory will move into the new Institute for Regeneration and Repair at University of Edinburgh. Dr Hansen's webpage: https:// gramhansenlab.com/. The Gram Hansen Lab can be followed at @gram_lab.

cells   **MDPI**

Editorial

Special Issue on "Disease and the Hippo Pathway"

Carsten Gram Hansen [1,2]

[1] University of Edinburgh Centre for Inflammation Research, Queen's Medical Research Institute, Edinburgh bioQuarter, 47 Little France Crescent, Edinburgh EH16 4TJ, UK; Carsten.G.Hansen@ed.ac.uk

[2] Institute for Regeneration and Repair, University of Edinburgh, Edinburgh bioQuarter, 5 Little France Drive, Edinburgh EH16 4UU, UK

Received: 25 September 2019; Accepted: 27 September 2019; Published: 30 September 2019

The Hippo pathway is a cellular signalling network, which plays major roles in organ homeostasis and development [1–5]. However, when this cellular signalling pathway is perturbed, diseases such as cancer, excessive fibrosis, metabolic disorders and impaired immune responses occur [1–6]. The current significant interest in the pathway continues to reveal new links between diseases and the Hippo pathway, as well as to provide new insights into what role the Hippo pathway plays in normal development, regenerative processes, organ size control [2–5], and cellular homeostasis, including fundamental processes such as cell size regulation [7,8]. This Special Issue provides an up-to-date overview of this exciting cellular signalling pathway.

The Hippo pathway is a serine/threonine kinase cascade that mediates the phosphorylation and, thereby, inactivation of the transcriptional co-regulators YAP and TAZ. A range of scaffolding proteins plays central roles in the dynamic regulation of this pathway. YAP/TAZ does not bind DNA directly and, therefore, utilizes transcription factors to mediate its transcriptional response. Phosphorylated YAP/TAZ is sequestered in the cytoplasm and subsequently does not bind to nuclear localised transcription factors [1].

In this special issue, Gundogdu and Hergovich authoritatively highlight the Mps one Binder (MOB) scaffolding proteins as central players. They further emphasize not only the disease relevance of the precise regulation of the MOBs and their direct involvement in the kinase regulation of the Hippo pathway, but also the MOBs' important function in scaffolding other kinase complexes [9]. Next, Huh, Kim, Jeong, and Park analyse the regulation of the TEADs, the main transcription factors used by YAP/TAZ [10]. This is an exciting area of research that up until recently has been understudied. Targeting the transcription factors directly, instead of focusing on YAP/TAZ, could be a fertile avenue to pursue, especially considering the current challenges in targeting the Hippo pathway. Hillmer and Link then describe how YAP/TAZ modulates the chromatin through TEADs and a range of additional cofactors [11]. YAP/TAZ-mediated chromatin remodelling is a field that, due to recent technological advances, has revealed major mechanistic insights into how YAP/TAZ either activates or represses gene transcription. Hillmer and Link provide a concise overview of these recent developments and also summarise outstanding questions that need addressing in the years to come [11].

Luo and Yu then highlight the importance of G-protein-coupled receptors (GPCRs) as one of the main upstream regulators of the Hippo pathway [12]. Their discussion focuses on how mutations [13,14], as well as altered GPCR activity [13,15], might drive tumourigenesis in multiple tissues via elevated YAP/TAZ activity. Luo and Yu examine how this GPCR–Hippo axis can be targeted in cancer [12].

Rognoni and Walko (featured on the front page) discuss the importance of YAP/TAZ in skin physiology, including in wound healing processes [16]. The mammalian skin is a well-structured organ with distinct cell layers. The skin is therefore an intriguing model organ to study in the context of the Hippo pathway, as it highlights the importance of the mesoscale organisation and the context-specific temporal and spatial regulation of YAP/TAZ [16]. Wound healing that is not resolved causes excessive fibrotic scarring. Kim, Choi, and Mo follow on and continue the discussion on how dysregulated YAP/TAZ drives fibrosis in multiple organs and also elaborate on its consequences for cancer development and its therapeutic implications [17]. Most types of solid tumours have high YAP/TAZ activity, and the majority of these cancers appear addicted to YAP/TAZ hyperactivity [18,19]. Advanced prostate cancer is one of the leading cancers killing men worldwide. This prevalent cancer therefore urgently needs improved therapeutics [20]. Omar Salem and I detail the role of the Hippo pathway in prostate cancer. We interrogate how impaired Hippo pathway activity contributes to this deadly disease [21].

Tumour growth needs additional blood supply, as cancer cells require both nutrients and oxygen. Malignant progression is consequently often paralleled by an angiogenic switch, where the vasculature transitions from a quiescent to a proliferative state [22]. In addition, angiogenesis also facilitates metastasis. The role and importance of angiogenesis in cancer development and growth is therefore well established [22]. Angiogenesis is also important in a range of healthy processes, such as embryonic development and wound healing. Azad, Ghahremani, and Yang highlight YAP/TAZ's critical roles in endothelial cells during angiogenesis not only in healthy but also in pathological processes, such as tumour vascular mimicry [23]. Brandt, North, and Link take advantage of the recent technical developments using CrispR to generate *lats2* knockout zebrafish. Interestingly, fish with somatic loss of function mutations of *lats2*, but not *lats1*, develop peripheral nerve sheath tumours [24]. The comparative low cost of maintaining zebrafish, the relative ease of genetic manipulations, the fast embryonic development, and the ability to carry out robust drug screens [25] will likely continue to make this a powerful model organism for research aiming at obtaining further fundamental insights into hyperactive-YAP/TAZ-driven human disease. Finally, Yamauchi and Moroishi [26] give an up-to-date and commanding overview of the Hippo pathway's role in the adaptive immune system [27]. Upstream core kinase components play pivotal roles in adaptive immunity and in both pro- and anticancer immune responses. Interestingly, these processes occur both independently of as well as through YAP/TAZ regulation [27–30].

The pioneering discovery of the Hippo pathway in *Drosophila melanogaster* [31] and the subsequent recognition that the major pathway components are greatly conserved in mammals have firmly established the need for both powerful model organisms and in vitro cellular model systems as means to obtain detailed insights into fundamental biological processes driven by the Hippo pathway [1–6]. Recent discoveries reveal that a wide range of diseases are driven by dysfunctional Hippo pathway activity and drive the continued interest in the pathway [1–6]. The articles in this Special Issue written by leading experts cover a wide range of diseases that are driven by perturbed Hippo pathway activity. Research into the Hippo pathway is a fascinating field that continues to hold great promise, and which will undoubtedly produce further major discoveries in years to come.

Acknowledgments: I want to especially thank the contributors to the Special Issue as well as to acknowledge the expert reviewers who reviewed submissions in a timely, fair, and constructive manner. Ongoing work in the Gram Hansen Lab is funded by a University of Edinburgh Chancellor's Fellowship and the Worldwide Cancer Research charity.

References

1. Hansen, C.G.; Moroishi, T.; Guan, K.-L. YAP and TAZ: A nexus for Hippo signaling and beyond. *Trends Cell Boil.* **2015**, *25*, 499–513. [CrossRef] [PubMed]

2. Moya, I.M.; Halder, G. Hippo-YAP/TAZ signalling in organ regeneration and regenerative medicine. *Nat. Rev. Mol. Cell Biol.* **2019**, *20*, 211–226. [CrossRef] [PubMed]

3. Davis, J.R.; Tapon, N. Hippo signalling during development. *Development* **2019**, *146*, dev167106. [CrossRef] [PubMed]

4. Zheng, Y.; Pan, D. The Hippo Signaling Pathway in Development and Disease. *Dev. Cell* **2019**, *50*, 264–282. [CrossRef]

5. Ma, S.; Meng, Z.; Chen, R.; Guan, K.-L. The Hippo Pathway: Biology and Pathophysiology. *Annu. Rev. Biochem.* **2019**, *88*, 577–604. [CrossRef]

6. Koo, J.H.; Guan, K.L. Interplay between YAP/TAZ and Metabolism. *Cell Metab.* **2018**, *28*, 196–206. [CrossRef]
7. Hansen, C.G.; Ng, Y.L.D.; Lam, W.-L.M.; Plouffe, S.W.; Guan, K.-L. The Hippo pathway effectors YAP and TAZ promote cell growth by modulating amino acid signaling to mTORC1. *Cell Res.* **2015**, *25*, 1299–1313. [CrossRef]
8. Tumaneng, K.; Schlegelmilch, K.; Russell, R.C.; Yimlamai, D.; Basnet, H.; Mahadevan, N.; Fitamant, J.; Bardeesy, N.; Camargo, F.D.; Guan, K.-L. YAP mediates crosstalk between the Hippo and PI(3)K–TOR pathways by suppressing PTEN via miR-29. *Nat. Cell Biol.* **2012**, *14*, 1322–1329. [CrossRef]
9. Gundogdu, R.; Hergovich, A. MOB (Mps one Binder) Proteins in the Hippo Pathway and Cancer. *Cells* **2019**, *8*, 569. [CrossRef]
10. Huh, H.D.; Kim, D.H.; Jeong, H.-S.; Park, H.W. Regulation of TEAD Transcription Factors in Cancer Biology. *Cells* **2019**, *8*, 600. [CrossRef]
11. Hillmer, R.E.; Link, B.A. The Roles of Hippo Signaling Transducers Yap and Taz in Chromatin Remodeling. *Cells* **2019**, *8*, 502. [CrossRef] [PubMed]
12. Luo, J.; Yu, F.-X. GPCR-Hippo Signaling in Cancer. *Cells* **2019**, *8*, 426. [CrossRef] [PubMed]
13. Yu, F.-X.; Zhao, B.; Panupinthu, N.; Jewell, J.L.; Lian, I.; Wang, L.H.; Zhao, J.; Yuan, H.; Tumaneng, K.; Li, H.; et al. Regulation of the Hippo-YAP pathway by G-protein-coupled receptor signaling. *Cell* **2012**, *150*, 780–791. [CrossRef] [PubMed]
14. Yu, F.-X.; Luo, J.; Mo, J.-S.; Liu, G.; Kim, Y.C.; Meng, Z.; Zhao, L.; Peyman, G.; Ouyang, H.; Jiang, W.; et al. Mutant Gq/11 promote uveal melanoma tumorigenesis by activating YAP. *Cancer Cell* **2014**, *25*, 822–830. [CrossRef] [PubMed]
15. Liu, G.; Yu, F.-X.; Kim, Y.C.; Meng, Z.; Naipauer, J.; Looney, D.J.; Liu, X.; Gutkind, J.S.; Mesri, E.A.; Guan, K.-L. Kaposi sarcoma-associated herpesvirus promotes tumorigenesis by modulating the Hippo pathway. *Oncogene* **2015**, *34*, 3536–3546. [CrossRef] [PubMed]
16. Rognoni, E.; Walko, G. The Roles of YAP/TAZ and the Hippo Pathway in Healthy and Diseased Skin. *Cells* **2019**, *8*, 411. [CrossRef] [PubMed]
17. Kim, C.-L.; Choi, S.-H.; Mo, J.-S. Role of the Hippo Pathway in Fibrosis and Cancer. *Cells* **2019**, *8*, 468. [CrossRef]
18. Zanconato, F.; Cordenonsi, M.; Piccolo, S. YAP and TAZ: a signalling hub of the tumour microenvironment. *Nat. Rev. Cancer* **2019**, *19*, 454–464. [CrossRef]
19. Moroishi, T.; Hansen, C.G.; Guan, K.-L. The emerging roles of YAP and TAZ in cancer. *Nat. Rev. Cancer* **2015**, *15*, 73–79. [CrossRef]
20. Wang, G.; Zhao, D.; Spring, D.J.; Depinho, R.A. Genetics and biology of prostate cancer. *Genome Res.* **2018**, *32*, 1105–1140. [CrossRef]
21. Salem, O.; Hansen, C.G. The Hippo Pathway in Prostate Cancer. *Cells* **2019**, *8*, 370. [CrossRef] [PubMed]
22. De Palma, M.; Biziato, D.; Petrova, T.V. Microenvironmental regulation of tumour angiogenesis. *Nat. Rev. Cancer* **2017**, *17*, 457–474. [CrossRef] [PubMed]
23. Azad, T.; Ghahremani, M.; Yang, X. The Role of YAP and TAZ in Angiogenesis and Vascular Mimicry. *Cells* **2019**, *8*, 407. [CrossRef] [PubMed]
24. Brandt, Z.J.; North, P.N.; Link, B.A. Somatic Mutations of lats2 Cause Peripheral Nerve Sheath Tumors in Zebrafish. *Cells* **2019**, *8*, 972. [CrossRef]
25. Wiley, D.S.; Redfield, S.E.; Zon, L.I. Chemical screening in zebrafish for novel biological and therapeutic discovery. *Methods Cell Biol.* **2017**, *138*, 651–679.
26. Spotlight on early-career researchers: an interview with Toshiro Moroishi. *Commun. Boil.* **2019**, *2*, 1–2.
27. Yamauchi, T.; Moroishi, T. Hippo Pathway in Mammalian Adaptive Immune System. *Cells* **2019**, *8*, 398. [CrossRef] [PubMed]
28. Moroishi, T.; Hayashi, T.; Pan, W.-W.; Fujita, Y.; Holt, M.V.; Qin, J.; Carson, D.A.; Guan, K.-L. The Hippo Pathway Kinases LATS1/2 Suppress Cancer Immunity. *Cell* **2016**, *167*, 1525–1539.e17. [CrossRef]
29. Wu, A.; Deng, Y.; Liu, Y.; Lu, J.; Liu, L.; Li, X.; Liao, C.; Zhao, B.; Song, H. Loss of VGLL4 suppresses tumor PD-L1 expression and immune evasion. *Embo J.* **2019**, *38*, e99506. [CrossRef]
30. Liu, H.; Dai, X.; Cao, X.; Yan, H.; Ji, X.; Zhang, H.; Shen, S.; Si, Y.; Zhang, H.; Chen, J.; et al. PRDM4 mediates YAP-induced cell invasion by activating leukocyte-specific integrin beta2 expression. *EMBO Rep.* **2018**, *19*, e45180. [CrossRef]

31. Gokhale, R.; Pfleger, C.M. The Power of Drosophila Genetics: The Discovery of the Hippo Pathway. *Methods Mol. Biol.* **2019**, *1893*, 3–26. [PubMed]

cells

MDPI

Review

MOB (Mps one Binder) Proteins in the Hippo Pathway and Cancer

Ramazan Gundogdu [1] and Alexander Hergovich [2,*

[1] Vocational School of Health Services, Bingol University, 12000 Bingol, Turkey; rgundogdu@bingol.edu.tr
[2] UCL Cancer Institute, University College London, WC1E 6BT London, UK
* Correspondence: a.hergovich@ucl.ac.uk; Tel.: +44-20-7679-2000; Fax: +44-20-7679-6817

Received: 1 May 2019; Accepted: 4 June 2019; Published: 10 June 2019

Abstract: The family of MOBs (monopolar spindle-one-binder proteins) is highly conserved in the eukaryotic kingdom. MOBs represent globular scaffold proteins without any known enzymatic activities. They can act as signal transducers in essential intracellular pathways. MOBs have diverse cancer-associated cellular functions through regulatory interactions with members of the NDR/LATS kinase family. By forming additional complexes with serine/threonine protein kinases of the germinal centre kinase families, other enzymes and scaffolding factors, MOBs appear to be linked to an even broader disease spectrum. Here, we review our current understanding of this emerging protein family, with emphases on post-translational modifications, protein-protein interactions, and cellular processes that are possibly linked to cancer and other diseases. In particular, we summarise the roles of MOBs as core components of the Hippo tissue growth and regeneration pathway.

Keywords: Mps one binder; Hippo pathway; protein kinase; signal transduction; phosphorylation; protein-protein interactions; structure biology; STK38; NDR; LATS; MST; STRIPAK

1. Introduction

The family of MOBs (monopolar spindle-one-binder proteins) is highly conserved in eukaryotes [1–4]. To our knowledge, at least two different MOBs have been found in every eukaryote analysed so far. For example, in unicellular organisms such as yeast the MOB proteins Mob1p and Mob2p are expressed by two independent genes [5]. In multicellular organisms such as flies, at least four different MOBs, termed dMOBs (Drosophila MOB proteins), have been reported [4]. In humans, as many as seven different MOB proteins, termed hMOBs (human MOB proteins), are encoded by different gene loci. Since each hMOB has been given several names over the years, we have simplified the nomenclature of hMOBs as follows [1]: hMOB1A (UniProtKB: Q9H8S9; also termed MOBKL1B, MOBK1B, Mob1a, MOB1α, Mob4b, MATS1 and C2orf6), hMOB1B (UniProtKB: Q7L9L4; also termed MOBKL1A, Mob1b, Mob4a and MATS2), hMOB2 (UniProtKB: Q70IA6; also termed MOBKL2, Mob2 and HCCA2), hMOB3A (UniProtKB: Q96BX8; also termed MOBKL2A, Mob3A, MOB-LAK, MOB2A, Mob1C), hMOB3B (UniProtKB: Q86TA1; MOBKL2B, Mob3B, MOB2B, Mob1D and C9orf35), hMOB3C (UniProtKB: Q70IA8; also termed MOBKL2C, Mob3C, MOB2C and Mob1E) and hMOB4 (UniProtKB: Q9Y3A3; also termed MOBKL3, Class II mMOB1, MOB3, Phocein/PHOCN, PREI3, 2C4D, and CGI-95). Given the very high identity between hMOB1A and hMOB1B (Figure 1), we refer to hMOB1A/B also as hMOB1 in this review.

	vs. hMOB1A	vs. hMOB2	vs. hMOB3A	vs. hMOB4
hMOB1A	100 %			
hMOB1B	96 %			
hMOB2	37 %	100 %		
hMOB3A	51 %	29 %	100 %	
hMOB3B	51 %	29 %	80 %	
hMOB3C	48 %	28 %	72 %	
hMOB4	22 %	17 %	20 %	100 %
dMOB1	85 %	36 %	47 %	20 %
dMOB2	30 %	42 %	28 %	17 %
dMOB3	51 %	27 %	64 %	20 %
dMOB4	21 %	17 %	20 %	80 %

Figure 1. Primary sequence identities of the fly and human MOBs. The identities between primary sequences are displayed. Identities were defined using EMBOSS needle for pairwise alignments (https://www.ebi.ac.uk/Tools/psa/emboss_needle/). The UniProtKB nomenclature for hMOBs can be found in the introduction section of the main text. The UniProtKB names for dMOBs are as follows [4]: dMOB1 (aka Mats and CG13852)–Q95RA8, dMOB2 (aka CG11711)–Q8IQG1, dMOB3 (aka CG4946)–Q9VL13, and dMOB4 (aka CG3403)–Q7K0E3.

In comparing the sequence identities of MOB proteins expressed in fly and human cells (Figure 1), it becomes apparent that hMOB1 is highly conserved in flies. dMOB1 is 85% identical to hMOB1 and functionally conserved, since exogenous expression of hMOB1A rescues the phenotypes associated with dMOB1 loss-of-function [6,7]. Among other MOBs, hMOB1 is the most closely related to hMOB3A/B/C and dMOB3 (Figure 1). hMOB2 is not as well conserved as hMOB1, since hMOB2 aligns similarly with dMOB1 and dMOB2. hMOB3A is very similar to hMOB3B and hMOB3C, being also 64% identical to dMOB3. hMOB4 is 80% identical to dMOB4, and hMOB4 and dMOB4 do not show any significant identity overlaps with other MOBs (Figure 1). A phylogenetic tree analysis of the dMOBs and hMOBs showed that hMOB1A, hMOB1B and dMOB1 cluster together into one subgroup (Figure 2). Likewise, hMOB2 and dMOB2, as well as hMOB3A, hMOB3B, hMOB3C and dMOB3 form separate subgroups, respectively. It is noteworthy that hMOB4 and dMOB4 also form a subgroup of their own, representing the most distant subgroup of MOBs (Figure 2). However, in spite of these striking sequence similarities, it is currently unknown whether loss-of-function of dMOB2, dMOB3 or dMOB4 can be functionally compensated by the expression of the corresponding hMOB family member. Nevertheless, research covering the past two decades has uncovered important aspects of MOBs that seem to be universally valid for all members of the MOB family. In summary, it was found that MOBs mainly function as intracellular co-regulatory proteins. In particular MOBs can directly bind to protein kinases, and thereby influence the activities of their binding partners. As such, before discussing MOBs in flies and mammals, we will provide a brief introductory background on MOBs studied in unicellular organisms such as yeast.

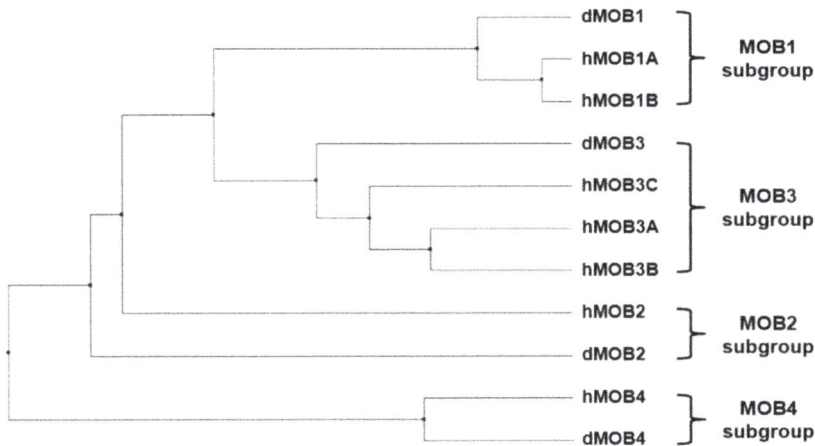

Figure 2. Phylogenetic relationships within the fly and human MOB protein family. The phylogenetic tree was defined using Clustal Omega (https://www.ebi.ac.uk/Tools/msa/clustalo/) together with the Jalview 2.10.5 software using phylogenetic calculation based on the neighbour-joining method. The UniProtKB nomenclature for the analysed proteins is defined in the legend of Figure 1.

More than two decades ago Mob1p, the first MOB (monopolar spindle-one-binder) protein, was described in budding yeast [1,5,8]. It turned out that Mob1p is a central component of the mitotic exit network (MEN) and the septation initiation network (SIN) of budding and fission yeast, respectively (summarised in refs. [8–12]). Mechanistically, budding yeast Mob1p in complex with the Dbf2p protein kinase (a budding yeast counterpart of mammalian NDR/LATS kinases [13]) regulates the release of phosphatase Cdc14p to promote exit from mitosis through a series of dephosphorylation events. Similarly, fission yeast Mob1p bound to Sid2p (a fission yeast counterpart of mammalian NDR/LATS kinases) controls the Clp1p phosphatase (the equivalent of Cdc14p) to support the SIN. Mob2p, another member of the MOB family [1], was uncovered together with Mob1p in budding yeast [5]. Subsequent research revealed that Mob2p coordinates morphogenesis networks in budding and fission yeast [1,13], as well as fungal species [14]. In budding yeast, Mob2p in complex with Cbk1p (the second NDR/LATS kinase expressed in budding yeast [13]) controls the RAM (regulation of Ace2p activity and cellular morphogenesis) signalling network. In fission yeast, Mob2p associated with Orb6p (the second NDR/LATS kinase in fission yeast) coordinates a similar morphogenesis network.

In summary, yeast cells express two different MOB proteins, Mob1p and Mob2p. While Mob1p controls mitotic exit, Mob2p is a regulator of cell morphogenesis and polarized growth. Mob1p and Mob2p function in complex with different members of the NDR/LATS kinase family. Depending on the yeast species Mob1p binds to Dbf2p or Sid2p whereas Mob2p associates with Cbk1p or Orb6p, respectively. Consequently, in yeast Mob1p and Mob2p act as part of independent and non-interchangeable complexes containing different NDR/LATS kinases in order to regulate diverse cellular processes (please consult refs. [1,13] for a more detailed discussion of these points).

2. An Overview of MOBs in *Drosophila melanogaster*

Since 2005, MOBs have been studied in multicellular organisms. In contrast to yeast, fly cells encode four different MOB proteins [4]. Lai and colleagues found that dMOB1 (also termed Mats for MOB as tumour suppressor) is an essential regulator of cell proliferation and cell death in fruit flies [7]. Follow up studies established dMOB1 as a core component of the Hippo tissue growth control pathway (for more details see subchapter 4 of this review). Briefly, as a core component of the Hippo pathway, dMOB1 acts as a co-activator of the Warts protein kinase [15–17], one of two NDR/LATS kinases in

D. melanogaster [13]. Intriguingly, dMOB1 also genetically interacts with Tricornered (Trc) [4], the second NDR/LATS kinase in flies [13]. Thus, dMOB1 does not bind specifically to a single NDR/LATS kinase, as observed for yeast Mob1p [1]. In addition, dMOB1 can form a complex with the Hippo (Hpo) protein kinase [18], with Hpo being able to phosphorylate dMOB1 [18] as well as Warts and Trc [19]. The importance of these protein-protein interactions and phosphorylation events is discussed in subchapters 4, 5 and 6. Noteworthy, dMOB1 can further play a role in mitosis (summarised in refs. [1,20,21]).

Unlike those of dMOB1, the biological functions of dMOB2, dMOB3 and dMOB4 are yet to be completely understood. Based on the work by the Adler laboratory, it seems that dMOB2 can play a role in wing hair morphogenesis, possibly by forming a complex with Trc [4]. However, the precise mechanism of action remains unknown. Moreover, dMOB2 supports the development of photoreceptor cells [22] and the growth of larval neuromuscular junctions in flies [23]. It is possible that these neurological roles are linked to the association of dMOB2 with Tricornered [4]. So far, only Trc has been established as a bona fide binding partner of dMOB2.

Interestingly, like dMOB1 and dMOB2, dMOB3 can also genetically interact with Trc [4], but any function or direct binding partners of dMOB3 have remained elusive. Similar to dMOB2, dMOB4 has been linked to a neurological function. More precisely, dMOB4 was found to play a role as a regulator of neurite branching [24]. Furthermore, dMOB4 depletion in fly cells results in defective focusing of kinetochore fibres in mitosis [25]. Possibly some of these roles of dMOB4 could be linked to dMOB4 being part of the Striatin-interacting phosphatase and kinase (STRIPAK) complex (see subchapters 6 and 7 for more details). Yet, these possible connections have yet to be experimentally explored.

In summary, based on current evidence it is tempting to conclude that, unlike in yeast, individual dMOBs can interact (at least genetically) with Warts and Trc, the two NDR/LATS kinases expressed in fly cells. Conversely, individual NDR/LATS kinases can interact with different dMOBs. Therefore, it appears that in multicellular organisms such as flies, the binding of MOBs is not restricted to a unique member of the NDR/LATS kinase family.

3. An Overview of MOBs in Human Cells

hMOB1A and hMOB1B are also known as hMOB1, since hMOB1B is 96% identical to hMOB1A (see Figure 1 and refs. [1,3,26]). hMOB1A and hMOB1B are mainly cytoplasmic proteins [27,28]. However, upon targeting to the plasma membrane of mammalian cells hMOB1 can trigger a binding-dependent activation of the human NDR/LATS kinases [27,29]. hMOB1 can form complexes with NDR1 (aka STK38), NDR2 (aka STK38L), LATS1 and LATS2, all four human NDR/LATS kinases [1,13]. These interactions are mediated through one unique and highly conserved domain in NDR/LATS kinases [29]. Like dMOB1 [18], hMOB1 can also bind to the MST1/2 kinases [30], the human counterparts of Hpo [31–34]. The significance and regulation of these interactions is discussed in much detail in subchapters 5 and 6. hMOB1 can further associate with other proteins, although the importance of these additional interactions is yet to be defined (see subchapter 6). In cases whereby a new hMOB1 binding partner has been validated by conventional interaction assays, we have included the discussion of this novel aspect in the appropriate subchapters. Nevertheless, the protein-protein interactions of hMOB1 with NDR/LATS kinases is the best understood. For a summary of the cellular roles of hMOB1 please consult subchapter 7 and ref. [1].

In contrast to hMOB1, a significant portion of hMOB2 is nuclear [27,28]. Intriguingly, hMOB2 forms a complex with NDR1/2, while hMOB2 neither associates with LATS1/2 [1,35] nor MST1/2 [36]. hMOB2 can compete with hMOB1 for NDR1/2 binding, since hMOB2 interacts with the same domain on NDR1/2 as hMOB1 [27,28]. However, the binding modes of hMOB1 and hMOB2 to NDR1/2 seem to differ [1,35]. hMOB2 can also interact with RAD50, a core component of the MRE11-RAD50-NBS1 (MRN) DNA damage sensor complex, thereby playing a role in cell cycle-related DNA damage signalling [37]. Whether the hMOB2/RAD50 complex is linked to NDR1/2 signalling is yet to be defined [38]. For a summary of the cellular roles of hMOB2, see subchapter 7 and refs. [1,38].

Based on their similarities hMOB3A, hMOB3B and hMOB3C are sometimes also referred to as hMOB3 [39]. Like hMOB1, hMOB3s are mainly cytoplasmic proteins [28,40]. The best-understood binding partner of hMOB3s is the MST1 kinase [39]. Given the high sequence similarities between hMOB3s and hMOB1 (Figure 1), it was rather surprising that hMOB3s do not bind to any NDR/LATS kinase [35]. Nonetheless, large-scale interactome studies suggest that hMOB3s may have other binding partners in addition to MST1/2 (see ref. [1] and this review). Our current understanding of the cellular roles of hMOB3s is summarised in subchapter 7.

hMOB4 (aka Phocein) is best known as part of the multi-component STRIPAK (Striatin-interacting phosphatase and kinase) complex [41–44]. hMOB4 can bind to two distinct regions of Striatin [45]. Besides the Striatin-associated hMOB4, the core STRIPAK complex contains serine/threonine protein phosphatase 2 (PP2A) subunits such as PP2Ac and PP2A-A, as well as the Striatins as PP2A regulatory subunits. STRIPAK also encompasses CCM3 (cerebral cavernous malformation 3; also known as PDCD10) as well as the MST3 (aka STK24), MST4 (aka MASK or STK26) and STK25 (aka YSK1 or SOK1) protein kinases, the three members of the GCKIII (germinal centre kinase III) subfamily of Sterile 20 kinases [46]. In addition, the core complex contains STRIP1/2 (aka FAM40A/B). Two mutually exclusive STRIPAK complexes have been defined based on additional components such as CTTNBP2-like adaptors, SLMAP (sarcolemmal membrane-associated protein) or others [47]. However, although the STRIPAK complex has been linked to the Hippo pathway (see subchapter 4), the precise role(s) of hMOB4 in the STRIPAK complex has remained of a more speculative nature. In this regard, hMOB4 has been reported to form a complex with MST4 in a phosphorylation-dependent manner [48], an aspect that is discussed briefly in subchapter 6. The cellular functions of hMOB4 are discussed in subchapter 7.

4. An Overview of MOBs and the Hippo Pathway

The Hippo tissue growth control and regeneration pathway represents a key signalling cascade in flies and mammals [16,49–51]. By co-ordinating death, growth, proliferation, and differentiation on the cellular level the Hippo pathway controls organ growth, tissue homeostasis and regeneration. Hence, the deregulation of Hippo signalling has been linked to serious human diseases, such as cancers [50,52,53]. At the molecular level, a conserved Hippo core cassette is fundamental for the regulation of the co-transcriptional regulators YAP and TAZ, two main effectors of the Hippo pathway [51,54]. To date, the MST1/2 (aka STK4/3; Hpo in *D. melanogaster*) and LATS1/2 (aka Warts in flies) protein kinases are the best understood members of the Hippo core cassette. In this regard, MOB1 acts as central signal transducers in the Hippo core cassette. Upon activation of the Hippo pathway MST1/2 phosphorylate LATS1/2 and MOB1, thereby supporting the formation of an active MOB1/LATS complex (see subchapter 6 for molecular details), which is essential for development and tissue growth control [6]. Activated LATS1/2 in complex with MOB1 then phosphorylate YAP/TAZ on different Ser/Thr residues, thereby inhibiting nuclear activities of YAP/TAZ through cytoplasmic retention and/or degradation of phosphorylated YAP/TAZ [51,54]. Markedly, the NDR1/2 protein kinases (aka STK38 and STK38L; the closest relatives of LATS1/2 [13]), can also phosphorylate and thereby constrain YAP1 to the cytoplasm [55,56]. Similar to LATS1/2, NDR1/2 can stably associate with MOB1 [1,6] and are direct effector substrates of MST1/2 [55,57]. As already mentioned, MST1/2 also phosphorylate MOB1 [30]. Hence, LATS1/2, NDR1/2 and MOB1 represent well-established substrates of MST1/2 [1,55,57,58]. Taken together, the core of the Hippo pathway comprises distinct kinases, such as MST1/2, LATS1/2 and NDR1/2, which act together with the fundamental signal adaptor MOB1.

By directly interacting with the MST1/2, LATS1/2 and NDR1/2 kinases (aka Hpo, Warts and Trc in flies) MOB1 functions as a key signal transducer in the Hippo pathway. However, although dMOB1 and MOB1 are essential in flies and mice [6,7,59], hMOB1 appears to be dispensable in human cells, at least in transformed HEK293A cells [60]. hMOB1A/B double-knockout (DKO) cells were viable in spite of drastically impaired LATS1/2-mediated YAP/TAZ phosphorylation upon serum deprivation [60]. Very similar to LATS1/2 DKO cells, YAP/TAZ remained nuclear and co-transcriptionally active in hMOB1A/B DKO cells. Even more puzzling, hMOB1 phosphorylation by MST1/2 does not seem to be required for LATS1/2 activation in HEK293A cells [60], although biochemical evidence strongly suggests that MST1/2-mediated phosphorylation of hMOB1 is needed for LATS1/2 binding (see subchapters 5 and 6). To make things even more complicated, hMOB1 can associate with other intracellular regulatory proteins besides binding to kinases of the Hippo core cassette (see subchapter 6). Therefore, it is quite possible that hMOB1 as a central Hippo component is acting in diverse cancer-associated cellular processes.

Until recently [37], NDR1/2 were the only reported binding partners of hMOB2 [1,35]. More specifically, it was documented that hMOB2 could compete with hMOB1 for NDR1/2 binding, with hMOB2 counteracting hMOB1 as a co-activator of NDR1/2 [1,35,61]. In this context, Zhang et al. studied Hippo signalling in a hMOB2 knockout hepatocellular carcinoma (HCC) cell line [62]. They found that MST1/2-mediated phosphorylation of NDR1/2 was induced, while MST1/2-mediated phosphorylations of hMOB1 and LATS1/2 were decreased upon hMOB2 knockout [62]. LATS1/2-mediated phosphorylation of YAP was also reduced in hMOB2 knockout cells, while MOB2 overexpression resulted in opposite effects [62]. Subsequently, the authors concluded that loss of hMOB2 can favour hMOB1 binding to NDR1/2, thereby reducing the pool of hMOB1 available for LATS1/2 binding. Conversely, hMOB2 overexpression may reduce hMOB1 binding to NDR1/2, thereby freeing hMOB1 to bind to LATS, resulting in the activation of Hippo signalling upstream of YAP [62]. However, this study [62] did not examine whether this molecular reprogramming of binding partners could be a possible consequence of cell cycle effects upon hMOB2 loss-of-function. Given that loss of hMOB2 can trigger a p53-dependent G1/S cell cycle arrest [37] and that NDR/LATS are associated with cell cycle progression [20,21], it will certainly be necessary to re-examine these HCC-based hMOB2 knockout cells to ensure the changes in NDR1/2, LATS1/2, and hMOB1 phosphorylation upon hMOB2 loss [62] are not merely reflecting indirect effects triggered by an underlying cell cycle arrest/delay.

Like hMOB1 and hMOB2, the group of hMOB3 signal transducers has been linked to the Hippo pathway. It was found that hMOB3s can directly bind to MST1, while they do not bind to any member of the NDR/LATS kinase family [35,39]. Considering that hMOB3s share high sequence similarities with hMOB1 (Figure 1) this was unexpected. Nevertheless, hMOB3s have at least one aspect in common with hMOB1, namely the direct binding to the MST1/2 kinases. More specifically, hMOB3s require two conserved positively charged residues to bind to MST1 [39], with both positively charged residues also being essential for the formation of a stable complex between hMOB1 and MST1/2 [6]. However, current evidence suggests that hMOB3 binding to MST1 is inhibitory and hMOB3s can act upstream of apoptotic MST1 signalling [39], while hMOB1 seems to mainly function downstream of MST1/2 [1,30,51]. Therefore, it is possible that hMOB3 competes with hMOB1 for MST1/2 binding, like hMOB2 is competing with hMOB1 for NDR1/2 binding, but these competitive interactions are likely to involve different interaction domains (see subchapter 6).

Intriguingly, hMOB4 has likewise been connected with the Hippo pathway. In fly and human cells, the hMOB4-containing supramolecular STRIPAK complexes [43,63] have been implicated in Hippo signalling on diverse regulatory levels [48,64–69]. However, to our knowledge, the specific contributions of hMOB4 to the crosstalk between Hippo and STRIPAK signalling remain undetermined. Nevertheless, the MST4 kinase, a relative of MST1/2 [46], was shown to form a complex with hMOB4, with the overall structure of the hMOB4/MST4 complex resembling the hMOB1/MST1 complex [48]. But one of the key residues (Lys105) that promotes MOB1 binding to MST1 [6] is not conserved in hMOB4 (Figure 3). Notably, the phospho-Ser/Thr binding motif of hMOB1 [70] seems to be present in hMOB4, while key residues such as Glu51 of MOB1 that mediate NDR/LATS binding [6,71,72] are not conserved in hMOB4 (Figure 3). Therefore, the binding mode of MST4 and other GCKIIIs to hMOB4 is yet to be completely understood [73].

Different members of the Ste20-like kinase family can phosphorylate, and thereby activate, NDR/LATS kinases (summarised in ref. [57]). Briefly, MST1/2 phosphorylate NDR/LATS in human cells. Moreover, MAP4K (mitogen-activated protein kinase kinase kinase kinase)-type kinases can regulate

NDR/LATS kinases through phosphorylation. In addition, MST3, MST4 and STK25, all members of the GCKIII subfamily of Ste20-like kinases [46], can phosphorylate NDR kinases in human and fly cells [57,74]. Strikingly, hMOB1 can bind to MST1/2, while hMOB4 is part of the GCKIII-containing STRIPAK complex. In this regard, the MAP4K-type kinase MINK1 was also identified as a component of the STRIPAK complex [75]. Consequently, it will be very interesting to properly map the binding patterns of hMOBs across the Ste20-like kinase family. Potentially, an array of different Ste20-like kinases is regulated by and/or regulates diverse hMOBs upstream of NDR/LATS signalling.

5. Post-Translational Modifications (PTMs) of MOBs

As already exemplified by the description of the Hippo pathway, signal transduction cascades transmit extra- and intra- cellular inputs through a broad array of protein kinases [76]. Thus, the phosphorylation status of signal transducers and effectors is finely orchestrated by interactions between kinases and phosphatases to achieve pathway regulation. Consequently, it is not surprising that MOBs can be regulated by phosphorylation events. In this subchapter, we summarise current knowledge on specific phosphorylation events of MOBs. Where appropriate, we also mention PTMs other than phosphorylations.

About two decades ago, yeast Mob1p was already described as a phospho-protein [5] that can be phosphorylated by Cdc15p (the yeast counterpart of the fly Hippo kinase) [77]. The phosphorylation of dMOB1 by Hippo was subsequently reported in fly cells [18]. However, Avruch and colleagues were the first to define the specific phosphorylation events that linked Hippo kinases such as MST1/2 with the regulation of hMOB1 [30,58]. More precisely, hMOB1 is specifically phosphorylated on Thr12 and Thr35 by MST1/2, thereby supporting complex formation between hMOB1 and NDR/LATS kinases [30]. Considering that the then available crystal structure of hMOB1A did not cover the N-terminal 32 residues [3], the precise molecular mechanism remained poorly understood for a long time [1]. However, Kim et al. recently deciphered the high-resolution crystal structure of full-length hMOB1 in its auto-inhibited form, and further reported the structural composition of a complex between hMOB1 and the N-terminal regulatory domain (NTR) of LATS1 [71]. These structures enabled the Hakoshima laboratory to define how hMOB1 is released from its auto-inhibitory conformation [71]. They found that hMOB1 phosphorylation on Thr12 and Thr35 accelerates the dissociation of the Switch helix, a long flexible positively-charged linker, from the LATS1-binding surface within hMOB1. Consequently, they uncovered a "pull-the-string" mechanism through which hMOB1 is rendered accessible for LATS1 binding [71], a model supported by an independent recent study [78]. This discovery was significant, since it helped to understand an important aspect of the molecular regulation of the core of the Hippo pathway [51,72].

Of note: in addition to the phosphorylation of Thr12 and Thr35 by MST1/2 [30], other PTMs of hMOB1, such as the phosphorylation of Tyr26 and Ser38, have been reported [79]. While the function of Ser38 phosphorylation has remained unclear, Tyr26 phosphorylation of hMOB1 has very recently been attributed an important role in tyrosine kinase signalling [80]. Gutkind and colleagues found that focal adhesion kinase (FAK) phosphorylates hMOB1 on Tyr26, thereby preventing the formation of a functional hMOB1/LATS complex and thus causing Hippo signalling to remain inactive [80]. It remains to be tested whether this new exciting regulatory mechanism of hMOB1 by Tyr26 phosphorylation also affects hMOB1/NDR complex formation. In addition, one should note that Ser23 of hMOB1 has also been observed [79], with Ser23 of hMOB1 representing a possible ATM target site [81], suggesting a possible link of hMOB1 to the DNA damage response (DDR). Last, but not least, it has been reported that GSK3β can destabilize hMOB1 by phosphorylating serine 146 of hMOB1 in the context of neurite outgrowth downstream of the PTEN-GSK3β axis [82]. Therefore, hMOB1 can be regulated by different phosphorylation events performed by diverse kinases.

In contrast to hMOB1, the only reported PTMs for hMOB2 are modifications with ubiquitin (or NEDD8) at Lys23, Lys32 and Lys131 [83,84]. The precise nature and functions of these three PTMs are currently unknown. In addition, the hMOB2 protein contains several putative ATM phosphorylation sites, which possibly are linked to the DDR function of MOB2 [81]. However, phosphorylation of hMOB2 has yet to be documented.

Similar to hMOB1, several phosphorylations of hMOB3s have been observed. Phosphorylations of hMOB3A on Thr15, Thr26 and Ser38, hMOB3B on Thr25 and Thr77, and hMOB3C on Thr14, Thr25 and Ser37 have been documented [79]. Ser37 of hMOB3C may represent an ATM targeting site [81]. Of note, the sequence motifs surrounding Thr15 and Ser38 of hMOB3A are quite similar to Thr12 and Thr35 of hMOB1 (Figure 3), hence raising the possibility that hMOB3A is possibly phosphorylated on Thr15 and/or Ser38 by MST1/2 in a similar fashion as reported for hMOB1 [30]. However, regulatory roles for any of these putative phosphorylation events are yet to be reported.

Last but not least, specific PTMs of hMOB4 are also of potential interest. In this regard, Ser147 and Tyr141 phosphorylations of hMOB4 have been observed [79]. Ser147 phosphorylation of hMOB4 is possibly performed by the DDR-linked ATM kinase [81], proposing that hMOB4 is linked to DNA damage signalling. However, no regulatory roles for any of these phosphorylation events have been described so far.

6. Protein-Protein Interactions of MOBs

Given that MOBs do not have any known enzymatic activities, one wonders how they contribute to intracellular signalling. Based on the currently available knowledge, the answer to this question is that MOBs are globular scaffold proteins that can regulate other factors such as protein kinases by direct protein-protein interactions [1]. Therefore, we will focus in this subchapter on summarising how MOBs form protein complexes and thereby help to regulate molecular machineries.

As already mentioned, hMOB1 can bind directly to MST1/2, LATS1/2 and NDR1/2. Interestingly, these interactions can occur in a phosphorylation-dependent manner. However, before discussing this important aspect in much detail, one should first note that hMOB1 can also associate with other signalling complexes. Current evidence suggests that hMOB1 can bind to at least seven apparently mutually exclusive and independent proteins in a sometimes phosphorylation-dependent fashion: (1) the NDR1/2 kinases, (2) the LATS1/2 kinases, (3) the MST1/2 kinases, (4) the PP6 phosphatase module, (5) the atypical guanine nucleotide exchange factors 6 to 8 (DOCK6-8), (6) the ubiquitin ligase Praja2, and (7) yet poorly characterized proteins, including leucine rich repeats and calponin homology domain containing (LRCH1-4) proteins and cytokine receptor like factor 3 (CRLF3) [1,6,30,65,78,85,86]. Notably, Xiong et al. found that the phosphorylation-dependent binding of hMOB1 to the PP6 and DOCK6-8 complexes appears to differ from that elucidated for protein kinase binding, with hMOB1 preferentially interacting with MST1/2 and NDR/LATS, and not the DOCK6-8 and PP6 modules [78]. Future investigations are needed to properly understand the underlying cellular functions of complexes containing hMOB1 and non-kinase binding partners such as PP6, DOCK6-8 and others.

While hMOB1 can bind to a variety of components, the complex formations with MST1/2 and NDR/LATS kinases are the only protein-protein interactions understood in detail. In Hippo core signalling, hMOB1 acts as a central signal adaptor helping to regulate NDR/LATS activation [1,6,13,71,72,87,88]. On the one hand, a stretch composed of highly conserved hydrophobic and positively charged residues N-terminal of the catalytic domains of NDR/LATS kinases, termed N-terminal regulatory domain (NTR), is essential for hMOB1 binding to NDR/LATS kinases to promote their activities. On the other hand, a conserved cluster of negatively charged residues on hMOB1 is needed for the interaction with NDR/LATS [1,3,6,13,28,71,72,87,88]. MST1/2 phosphorylation of hMOB1 on Thr12 and Thr35 regulates hMOB1 binding to the NTRs of NDR/LATS [6,30,71,72], although hMOB1 can also from a complex with NDR1/2 in a phosphorylation-independent manner [6]. In summary, hMOB1 binds to NDR/LATS kinases through highly conserved residues on hMOB1 and NDR/LATS kinases in a manner that can be regulated by MST1/2 phosphorylation of hMOB1.

hMOB1 binding to NDR/LATS has multiple functions. First, hMOB1 binding helps to stimulate auto-phosphorylation on a Ser residue located in the T-loop (also known as activation segment) of NDR/LATS kinases (summarised in [1,13,87,89]). Second, MOB1 also acts as an allosteric activator of NDR/LATS by mediating MST1/2-mediated trans-phosphorylation of a Thr residue located in the hydrophobic motif (HM) outside of the catalytic domain of NDR/LATS [6,28,78,87–91]. Third, it was biochemically shown that stable hMOB1 binding to MST2 can function as an important step in activating the MST1/2-LATS1/2 kinase cascade [72,92], although stable MOB1/MST1/2 binding seems dispensable for development and tissue growth control [6,90] and a stable hMOB1/MST1/2 complex is not required for efficient MST1/2-mediated phosphorylation of hMOB1 [6]. Since some of these hMOB1 functions appear to contradict each other, Manning and Harvey [93] proposed a unifying model, wherein hMOB1 acts before and after MST1/2-mediated phosphorylation of LATS1/2 and hMOB1. Notably, this model is similar, but still different, from the model proposed for NDR1/2 activation by MST1/2 and hMOB1 [6,88]. In this regard, somewhat surprisingly, T-loop and HM phosphorylations of LATS1/2 can also occur without increased MST1/2-mediated phosphorylation of hMOB1 on Thr12/Thr35 [94], suggesting that hMOB1/LATS complex formation can be dispensable for efficient LATS1/2 phosphorylation in certain settings. Nevertheless, hMOB1/LATS complex formation appears to be essential for development and tissue growth control [6].

Taken together, these data show collectively that hMOB1/NDR and hMOB1/LATS complexes are not equal, thereby allowing cells to process diverse biological inputs through diverse hMOB1 signalling. hMOB1 as an adapter protein can help to differentially regulate NDR/LATS kinases by concurrently forming complexes with NDR/LATS and their upstream activator kinases such as MST1/2 and by allosterically activating the auto- and trans-phosphorylation activities of NDR/LATS.

Interestingly, recent structural and biochemical analyses of the catalytic domain of NDR1, lacking the entire NTR domain, have suggested that the regulation of NDR1 activity by the elongated activation segment and by hMOB1 binding may represent mechanistically distinct events [89]. This notion is supported by the biochemical and structural dissection of the yeast Cbk1p-Mob2p complex [95]. However, much more work is needed to fully understand how the regulation of NDR/LATS kinases is orchestrated by hMOB1 binding to the NTR, activation segment and HM phosphorylations–three molecular events that possibly represent three distinct regulatory levels. In particular, the role of MST1/2-mediated phosphorylation of hMOB1 is yet to be completely understood with regard to the regulation of NDR1/2, in particular when considering that phospho-hMOB1 displays a much higher affinity for NDR1/2 than unphosphorylated hMOB1 [6,78], while LATS1/2 can only bind to phospho-MOB1 [6].

In contrast to NDR1/2 kinases, hMOB1 seems to associate with LATS1/2 and MST1/2 in a purely phospho-dependent manner [6,65,69–72,78,92]. In the case of MST1/2, the direct interactions involve several auto-phosphorylation sites in MST1/2 that help to recruit hMOB1 [51,65,69,72,78,92]. Noteworthy, these interactions involve a phospho-Ser/Thr binding pocket of hMOB1, encompassing the positively charged residues Lys153, Arg154 and Arg157 of hMOB1 [70]. These three key residues appear to be conserved in hMOB1 and hMOB4, while hMOB3s have only two conserved positively charged residues (Figure 3). However, phospho-peptide binding properties of hMOB2, hMOB3s and hMOB4 are yet to be reported.

Based on the structure of a short fragment of MST2 (371 to 400) bound to hMOB1 lacking the first 50 amino acids, Ni et al. [72] defined two MOB1/MST2 interfaces, namely one supportive binding site (residues 390 to 398 of MST2) and one main phospho-Thr binding site, centring around phosphorylated Thr378 of MST2 and the phospho-Ser/Thr binding pocket of hMOB1. However, other important sites involved in hMOB1 binding on MST2, such as phospho-Thr349, phospho-Thr356, and phospho-Thr364, were not examined on the structural level [72]. This notion [72] that Thr378 of MST2 (Thr380 of MST1, respectively) is not the sole main interaction site on hMOB1 is supported by other studies [78,92]. At least six different phospho-threonine sites on MST1 (Thr329, Thr340, Thr353, Thr367, Thr380 and Thr387) contribute to hMOB1 binding [78,92]. Intriguingly, Pro106 of hMOB1 contributes to phospho-Thr binding on MST1 [72,78,92], and the positively charged Lys104 and Lys105 of hMOB1 are crucial for MST1/2 binding, while being expendable for NDR/LATS binding [6]. Thus, it is quite possible that Lys104/Lys105 together with Pro106 of hMOB1 may bond with multiple negatively charged phosphorylated Thr residues on MST1/2. Notably, the Lys104/Lys105/Pro106 motif of hMOB1 is conserved in dMOB1, hMOB3s and dMOB3 (Figure 3). Alterations of Arg108 and Lys109 in hMOB3A, corresponding to Lys104 and Lys105 in hMOB1 (Figure 3), abolish MST1 binding [39], suggesting that the Lys104/Lys105/Pro106 binding motif of hMOB1 is functionally conserved.

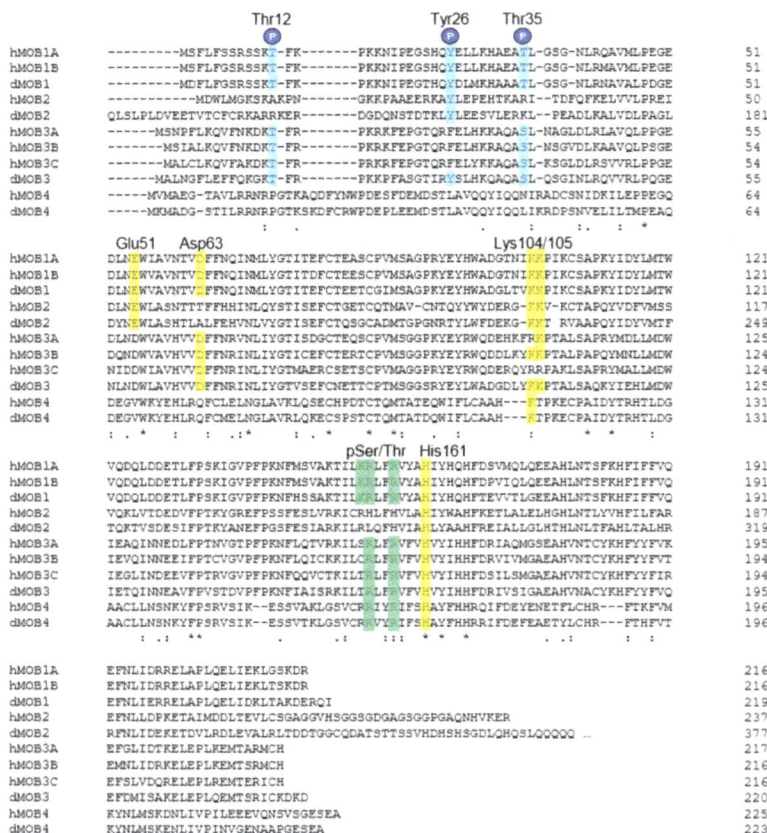

```
                Thr12             Tyr26    Thr35
hMOB1A  ---------MSFLFSSRSSKT-FK-------PKKNIPEGSHQYELLKHAEATL-GSG-NLRQAVMLPEGE  51
hMOB1B  ---------MSFLFGSRSSKT-FK-------PKKNIPEGSHQYELLKHAEATL-GSG-NLRMAVMLPEGE  51
dMOB1   ---------MDFLFGSRSSKT-FK-------PKKNIPEGTHQYDLMKHAAATL-GSG-NLRNAVALPDGE  51
hMOB2   ------------MDWLMGKSRAKPN-------GKKPAAEERKAYLEPEHTKARI--TDFQFKELVVLPREI  50
dMOB2   QLSLPLDVEETVTCFCRKARRKER-------DGDQNSTDTKLYLEESVLRERKL--PEADLKALVDLPAGL  181
hMOB3A  ------MSNPFLKQVFNKDKT-FR-------PKRKFEPGTQRFELHKKAQASL-NAGLDLRLAVQLPPGE  55
hMOB3B  -------MSIALKQVFNKDKT-FR-------PKRKFEPGTQRFELHKRAQASL-NSGVDLKAAVQLPSGE  54
hMOB3C  -------MALCLKQVFAKDKT-FR-------PKRKFEPGTQRFELYKRAQASL-KSGLDLRSVVRLPPGE  54
dMOB3   ------MALNGFLEFFQKGKT-FR-------PKKPFASGTIRYSLHKQAQASL-QSGINLRQVVRLPQGE  55
hMOB4   -----MVMAEG-TAVLRRNRPGTKAQDFYNWPDESFDEMDSTLAVQQYIQQNIRADCSNIDKILEPPEGQ  64
dMOB4   -----MKMADG-STILRRNRPGTKSKDFCRWPDEPLEEMDSTLAVQQYIQQLIKRDPSNVELILTMPEAQ  64
                  :                          :            :   :.   :   *

        Glu51   Asp63                              Lys104/105
hMOB1A  DLNEWIAVNTVDFFNQINMLYGTITEFCTEASCPVMSAGPRYEYHWADGTNIKFPIKCSAPKYIDYLMTW  121
hMOB1B  DLNEWVAVNTVDFFNQINMLYGTITDFCTEESCPVMSAGPRYEYHWADGTNIKFPIKCSAPKYIDYLMTW  121
dMOB1   DLNEWVAVNTVDFFNQINMLYGTITEFCTEETCGIMSAGPRYEYHWADGLTVKFPIKCSAPKYIDYLMTW  121
hMOB2   DLNEWLASNTTTFFHHINLQYSTISEFCTGETCQTMAV-CNTQYVWYDERG-YKV-KCTAPQYVDFVMSS  117
dMOB2   DYNEWLASHTLALFEHVNLVYGTISEFCTQSGCADMTGPGNRTYLWFDEKG-YKT RVAAPQYIDVVMTF  249
hMOB3A  DLNDWVAVHVVDFFNRVNLIYGTISDGCTEQSCPVMSGGPKYEYRWQDEHKFRKPTALSAPRYMDLLMDW  125
hMOB3B  DQNDWVAVHVVDFFNRINLIYGTICEFCTERTCPVMSGGPKYEYRWQDDLKYRTPTALPAPQYMNLLMDW  124
hMOB3C  NIDDWIAVHVVDFFNRINLIYGTMAERCSETSCPVMAGGPRYEYRWQDERQYRRFAKLSAPRYMALLMDW  124
dMOB3   NLNDWLAVHVVDFFNRINLIYGTVSEFCNETTCPTMSGGSRYEYLWADGDLYPKPTALSAQKYIEHLMDW  125
hMOB4   DEGVWKYEHLRQFCLELNGLAVKLQSECHPDTCTQMTATEQWIFLCAAH---PTPKECPAIDYTRHTLDG  131
dMOB4   DEGVWKYEHLRQFCMELNGLAVRLQKECSPSTCTQMTATDQWIFLCAAH---PTPKECPAIDYTRHTLDG  131
         :  .  *    ..:*     ::     *   .         :     *        *    *

                            pSer/Thr  His161
hMOB1A  VQDQLDDETLFPSKIGVPVFPFKNFMSVAKTILKRLFPVYAHIYHQHFDSVMQLQEEAHLNTSFKHFIFFVQ  191
hMOB1B  VQDQLDDETLFPSKIGVPVFPFKNFMSVAKTILKRLFPVYAHIYHQHFDPVIQLQEEAHLNTSFKHFIFFVQ  191
dMOB1   VQDQLDDETLFPSKIGVPVFPFKNFHSSAKTILKRLFPVYAHIYHQHFTEVVTLGEEAHLNTSFKHFIFFVQ  191
hMOB2   VQKLVTDEDVFPTKYGREFPSSFESLVRKICRHLFHVLAHIYWAAHFKETLALELHGHLNTLYVHFILFAR  187
dMOB2   TQKTVSDESIFPTKYANEFPGSFESIARKILRLQFHVIAHLYAAHFREIALLGLHTHLNLTFAHLTALHR  319
hMOB3A  IEAQINNEDLFPTNVGTPFFRNFLQTVFRILSHLFVFVHVYIHHFDRIAQMGSEAHVNTCYKHFYFVK  195
hMOB3B  IEVQINNEEIFPTCVGVPFFKNFLQICKRILRCHLFPVFVHVYIHHFDSILSMGAEAHVNTCYKHFYFVT  194
hMOB3C  IEGLINDEEVFPTRVGVPFFKNFPQQVCTKILTHLFPVFVHVYIHHFDRIAQMGSEAHVNTCYKHFYFIR  194
dMOB3   IETQINNEAVFPVSTDVPFFKNFIAISRKILTHLFPVFVHVYIHHFDRIVSIGAEAHVNACYKHFYFVQ  195
hMOB4   AACLLNSNKYFPSRVSIK--ESSVAKLGSVCRAIYHIFSHAYFHHRQIFDEYENETFLCHR---FTKFVM  196
dMOB4   AACLLNSNKYFPSRVSIK--ESSVTKLGSVCRAIYHIFSHAYFHHRRIFDEFEAETYLCHR---FTHFVT  196
         :  .:   **      .     .:    :::  *  *   .     ..:    .

hMOB1A  EFNLIDRRELAPLQELIEKLGSKDR                                  216
hMOB1B  EFNLIDRRELAPLQELIEKLTSKDR                                  216
dMOB1   EFNLIERRELAPLQELIDKLTAKDERQI                               219
hMOB2   EFNLLDPKETAIMDDLTEVLCSGAGGVHSGGSGDGAGSGGPGAQNHVKER         237
dMOB2   RFNLIDEKETDVLRDLEVALRLTDDTGGCQDATSTTSSVHDHSHSGDLQHQSLQQQQQ ...  377
hMOB3A  EFGLIDTKELEPLKEMTARMCH                                     217
hMOB3B  EMNLIDRKELEPLKEMTSRMCH                                     216
hMOB3C  EFSLVDQRELEPLREMTERICH                                     216
dMOB3   EFDMISAKELEPLQEMTSRICKDKD                                  220
hMOB4   KYNLMSKDNLIVPILEEEVQNSVSGESEA                              225
dMOB4   KYNLMSKENLIVPINVGENAAPGESEA                                223
         .  .:
```

Figure 3. Comparison of the primary protein sequences of human and fly MOBs. Primary protein sequences were aligned using Clustal Omega as defined in the legend of Figure 2. The sites of Thr12, Tyr26 and Thr35 phosphorylations of hMOB1A are highlighted in light blue. Important interaction sites of hMOB1A that were verified by different experimental approaches are displayed in yellow. The key residues of the phospho-Ser/Thr (pSer/Thr) binding pocket of hMOB1A are shown in green. For more details on key residues of hMOB1 that are involved in kinase binding please consult refs. [3,6,70–72,78,92,95]. The UniProtKB nomenclature for the analysed proteins is defined in the legends of Figures 1 and 2.

Notably, phospho-Thr378 of MST2 can also bind to SLMAP [68], in addition to hMOB1 [69], and a T-loop mutation of MST2 reduces the MST2/SLMAP interaction [65] in the context of STRIPAK-Hippo signalling. Thus, the multisite auto-phosphorylation of the MST1/2 linkers appears to function as a molecular platform able to integrate different Hippo signalling events [68,69].

In contrast to MST1/2, the LATS1/2 kinases must not be phosphorylated to bind to phospho-MOB1 [6,71,72]. Nevertheless, the phosphorylation of LATS1/2 on serine 690 and 653, respectively (corresponding to threonine 74 and 75 in NDR1 and NDR2), are very likely to represent another regulatory layer. LATS1 S690A, LATS2 S653A, NDR1 T74A and NDR2 T75A mutants display diminished binding to MOB1 [27–29,87,88,96,97] and the crystal structure of phospho-MOB1 bound to a LATS1 fragment highlighted the importance of serine 690 of LATS1 for MOB1 binding [72]. In this regard, hMOB1 must be phosphorylated to bind to LATS1/2 [6,71,72], with Asp63 of hMOB1 being essential and providing the selectivity for LATS1/2 binding [6]. In this regard, the analysis of hMOB1 mutants with alterations of Lys153 and Arg154 revealed that the phospho-Ser/Thr binding interface of hMOB1 is important for MST1/2 as well as LATS1/2 binding [6]. Consequently, Lys153 and Arg154 of hMOB1 are very likely to represent central residues in a more general phospho-Ser/Thr binding domain as already proposed by Rock et al. [70]. Notably, Lys153 and Arg154 of hMOB1 can also contribute to Praja2 binding in the context of hMOB1 proteolysis [86].

In addition, the interactions of hMOB1 with NDR/LATS kinases can be modulated by Tyr26 phosphorylation of hMOB1 [80] and hMOB2 counteracting hMOB1 binding to NDR1/2 [35,62]. Comparable to hMOB2, the LIM domain protein TRIP6 can compete with MOB1 for binding to the NTR of LATS1/2, thereby negatively interfering with LATS1/2 activation by hMOB1 [98]. Similarly, phosphatidic acid-related lipid signalling can regulate the Hippo pathway by directly interacting with LATS1/2, thus disrupting hMOB1/LATS complex formation [99]. In this context, one should note that inactivation of the PTEN lipid phosphatase can interfere with the interaction of hMOB1 and LATS1/2 in gastric cancer [100]. Moreover, the Angiomotin scaffolds can support hMOB1-mediated LATS1/2 auto-phosphorylation [101].

The best understood binding partners of hMOB2, hMOB3s and hMOB4 were already mentioned in subchapters 3 and 4. Briefly, two binding partners of hMOB2 have been established: the NDR1/2 kinases [1,35] and the RAD50 scaffold [37]. However, the physiological significance of these interactions is yet to be fully defined [38], with hMOB2 possibly also having indirect effects on LATS1/2 signalling by competing with hMOB1 binding to LATS1/2 [35,62].

hMOB3s have one established binding partner: the MST1 kinase [36]. In response to apoptotic stimuli, hMOB3s bind to MST1 through conserved positively charged residues, thereby negatively regulating apoptotic MST1 signalling in glioblastoma multiforme cells by inhibiting the MST1 cleavage-based activation process [36]. In this regard, it is crucial to note that hMOB3s do not function like hMOB1 in the Hippo pathway as proposed recently [102]. While hMOB3s and hMOB1 share similar primary sequences (Figures 1 and 3), they differ significantly in their binding partners, and hence their potential to regulate the core of the Hippo pathway. Nevertheless, one cannot rule out the possibility that hMOB3 and hMOB1 may compete for MST1/2 binding, similar as already reported for the competition of hMOB2 and hMOB1 for NDR1/2 binding [1,35].

As mentioned, hMOB4 is a core member of the STRIPAK complex, known to function as a regulator of Hippo signalling [43,63]. Recently, it was reported that hMOB4 forms a complex with MST4 [48], a distant relative of MST1/2 [46] and a known component of the STRIPAK complex [73]. It was proposed that MST4 and hMOB4 may disrupt assembly of the hMOB1/MST1 complex through alternative pairing [48]; however, this model has yet to be explored in more detail.

Taken together, we have started to understand the mechanistic importance of protein-protein interactions involving hMOBs. However, some of our current working models have yet to be consolidated by future studies, ranging from more structure biology to the testing of the significance of selected protein-protein interactions in adequate biological model systems. In particular, we need to comprehend how phosphorylation-triggered interactions of hMOBs help to coordinate the specificity and robustness of essential signalling cascades. In this context, it will be vital to understand the potential functional contribution of the abilities of hMOBs to engage with multiple ligands.

7. Cancer-Associated Cellular Functions of MOBs

7.1. Roles in Mitosis and Cell Cycle Progression

Since we are only briefly summarising here the mitotic functions of hMOB1, for a detailed overview of the cell cycle roles of hMOB1 see refs. [20,21]. Initially, a kinase-targeting RNAi screen found that the NDR1 kinase, a binding partner of hMOB1 and hMOB2 [1], is possibly involved in regulating spindle orientation [103]. Yao and colleagues subsequently discovered a new level of hMOB-mediated NDR1 regulation in mitosis [61]. More precisely, PLK1 phosphorylates NDR1 at Thr7, Thr183 and Thr407 upon mitotic entry, thereby eliciting PLK1-dependent suppression of NDR1 activity to ensure correct spindle orientation in mitosis [61]. Mechanistically, PLK1-mediated phosphorylation of NDR1 switches NDR1 binding from hMOB1 to hMOB2, thereby interfering with the binding of hMOB1 to NDR1 and subsequent NDR1 activation [61]. The involvement of hMOB1 in mitosis is further supported by the finding that hMOB1 knockdown displays a negative genetic interaction with the *PTTG1* gene [104], with *PTTG1* encoding Securin, a well-established regulator of chromosome segregation [105]. Noteworthy, hMOB1 can localise to kinetochore structures of mitotic cells [106] and contributes to cell abscission and centriole re-joining after telophase and cytokinesis [107].

Knockdown of hMOB4 can also cause mitotic spindle defects in human cells [108]. Similarly, dMOB4 depletion results in mitotic defects in fly cells [25]. However, it is currently not known whether this function is related to the MOB4-containing STRIPAK complex. In contrast to hMOB1 and hMOB4, the depletion of hMOB2 results in a G1/S cell cycle arrest [37,109]. More specifically, hMOB2 knockdown triggers a p53-dependent G1/S cell cycle arrest [37] that is discussed in more detail in subchapter 7.2. hMOB3s have no known role in mitosis.

7.2. Roles in the DNA Damage Response (DDR)

Considering the possible involvement of NDR1/2 in the DDR [38], one would predict that regulators of NDR1/2, such as hMOB1 [1], may also play a role in the DDR. Interestingly, this seems to be the case, since hMOB1 knockdown appears to be sufficient to cause spontaneous DNA double-strand break formation [110] and increased γH2AX phosphorylation in human cells [111]. The single knockdown of hMOB3A or hMOB4 can also result in increased γH2AX phosphorylation in cells [111], while hMOB3C knockdown cells seem to display an increased homologous recombination repair frequency [112]. Collectively, these reports propose that hMOB1, hMOB3 and hMOB4 are possibly linked to DNA damage signalling. However, it is yet to be established whether any of these possible DDR functions are linked to binding to kinases of the Hippo core, such as NDR/LATS, or components of the STRIPAK complex.

Regarding hMOB2 we have a somewhat clearer understanding regarding the DDR [38]. Given that a genome wide screen for novel putative DDR factors identified hMOB2 (aka HCCA2) as a potential candidate [113], we sought to understand how hMOB2 might function as part of the DDR. Subsequent experiments uncovered that loss of hMOB2 causes the accumulation of DNA damage, which activates the DDR kinases ATM and CHK2, consequently activating a p53 dependent G1/S cell cycle checkpoint in the absence of any exogenously induced DNA damage [37]. Complementary experiments showed that hMOB2 promotes cell survival and G1/S cell cycle arrest upon exposure to diverse DNA damaging agents [37]. Unexpectedly, hMOB2 seems to act in the DDR independent of NDR signalling [37], therefore, the biological significance of the hMOB2/NDR complex remains poorly understood [38]. Mechanistically, hMOB2 can interact with RAD50, facilitating the recruitment of the MRE11-RAD50-NBS1 (MRN) DNA damage sensor complex and activated ATM to DNA damaged chromatin [37]. However, this mechanistic link does not appear to be relevant in all DDR settings, suggesting that additional mechanisms should be considered [38].

Of note, other independent studies support the notion that hMOB2 represents a new player in the DDR. hMOB2 knockdown cells displayed an increased sensitivity to exogenously induced DNA-damage and a defective G2/M cell cycle checkpoint in response to ionizing radiation (IR) [113].

Furthermore, patient-derived fibroblasts with MOB2 loss-of-function exhibited increased susceptibility to exogenously induced DNA-damage [114]. hMOB2 knockdown in non-small cell lung cancer (NSCLC) cell lines is also a promising strategy to increase the susceptibility to IR in the context of radio-sensitization [115]. In addition, the search for mutator alleles which increase the rate of germline mutations has revealed the *hMob2* gene as one of the top hits [116]. Thus, current evidence collectively suggests that the involvement of MOB2 in the response to DNA damage is important to maintain genomic stability. hMOB2 can be part of the DDR by playing roles in DDR kinase signalling, cell survival and cell cycle checkpoints upon exposure to DNA damage. However, we have yet to understand how hMOB2 operates as a DDR protein on the molecular level [38]. Furthermore, we must comprehend which types of DNA damage repair mechanism(s) are linked to hMOB2 and whether this knowledge can be clinically exploited for precision medicine.

7.3. Apoptosis and Autophagy

hMOB1 has been linked to apoptotic signalling [1]. hMOB1 as a co-activator of human NDR kinases plays an important role in the formation of a MST1-NDR-hMOB1 complex, thereby supporting MST1-mediated phosphorylation of NDR1/2 upon the induction of apoptosis [91,117]. Interestingly, the apoptotic role of hMOB1 was further linked to cell fate decisions in response to autophagic stress conditions. Precisely, upon investigating the activation of NDR1 in macroautophagy, we observed that NDR1 is stimulated in a hMOB1-dependent manner upon autophagy induction [118]. Considering further that NDR1's kinase activity also plays an important role in mitophagy, the removal of damaged mitochondria by selective autophagy [119], it is likely that hMOB1 as a co-activator of NDR1/2 possibly functions in different autophagic processes. Taken together, hMOB1 seems to assist NDR1/2 to coordinate autophagic and apoptotic events.

7.4. Centrosome Biology

By functioning as a co-activator of human NDR kinases, hMOB1 can also play a role in centrosome duplication [28]. However, although hMOB1 can localise to centrosomes [106,120], we do not know whether the centrosomal pool of hMOB1 is the main supporter of centrosome duplication. Based on data from a large-scale screen hMOB1 appears to bind to DIPA [121,122], which is of potential interest, considering that DIPA was found on centrosomes [123]. Moreover, the NDR2 kinase was recently connected with ciliogenesis, a cellular process that utilises the mature mother centriole at the base of cilia extrusions [124,125]. Considering further that active (phosphorylated) NDR1/2 can localise to the mother centriole in a cell cycle dependent fashion [28], it is quite tempting to speculate that hMOB1 as a main regulator of NDR1/2 is linked to the ciliogenesis. This notion is supported by additional lines of evidence:

First, upon studying different canine models for retinal degeneration, it was found that hMOB1 expression was specifically elevated in the early retinal degeneration model [126], an autosomal recessive disorder caused by a NDR2 mutation [127]. Possibly this reflects the cellular attempt to compensate for NDR2 loss-of-function. Second, murine NDR1/2 kinases are important for regulation of proliferation of terminally differentiated cells in the retina [128]. Third, increased levels of phospho-Thr35 MOB1 were observed in a fish model for photoreceptor ciliogenesis and retinal development [129]. Fourth, MOB1 accumulates at the basal bodies and MOB1 depletion delays ciliogenesis in *Tetrahymena* [130]. Therefore, MOB1 appears to represent a Hippo component that provides a link between the Hippo pathway and ciliogenesis in the context of diseases such as retinal degeneration and possibly also polycystic kidney disease [131–139].

7.5. MOBs and RAS Onco-Proteins

Oncogenic Ras signalling occurs frequently in many human cancers. However, no effective targeted therapies are currently available to treat patients suffering from Ras-driven tumours. In this regard, a systematic screen uncovered YAP1, a key effector of Hippo signalling [17,140], as a significant component of oncogenic RAS signalling [141]. Specifically, YAP1 overexpression is sufficient to rescue cell viability in a KRAS-dependent fashion [141]. Interestingly, the same screen also identified hMOB2 and hMOB3C as possible factors in oncogenic RAS signalling [141]. hMOB2 overexpression could decrease cell viability in a KRAS-dependent manner, while hMOB3C overexpression restored cell viability without any obvious effects on MEK and PI3K signalling [141]. Another genome-wide RNAi screen identified a synthetic lethal interaction between hMOB3A loss-of-function and oncogenic KRAS [142]. Perhaps the RAS-related function of hMOB2 is linked to hMOB2/NDR complex formation, since NDR1/2 were recently linked to oncogenic RAS signalling [119,143]. Future research is warranted to probe these interesting links.

8. MOBs and Cancer

8.1. Lung cancer

hMOB1A mRNA levels are low in human lung cancer samples [144]. Consequently, Suzuki and colleagues studied inducible conditional MOB1 knockout mice in the context of lung physiology and cancer, revealing that mice with postnatally MOB1 loss-of-function did not develop spontaneous lung adenocarcinomas [145]. Unexpectedly, urethane treatment-induced lung tumour formation was decreased upon postnatal MOB1 deletion [145], suggesting that MOB1 rather functions as a cancer-promoting factor in this setting. Thus, it will be very interesting to elucidate which functional protein-protein interaction(s) involving MOB1 perform this pro-cancer role. Noteworthy, hMOB3B seems to interact with NT5C2 [121]. Given that NT5C2 has been used as a prognostic marker in lung cancer [146], it is tempting to speculate that hMOB3s might be of clinical relevance in lung cancer.

8.2. Pancreatic Cancer

Chen et al. found that miR-181c can directly and selectively repress hMOB1, MST1, LATS2, and SAV1 expression in human pancreatic cancer cells [147]. Inhibition of miR-181c increased the protein levels of hMOB1 and other Hippo components. In contrast, overexpression of miR-181c suppressed them, resulting in increased pancreatic cancer cell survival and chemoresistance [147]. Moreover, deregulated expression of the long non-coding RNA (lncRNA) UCA1 has been implicated in diverse human cancers, such as pancreatic cancer. UCA1 can interact with hMOB1, LATS1, and YAP to form shielding composites, thereby allowing YAP activation in pancreatic cancer cells [148]. Thus, deregulation of hMOB1 as part of the Hippo pathway can play a significant role in the progression of pancreatic cancer. Of note, expression levels of hMOB4 were found elevated in pancreatic cancer samples [48]. Thus, diverse hMOBs are of potential interest in clinical studies of pancreatic cancer.

8.3. Liver Cancer

The Thr12 phosphorylation levels of hMOB1 were significantly decreased in human liver cancer samples [94]. Mice with liver-specific *Mob1* null alleles are prone to develop liver cancers [149]. Given that this phenotype is the most severe among mutant mice lacking a Hippo signalling component in the liver [150], it is fair to conclude that MOB1 constitutes a critical hub of Hippo signalling in livers and possibly also other tissues. In this regard, it is noteworthy that increased expression of hMOB2 mRNA was associated with hepatocellular carcinoma development and progression [151]. However, due to the lack of MOB2 loss-of-function animal models it is currently unclear whether deregulated MOB2 expression can drive liver cancer development.

Given that co-infections with hepatitis delta virus (HDV) and hepatitis B virus (HBV) accelerate the development of hepatocellular carcinoma (HCC) [152], one should also note that hMOB1 can associate with CCDC85B (aka DIPA) [121,122], which is known to play a role in HDV replication [153]. hMOB1 also interacts with the hepatitis C virus (HCV) non-structural protein NS5A, although no relationship between the hMOB1A-NS5A interaction and HCV replication has been observed [154].

Nevertheless, the hMOB1-binding partners LATS1/2 represent host kinases that can phosphorylate NS5A at a highly conserved residue required for optimal HCV replication [155].

8.4. Haematological Malignancies

By studying the effects of 3-deazaneplanocin A (DZNep), a histone methyltransferase EZH2 inhibitor, on human T-cell acute lymphoid leukaemia (T-ALL) cells, Shen et al. found that DZNep treatment reduced histone methylation in hMOB1 promoters, resulting in an upregulation of hMOB1 expression associated with inhibited growth of T-ALL cells [156]. Thus, MOB1 levels may represent an indicative marker for DZNep treatments of T-ALL patients.

As mentioned, hMOB3 may interact with NT5C2 [121]. Given that NT5C2 expression levels can serve as a prognostic marker in haematological malignancies [157,158] and NT5C2 mutations were linked to ALL [159], it is therefore tempting to deduce that a link between hMOB3s and NT5C2 might be of clinical significance with regard to haematological cancers. In this regard, one should further note that altered expression levels of hMOB3A and hMOB3B were observed in mantle cell lymphoma [160,161]. However, animal studies that functionally link the deregulation of MOB3 with haematological malignancies are missing to date.

8.5. Breast and Ovarian Cancers

Kim et al. found that hMOB1 knockdown decreased the viability of breast cancer cell lines [162], suggesting that hMOB1 is supporting the survival of breast cancer cells. Changes in the methylation of the hMOB1 promoter were observed in triple negative breast cancer cells [163]. Likewise, a systematic study across cancer cell lines revealed that MOB1 knockdown results in decreased viability of ovarian cancer cells [164], suggesting that hMOB1 is supporting the survival of ovarian cancer cells. Of note, single knockdown of hMOB3A, hMOB3B or hMOB3C also decreased the viability of breast cancer cells [162].

8.6. Colon Cancer

The mRNA levels of hMOB1A are low in human colorectal cancer samples [165]. Recently, a study of mice with intestinal epithelial cell-specific depletion of MOB1 uncovered a link between MOB1 and intestinal homeostasis through the regulation of Wnt signalling [166]. Considering that the disruption of intestinal homeostasis by the deregulation of other Hippo components can result in colon cancer [167], it is quite likely that loss of MOB1 as a central component of the Hippo pathway is also associated with colon cancer development and/or progression. Of note, single knockdown of hMOB3A, hMOB3B or hMOB3C in human colon cancer cells displays a negative genetic interaction with the PTEN phosphatase [104].

8.7. Prostate Cancer

A single nucleotide polymorphisms (SNP) in the *hMob2* gene has been revealed to be a prostate cancer susceptibility loci [168]. The mRNA levels of hMOB3B were significantly decreased in prostate cancer, suggesting that hMOB3B may act as a tumour suppressor in prostate cancer [169]. Moreover, the hMOB3B promoter region is also hypermethylated in a significant number of prostate cancer samples [170]. Thus, loss of hMOB3B might promote prostate cancer development and/or progression.

8.8. Glioblastoma

On the one hand, it was found that decreased hMOB1 levels help to sustain glioblastoma growth [86]. On the other hand, the protein levels of hMOB3s are significantly upregulated in samples derived from patients suffering from glioblastoma multiforme [36]. Therefore, it appears that hMOB1 and hMOB3s may be considered in clinical assessments of glioblastoma samples.

8.9. Human Cancer Cell Lines

Based on genome-scale loss-of-functions screens using a broad spectrum of diverse human cancer cell lines, one is tempted to conclude that the single knockdown or knockout of hMOB1A, hMOB1B, hMOB3A, hMOB3B, hMOB3C or hMOB4 can decrease the survival of some cancer cells [171–174]. However, these findings have yet to be validated, and even more importantly, are most likely limited in their nature, since possible compensatory redundancies in the MOB1 and MOB3 groups were not examined. Of note, the status of hMOB3B together with a few other markers can serve as predictor of acquired and intrinsic resistance to epidermal growth factor receptor tyrosine kinase inhibitors, including gefitinib, erlotinib, and afatinib [175].

9. MOBs and Other Diseases

9.1. Neurobiology

MOB2 loss-of-function impairs the correct neuronal positioning within the developing cortex, thereby associating MOB2 inactivation with a disorder such as periventricular nodular heterotopia [114]. Moreover, MOB2 plays a part in PKA-mediated signalling in astrocytes [176]. Similarly, MOB2 plays a neurological function in invertebrates, since dMOB2 expression is necessary and sufficient to regulate the growth of larval neuromuscular junctions in flies [23].

Chromatin immunoprecipitation followed by deep sequencing identified the hMOB3A gene as a target for the NRF1 (Nuclear respiratory factor 1), a transcription factor linked to neurodegenerative diseases [177]. Recombinant hMOB3A and hMOB3B can bind to β-amyloid, suggesting that hMOB3s are possibly linked to Alzheimer disease and other atrophies [178]. Interestingly, the same study found that hMOB1 can also associate with β-amyloid [178].

hMOB4 is mainly expressed in the central and peripheral nervous system [179]. Striatin-1, a core component of the STRIPAK complex, appears to require hMOB4 binding to fulfil its neurological function [180]. This notion is supported by the observation that dMOB4 also plays a role in the regulation of axonal transport, neurite elongation, and synapse formation in flies [24].

Taken together, each member of the MOB family seems to have at least one neurological function, with the neurological roles of MOB2 and MOB4 being the best understood.

9.2. Immunity and Virology

Quite likely hMOB1, based on its associations with NDR/LATS and MST1/2 kinases [1] is linked to immunological functions as summarised in ref. [57]. In addition, it was reported that the conserved *Legionella pneumophila* effector kinase LegK7 can phosphorylate MOB1 and thereby mimic MST1/2 activation to shut down the Hippo effectors YAP/TAZ, resulting in the promotion of bacterial growth and infection [181]. In this regard, the other hMOBs are possibly also associated with immune signalling in health and disease. More precisely, a SNP in the *hMOB2* gene is associated with altered cytokine levels, in particular interleukin 9 (IL-9) levels [182]. SNPs in the *hMOB3B* gene are associated with rheumatoid arthritis [183], osteoarthritis [184] and asthma [185]. Moreover, a genome-wide RNAi screen identified hMOB4 as a possible regulator of cytokine secretion, since IL-8 levels were decreased upon hMOB4 knockdown [186].

9.3. Diverse Disease Spectrum

In addition to links to neurological and immune diseases, the deregulation of hMOBs has also been associated with other sicknesses as briefly summarised below. MOB1A/B-deficient mouse epidermis cannot be engrafted successfully onto donor mice, thereby influencing skin engraftment efficiency [187], suggesting a crucial role of MOB1 in the skin. Mice with chondrocyte-specific MOB1A/B-deficiency display impaired chondrocyte homeostasis, leading to chondrodysplasia, thereby linking MOB1 loss-of-function with a hereditary skeletal disorder [188]. Furthermore, an autosomal SNP of *hMob2* is associated with type 2 diabetes [189,190]. Last but not least, it is notable that the dysregulation of the hMOB4-containing STRIPAK complex correlates with a broad range of human diseases including, diabetes, autism, cardiac disease and cerebral cavernous malformation (CCM) in addition to cancer [43,44]. However, it remains to be determined the degree to which hMOB4 is essential to support fundamental cellular processes such as cell proliferation, growth, differentiation, migration and apoptosis.

10. Conclusions and Future Outlook

In summary, yeast cells express two different MOBs that act in independent and non-interchangeable complexes formed with members of the conserved NDR/LATS kinase family. In contrast, in multicellular organisms such as flies and humans, MOBs do not exclusively bind to a unique NDR/LATS kinase. Fly cells express at least four different dMOBs, while the human genome encodes at least seven different *hMOB* genes. hMOBs have been implicated in different crucial intracellular functions through forming complexes with diverse binding partners (Figure 4; see also subchapter 6). By associating with members of the NDR/LATS kinase family, hMOB1 has central roles in Hippo signalling (Figure 4, left panel). Importantly, hMOB1 also binds directly to MST1/2 as part of the Hippo core cassette (Figure 4, left panel). By forming a complex with the MRN DNA damage sensor complex, hMOB2 has been linked to DNA damage signalling (Figure 4, middle left panel). Similar to hMOB1, hMOB3s can associate with MST1 in the context of apoptotic signalling, but unlike hMOB1, hMOB3s do not interact with NDR/LATS kinases (Figure 4, middle right panel). Last but not least, hMOB4 has been documented as a core member of the STRIPAK complex (Figure 4, right panel).

Although the four branches of MOB signalling can be depicted as four separate signal transduction cascades, one should note that it is quite likely that they are interconnected (Figure 4, see dashed grey lines). For example, it is known that hMOB2 can oppose certain functions of the hMOB1/NDR complex by competing with hMOB1 for NDR1/2 binding. Hence, it could well be that hMOB2 is interlinked with the Hippo pathway through interfering with hMOB1-regulated processes. Another example is the possible connection of hMOB3s with the Hippo pathway. Perhaps hMOB3s can compete with hMOB1 for MST1/2 binding and thereby meddle with hMOB1-regulated Hippo signalling. Furthermore, hMOB4 as part of the STRIPAK complexes may help to connect the GCKIII upstream kinases with NDR1/2 as members of the Hippo pathway. As is apparent from these examples, it is conceivable that hMOB2, hMOB3s and hMOB4 are interconnected with hMOB1-regulated Hippo signalling to some degree (Figure 4, see dashed grey lines). Certainly, future research is warranted to address these predicted crosstalks between different hMOB-containing complexes.

Some characteristics of MOB proteins are conserved from yeast to man, but MOB signalling is surely more complex in multicellular organisms. Therefore, one future challenge is to fully comprehend and appreciate this complexity. Despite great progress in understanding how MOBs can help to regulate their binding partners, much more remains to be accomplished in this regard. For instance, more research is needed to fully understand how hMOBs are fine-tuning the activities of NDR/LATS kinases, with particular emphasis on delineating the in vivo implications of MST1/2-mediated phosphorylation of hMOB1. In this context, we also must understand which protein-protein interactions of hMOBs are essential on the cellular and organismal levels. For example, multiple binding partners of hMOB1 have been reported, but we have only begun to appreciate the physiological importance of a fraction of these hMOB1-focused complexes. The molecular and biological understanding of the significance of hMOB2/NDR, hMOB2/MRN, hMOB3/MST1 and hMOB4/STRIPAK complexes has also remained a pressing question that is yet to be dissected in much detail. Possibly, forthcoming studies on MOBs from organisms such as fungi, parasites, bacteria and plants [1,181,191,192] will help to provide breakthroughs in this regard; studies of yeast MOBs have initially led the way [1,5,8].

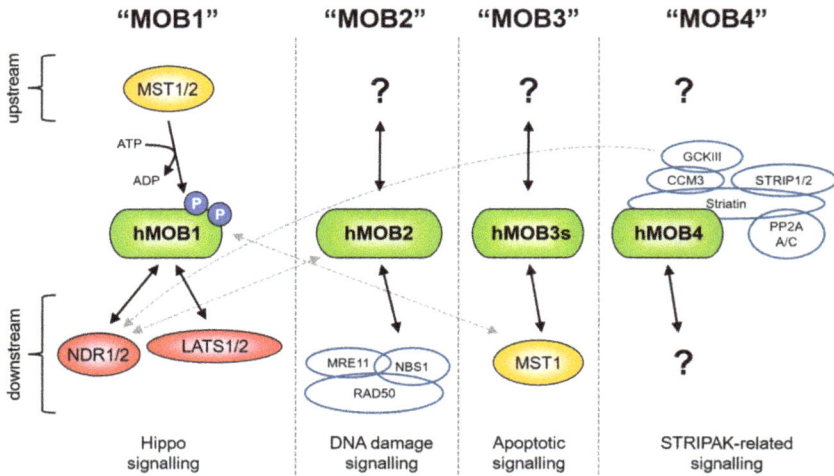

Figure 4. The four branches of MOB control. The "MOB1 branch": In human cells, hMOB1 phosphorylated by MST1/2 promotes the activation of NDR1/2 and LATS1/2 in the context of Hippo signalling and the regulation of other important cellular functions [1,55,57]. The "MOB2 branch": hMOB2 can interact with the MRN (Mre11-Rad50-Nbs1) DNA damage sensor complex, thereby supporting DNA damage and consequently cell cycle signalling [37,38]. The "MOB3 branch": hMOB3s form a complex with MST1, in so doing regulating apoptotic signalling [39]. The "MOB4 branch": hMOB4 as part of the STRIPAK (Striatin-interacting phosphatase and kinase) complex is likely to support the diverse cellular roles of the STRIPAK complex [43,44]. Notably, it is possible that these MOB branches do not signal in isolation, but rather may be interlinked as highlighted by dashed grey lines. For example, hMOB2 may connect with the "MOB1 branch" by competing with hMOB1 for NDR1/2 binding [35,62], the hMOB3s may link with the "MOB1 branch" through competing with hMOB1 for MST1/2 binding [6,39], or hMOB4 as part of the GCKIII-containing STRIPAK complex may connect with the "MOB1 branch" through GCKIII-mediated phosphorylation of NDR1/2 (see subchapter 4). Noteworthy, Hippo and STRIPAK signalling have already been found to be interconnected (see subchapter 4), although, to our knowledge, the specific roles of hMOB4 have not been defined. hMOBs are in green. MST1/2 and NDR/LATS kinases are in yellow and red, respectively. Established interactions are highlighted by black lines, while putative interconnections between "MOB branches" are shown by dashed grey lines. Phosphorylations are indicated by "P" in blue.

Taken together, future research is needed to improve the structural and biochemical understanding of hMOB-regulated signalling in order to fully elucidate the molecular mechanisms underlying human diseases, such as cancer, that are linked to the deregulation of hMOBs. In this regard, research into the cellular functions of MOBs is also timely and relevant. The biological roles of MOBs must likewise be addressed in appropriate animal models, such as transgenic mice. To date, only MOB1 knock-out animals have been reported. To our knowledge, transgenic mice allowing for studies of the physiological functions of MOB2, MOB3s and MOB4 are currently not available. Hopefully studies of disease-relevant cell biological roles of MOBs will soon be complemented by appropriate loss-of-function animal models. Many research avenues must be pursued to adequately describe the complexity of MOB signalling in multicellular organisms. Undoubtedly, exciting new discoveries related to MOBs are to be expected over the coming years.

Acknowledgments: We thank Y. Kulaberoglu and J. Lisztwan for their critical input on this manuscript. This research was funded by the Wellcome Trust grant number 090090/Z/09/Z, BBSRC grant number BB/I021248/1, Worldwide Cancer Research (formerly AICR) grant number 11-0634, UCL Cancer Research UK Centre funding, and the National Institute for Health Research University College London Hospitals Biomedical Research Centre. R.G. is supported by the Research Council of the Republic of Turkey (TUBITAK grant 119S007).

Conflicts of Interest: The authors declare no conflict of interest.

References

1. Hergovich, A. MOB control: Reviewing a conserved family of kinase regulators. *Cell Signal.* **2011**, *23*, 1433–1440. [CrossRef] [PubMed]

2. Chow, A.; Hao, Y.; Yang, X. Molecular characterization of human homologs of yeast MOB1. *Int. J. Cancer* **2010**, *126*, 2079–2089. [CrossRef] [PubMed]

3. Stavridi, E.S.; Harris, K.G.; Huyen, Y.; Bothos, J.; Verwoerd, P.M.; Stayrook, S.E.; Pavletich, N.P.; Jeffrey, P.D.; Luca, F.C. Crystal structure of a human Mob1 protein: Toward understanding Mob-regulated cell cycle pathways. *Structure* **2003**, *11*, 1163–1170. [CrossRef]

4. He, Y.; Emoto, K.; Fang, X.; Ren, N.; Tian, X.; Jan, Y.N.; Adler, P.N. Drosophila Mob family proteins interact with the related tricornered (Trc) and warts (Wts) kinases. *Mol. Biol. Cell* **2005**, *16*, 4139–4152. [CrossRef] [PubMed]

5. Luca, F.C.; Winey, M. MOB1, an essential yeast gene required for completion of mitosis and maintenance of ploidy. *Mol. Biol. Cell* **1998**, *9*, 29–46. [CrossRef] [PubMed]

6. Kulaberoglu, Y.; Lin, K.; Holder, M.; Gai, Z.; Gomez, M.; Assefa Shifa, B.; Mavis, M.; Hoa, L.; Sharif, A.A.D.; Lujan, C.; et al. Stable MOB1 interaction with Hippo/MST is not essential for development and tissue growth control. *Nat. Commun.* **2017**, *8*, 695. [CrossRef] [PubMed]

7. Lai, Z.C.; Wei, X.; Shimizu, T.; Ramos, E.; Rohrbaugh, M.; Nikolaidis, N.; Ho, L.L.; Li, Y. Control of cell proliferation and apoptosis by mob as tumor suppressor, mats. *Cell* **2005**, *120*, 675–685. [CrossRef]

8. Bardin, A.J.; Amon, A. Men and sin: What's the difference? *Nat. Rev. Mol. Cell Biol.* **2001**, *2*, 815–826. [CrossRef]

9. Meitinger, F.; Palani, S.; Pereira, G. The power of MEN in cytokinesis. *Cell Cycle* **2012**, *11*, 219–228. [CrossRef]

10. Baro, B.; Queralt, E.; Monje-Casas, F. Regulation of Mitotic Exit in Saccharomyces cerevisiae. *Methods Mol. Biol.* **2017**, *1505*, 3–17. [CrossRef]

11. Simanis, V. Pombe's thirteen–control of fission yeast cell division by the septation initiation network. *J. Cell Sci.* **2015**, *128*, 1465–1474. [CrossRef] [PubMed]

12. Hotz, M.; Barral, Y. The Mitotic Exit Network: New turns on old pathways. *Trends Cell Biol.* **2014**, *24*, 145–152. [CrossRef]

13. Hergovich, A.; Stegert, M.R.; Schmitz, D.; Hemmings, B.A. NDR kinases regulate essential cell processes from yeast to humans. *Nat. Rev. Mol. Cell Biol.* **2006**, *7*, 253–264. [CrossRef] [PubMed]

14. Saputo, S.; Chabrier-Rosello, Y.; Luca, F.C.; Kumar, A.; Krysan, D.J. The RAM network in pathogenic fungi. *Eukaryot Cell* **2012**, *11*, 708–717. [CrossRef] [PubMed]

15. Staley, B.K.; Irvine, K.D. Hippo signaling in Drosophila: Recent advances and insights. *Dev. Dyn.* **2012**, *241*, 3–15. [CrossRef] [PubMed]

16. Sun, S.; Irvine, K.D. Cellular Organization and Cytoskeletal Regulation of the Hippo Signaling Network. *Trends Cell Biol.* **2016**, *26*, 694–704. [CrossRef] [PubMed]

17. Misra, J.R.; Irvine, K.D. The Hippo Signaling Network and Its Biological Functions. *Annu. Rev. Genet.* **2018**, *52*, 65–87. [CrossRef] [PubMed]

18. Wei, X.; Shimizu, T.; Lai, Z.C. Mob as tumor suppressor is activated by Hippo kinase for growth inhibition in Drosophila. *EMBO J.* **2007**, *26*, 1772–1781. [CrossRef]

19. Emoto, K.; Parrish, J.Z.; Jan, L.Y.; Jan, Y.N. The tumour suppressor Hippo acts with the NDR kinases in dendritic tiling and maintenance. *Nature* **2006**, *443*, 210–213. [CrossRef]

20. Hergovich, A. Hippo Signaling in Mitosis: An Updated View in Light of the MEN Pathway. *Methods Mol. Biol.* **2017**, *1505*, 265–277. [CrossRef]

21. Hergovich, A.; Hemmings, B.A. Hippo signalling in the G2/M cell cycle phase: Lessons learned from the yeast MEN and SIN pathways. *Semin. Cell Dev. Biol.* **2012**, *23*, 794–802. [CrossRef] [PubMed]

22. Liu, L.Y.; Lin, C.H.; Fan, S.S. Function of Drosophila mob2 in photoreceptor morphogenesis. *Cell Tissue Res.* **2009**, *338*, 377–389. [CrossRef] [PubMed]
23. Campbell, M.; Ganetzky, B. Identification of Mob2, a novel regulator of larval neuromuscular junction morphology, in natural populations of Drosophila melanogaster. *Genetics* **2013**, *195*, 915–926. [CrossRef] [PubMed]
24. Schulte, J.; Sepp, K.J.; Jorquera, R.A.; Wu, C.; Song, Y.; Hong, P.; Littleton, J.T. DMob4/Phocein regulates synapse formation, axonal transport, and microtubule organization. *J. Neurosci.* **2010**, *30*, 5189–5203. [CrossRef] [PubMed]
25. Trammell, M.A.; Mahoney, N.M.; Agard, D.A.; Vale, R.D. Mob4 plays a role in spindle focusing in Drosophila S2 cells. *J. Cell Sci.* **2008**, *121*, 1284–1292. [CrossRef] [PubMed]
26. Devroe, E.; Erdjument-Bromage, H.; Tempst, P.; Silver, P.A. Human Mob proteins regulate the NDR1 and NDR2 serine-threonine kinases. *J. Biol. Chem.* **2004**, *279*, 24444–24451. [CrossRef] [PubMed]
27. Hergovich, A.; Bichsel, S.J.; Hemmings, B.A. Human NDR kinases are rapidly activated by MOB proteins through recruitment to the plasma membrane and phosphorylation. *Mol. Cell Biol.* **2005**, *25*, 8259–8272. [CrossRef]
28. Hergovich, A.; Kohler, R.S.; Schmitz, D.; Vichalkovski, A.; Cornils, H.; Hemmings, B.A. The MST1 and hMOB1 tumor suppressors control human centrosome duplication by regulating NDR kinase phosphorylation. *Curr. Biol.* **2009**, *19*, 1692–1702. [CrossRef]
29. Hergovich, A.; Schmitz, D.; Hemmings, B.A. The human tumour suppressor LATS1 is activated by human MOB1 at the membrane. *Biochem. Biophys. Res. Commun.* **2006**, *345*, 50–58. [CrossRef]
30. Praskova, M.; Xia, F.; Avruch, J. MOBKL1A/MOBKL1B phosphorylation by MST1 and MST2 inhibits cell proliferation. *Curr. Biol.* **2008**, *18*, 311–321. [CrossRef]
31. Harvey, K.; Tapon, N. The Salvador-Warts-Hippo pathway–an emerging tumour-suppressor network. *Nat. Rev. Cancer* **2007**, *7*, 182–191. [CrossRef] [PubMed]
32. Pan, D. Hippo signaling in organ size control. *Genes Dev.* **2007**, *21*, 886–897. [CrossRef] [PubMed]
33. Saucedo, L.J.; Edgar, B.A. Filling out the Hippo pathway. *Nat. Rev. Mol. Cell Biol.* **2007**, *8*, 613–621. [CrossRef] [PubMed]
34. Thompson, B.J.; Sahai, E. MST kinases in development and disease. *J. Cell Biol.* **2015**, *210*, 871–882. [CrossRef] [PubMed]
35. Kohler, R.S.; Schmitz, D.; Cornils, H.; Hemmings, B.A.; Hergovich, A. Differential NDR/LATS interactions with the human MOB family reveal a negative role for human MOB2 in the regulation of human NDR kinases. *Mol. Cell Biol.* **2010**, *30*, 4507–4520. [CrossRef] [PubMed]
36. Tang, F.; Gill, J.; Ficht, X.; Barthlott, T.; Cornils, H.; Schmitz-Rohmer, D.; Hynx, D.; Zhou, D.; Zhang, L.; Xue, G.; et al. The kinases NDR1/2 act downstream of the Hippo homolog MST1 to mediate both egress of thymocytes from the thymus and lymphocyte motility. *Sci. Signal.* **2015**, *8*, ra100. [CrossRef]
37. Gomez, V.; Gundogdu, R.; Gomez, M.; Hoa, L.; Panchal, N.; O'Driscoll, M.; Hergovich, A. Regulation of DNA damage responses and cell cycle progression by hMOB2. *Cell Signal.* **2015**, *27*, 326–339. [CrossRef] [PubMed]
38. Gundogdu, R.; Hergovich, A. The Possible Crosstalk of MOB2 With NDR1/2 Kinases in Cell Cycle and DNA Damage Signaling. *J. Cell Signal.* **2016**, *1*, 125. [CrossRef] [PubMed]
39. Tang, F.; Zhang, L.; Xue, G.; Hynx, D.; Wang, Y.; Cron, P.D.; Hundsrucker, C.; Hergovich, A.; Frank, S.; Hemmings, B.A.; et al. hMOB3 modulates MST1 apoptotic signaling and supports tumor growth in glioblastoma multiforme. *Cancer Res.* **2014**, *74*, 3779–3789. [CrossRef]
40. Simpson, J.C.; Wellenreuther, R.; Poustka, A.; Pepperkok, R.; Wiemann, S. Systematic subcellular localization of novel proteins identified by large-scale cDNA sequencing. *EMBO Rep.* **2000**, *1*, 287–292. [CrossRef]
41. Moreno, C.S.; Lane, W.S.; Pallas, D.C. A mammalian homolog of yeast MOB1 is both a member and a putative substrate of striatin family-protein phosphatase 2A complexes. *J. Biol. Chem.* **2001**, *276*, 24253–24260. [CrossRef] [PubMed]
42. Baillat, G.; Moqrich, A.; Castets, F.; Baude, A.; Bailly, Y.; Benmerah, A.; Monneron, A. Molecular cloning and characterization of phocein, a protein found from the Golgi complex to dendritic spines. *Mol. Biol. Cell* **2001**, *12*, 663–673. [CrossRef] [PubMed]
43. Shi, Z.; Jiao, S.; Zhou, Z. STRIPAK complexes in cell signaling and cancer. *Oncogene* **2016**, *35*, 4549–4557. [CrossRef] [PubMed]

44. Hwang, J.; Pallas, D.C. STRIPAK complexes: Structure, biological function, and involvement in human diseases. *Int. J. Biochem. Cell Biol.* **2014**, *47*, 118–148. [CrossRef] [PubMed]
45. Gordon, J.; Hwang, J.; Carrier, K.J.; Jones, C.A.; Kern, Q.L.; Moreno, C.S.; Karas, R.H.; Pallas, D.C. Protein phosphatase 2a (PP2A) binds within the oligomerization domain of striatin and regulates the phosphorylation and activation of the mammalian Ste20-Like kinase Mst3. *BMC Biochem.* **2011**, *12*, 54. [CrossRef] [PubMed]
46. Dan, I.; Watanabe, N.M.; Kusumi, A. The Ste20 group kinases as regulators of MAP kinase cascades. *Trends Cell Biol.* **2001**, *11*, 220–230. [CrossRef]
47. Goudreault, M.; D'Ambrosio, L.M.; Kean, M.J.; Mullin, M.J.; Larsen, B.G.; Sanchez, A.; Chaudhry, S.; Chen, G.I.; Sicheri, F.; Nesvizhskii, A.I.; et al. A PP2A phosphatase high density interaction network identifies a novel striatin-interacting phosphatase and kinase complex linked to the cerebral cavernous malformation 3 (CCM3) protein. *Mol. Cell Proteom.* **2009**, *8*, 157–171. [CrossRef]
48. Chen, M.; Zhang, H.; Shi, Z.; Li, Y.; Zhang, X.; Gao, Z.; Zhou, L.; Ma, J.; Xu, Q.; Guan, J.; et al. The MST4-MOB4 complex disrupts the MST1-MOB1 complex in the Hippo-YAP pathway and plays a pro-oncogenic role in pancreatic cancer. *J. Biol. Chem.* **2018**, *293*, 14455–14469. [CrossRef]
49. Johnson, R.; Halder, G. The two faces of Hippo: Targeting the Hippo pathway for regenerative medicine and cancer treatment. *Nat. Rev. Drug Discov.* **2014**, *13*, 63–79. [CrossRef]
50. Yu, F.X.; Zhao, B.; Guan, K.L. Hippo Pathway in Organ Size Control, Tissue Homeostasis, and Cancer. *Cell* **2015**, *163*, 811–828. [CrossRef]
51. Meng, Z.; Moroishi, T.; Guan, K.L. Mechanisms of Hippo pathway regulation. *Genes Dev.* **2016**, *30*, 1–17. [CrossRef] [PubMed]
52. Pan, D. The hippo signaling pathway in development and cancer. *Dev. Cell* **2010**, *19*, 491–505. [CrossRef] [PubMed]
53. Harvey, K.F.; Zhang, X.; Thomas, D.M. The Hippo pathway and human cancer. *Nat. Rev. Cancer* **2013**, *13*, 246–257. [CrossRef] [PubMed]
54. Zanconato, F.; Cordenonsi, M.; Piccolo, S. YAP/TAZ at the roots of cancer. *Cancer Cell* **2016**, *29*, 783–803. [CrossRef] [PubMed]
55. Hergovich, A. The Roles of NDR Protein Kinases in Hippo Signalling. *Genes* **2016**, *7*. [CrossRef] [PubMed]
56. Zhang, L.; Tang, F.; Terracciano, L.; Hynx, D.; Kohler, R.; Bichet, S.; Hess, D.; Cron, P.; Hemmings, B.A.; Hergovich, A.; et al. NDR functions as a physiological YAP1 kinase in the intestinal epithelium. *Curr. Biol.* **2015**, *25*, 296–305. [CrossRef] [PubMed]
57. Sharif, A.A.D.; Hergovich, A. The NDR/LATS protein kinases in immunology and cancer biology. *Semin. Cancer Biol.* **2018**, *48*, 104–114. [CrossRef]
58. Gomez, M.; Kulaberoglu, Y.; Hergovich, A. MST1/2 Kinase Assays Using Recombinant Proteins. *Methods Mol. Biol.* **2019**, *1893*, 319–331. [CrossRef]
59. Nishio, M.; Hamada, K.; Kawahara, K.; Sasaki, M.; Noguchi, F.; Chiba, S.; Mizuno, K.; Suzuki, S.O.; Dong, Y.; Tokuda, M.; et al. Cancer susceptibility and embryonic lethality in Mob1a/1b double-mutant mice. *J. Clin. Invest.* **2012**, *122*, 4505–4518. [CrossRef] [PubMed]
60. Plouffe, S.W.; Meng, Z.; Lin, K.C.; Lin, B.; Hong, A.W.; Chun, J.V.; Guan, K.L. Characterization of Hippo Pathway Components by Gene Inactivation. *Mol. Cell* **2016**, *64*, 993–1008. [CrossRef]
61. Yan, M.; Chu, L.; Qin, B.; Wang, Z.; Liu, X.; Jin, C.; Zhang, G.; Gomez, M.; Hergovich, A.; Chen, Z.; et al. Regulation of NDR1 activity by PLK1 ensures proper spindle orientation in mitosis. *Sci. Rep.* **2015**, *5*, 10449. [CrossRef] [PubMed]
62. Zhang, W.; Shen, J.; Gu, F.; Zhang, Y.; Wu, W.; Weng, J.; Liao, Y.; Deng, Z.; Yuan, Q.; Zheng, L.; et al. Monopolar spindle-one-binder protein 2 regulates the activity of large tumor suppressor/yes-associated protein to inhibit the motility of SMMC-7721 hepatocellular carcinoma cells. *Oncol. Lett.* **2018**, *15*, 5375–5383. [CrossRef] [PubMed]
63. Tang, Y.; Chen, M.; Zhou, L.; Ma, J.; Li, Y.; Zhang, H.; Shi, Z.; Xu, Q.; Zhang, X.; Gao, Z.; et al. Architecture, substructures, and dynamic assembly of STRIPAK complexes in Hippo signaling. *Cell Discov.* **2019**, *5*, 3. [CrossRef] [PubMed]
64. Ribeiro, P.S.; Josue, F.; Wepf, A.; Wehr, M.C.; Rinner, O.; Kelly, G.; Tapon, N.; Gstaiger, M. Combined functional genomic and proteomic approaches identify a PP2A complex as a negative regulator of Hippo signaling. *Mol. Cell* **2010**, *39*, 521–534. [CrossRef] [PubMed]

65. Couzens, A.L.; Knight, J.D.; Kean, M.J.; Teo, G.; Weiss, A.; Dunham, W.H.; Lin, Z.Y.; Bagshaw, R.D.; Sicheri, F.; Pawson, T.; et al. Protein interaction network of the mammalian Hippo pathway reveals mechanisms of kinase-phosphatase interactions. *Sci. Signal.* **2013**, *6*, rs15. [CrossRef] [PubMed]

66. Liu, B.; Zheng, Y.; Yin, F.; Yu, J.; Silverman, N.; Pan, D. Toll Receptor-Mediated Hippo Signaling Controls Innate Immunity in Drosophila. *Cell* **2016**, *164*, 406–419. [CrossRef] [PubMed]

67. Sakuma, C.; Saito, Y.; Umehara, T.; Kamimura, K.; Maeda, N.; Mosca, T.J.; Miura, M.; Chihara, T. The Strip-Hippo Pathway Regulates Synaptic Terminal Formation by Modulating Actin Organization at the Drosophila Neuromuscular Synapses. *Cell Rep.* **2016**, *16*, 2289–2297. [CrossRef]

68. Bae, S.J.; Ni, L.; Osinski, A.; Tomchick, D.R.; Brautigam, C.A.; Luo, X. SAV1 promotes Hippo kinase activation through antagonizing the PP2A phosphatase STRIPAK. *Elife* **2017**, *6*. [CrossRef]

69. Zheng, Y.; Liu, B.; Wang, L.; Lei, H.; Pulgar Prieto, K.D.; Pan, D. Homeostatic Control of Hpo/MST Kinase Activity through Autophosphorylation-Dependent Recruitment of the STRIPAK PP2A Phosphatase Complex. *Cell Rep.* **2017**, *21*, 3612–3623. [CrossRef]

70. Rock, J.M.; Lim, D.; Stach, L.; Ogrodowicz, R.W.; Keck, J.M.; Jones, M.H.; Wong, C.C.; Yates, J.R.; Winey, M.; Smerdon, S.J.; et al. Activation of the yeast Hippo pathway by phosphorylation-dependent assembly of signaling complexes. *Science* **2013**, *340*, 871–875. [CrossRef]

71. Kim, S.Y.; Tachioka, Y.; Mori, T.; Hakoshima, T. Structural basis for autoinhibition and its relief of MOB1 in the Hippo pathway. *Sci. Rep.* **2016**, *6*, 28488. [CrossRef] [PubMed]

72. Ni, L.; Zheng, Y.; Hara, M.; Pan, D.; Luo, X. Structural basis for Mob1-dependent activation of the core Mst-Lats kinase cascade in Hippo signaling. *Genes Dev.* **2015**, *29*, 1416–1431. [CrossRef] [PubMed]

73. Sugden, P.H.; McGuffin, L.J.; Clerk, A. SOcK, MiSTs, MASK and STicKs: The GCKIII (germinal centre kinase III) kinases and their heterologous protein-protein interactions. *Biochem. J.* **2013**, *454*, 13–30. [CrossRef] [PubMed]

74. Poon, C.L.C.; Liu, W.; Song, Y.; Gomez, M.; Kulaberoglu, Y.; Zhang, X.; Xu, W.; Veraksa, A.; Hergovich, A.; Ghabrial, A.; et al. A Hippo-like Signaling Pathway Controls Tracheal Morphogenesis in Drosophila melanogaster. *Dev. Cell* **2018**, *47*, 564–575. [CrossRef] [PubMed]

75. Hyodo, T.; Ito, S.; Hasegawa, H.; Asano, E.; Maeda, M.; Urano, T.; Takahashi, M.; Hamaguchi, M.; Senga, T. Misshapen-like kinase 1 (MINK1) is a novel component of striatin-interacting phosphatase and kinase (STRIPAK) and is required for the completion of cytokinesis. *J. Biol. Chem.* **2012**, *287*, 25019–25029. [CrossRef] [PubMed]

76. Manning, G.; Whyte, D.B.; Martinez, R.; Hunter, T.; Sudarsanam, S. The protein kinase complement of the human genome. *Science* **2002**, *298*, 1912–1934. [CrossRef] [PubMed]

77. Mah, A.S.; Jang, J.; Deshaies, R.J. Protein kinase Cdc15 activates the Dbf2-Mob1 kinase complex. *Proc. Natl. Acad. Sci. USA* **2001**, *98*, 7325–7330. [CrossRef] [PubMed]

78. Xiong, S.; Couzens, A.L.; Kean, M.J.; Mao, D.Y.; Guettler, S.; Kurinov, I.; Gingras, A.C.; Sicheri, F. Regulation of Protein Interactions by Mps One Binder (MOB1) Phosphorylation. *Mol. Cell Proteom.* **2017**, *16*, 1111–1125. [CrossRef]

79. Hornbeck, P.V.; Zhang, B.; Murray, B.; Kornhauser, J.M.; Latham, V.; Skrzypek, E. PhosphoSitePlus, 2014: Mutations, PTMs and recalibrations. *Nucleic Acids Res.* **2015**, *43*, D512–D520. [CrossRef]

80. Feng, X.; Arang, N.; Rigiracciolo, D.C.; Lee, J.S.; Yeerna, H.; Wang, Z.; Lubrano, S.; Kishore, A.; Pachter, J.A.; Konig, G.M.; et al. A Platform of Synthetic Lethal Gene Interaction Networks Reveals that the GNAQ Uveal Melanoma Oncogene Controls the Hippo Pathway through FAK. *Cancer Cell* **2019**, *35*, 457–472. [CrossRef]

81. Wong, Y.H.; Lee, T.Y.; Liang, H.K.; Huang, C.M.; Wang, T.Y.; Yang, Y.H.; Chu, C.H.; Huang, H.D.; Ko, M.T.; Hwang, J.K. KinasePhos 2.0: A web server for identifying protein kinase-specific phosphorylation sites based on sequences and coupling patterns. *Nucleic Acids Res.* **2007**, *35*, W588–W594. [CrossRef] [PubMed]

82. Song, Z.; Han, X.; Zou, H.; Zhang, B.; Ding, Y.; Xu, X.; Zeng, J.; Liu, J.; Gong, A. PTEN-GSK3beta-MOB1 axis controls neurite outgrowth in vitro and in vivo. *Cell Mol. Life Sci.* **2018**, *75*, 4445–4464. [CrossRef] [PubMed]

83. Wagner, S.A.; Beli, P.; Weinert, B.T.; Scholz, C.; Kelstrup, C.D.; Young, C.; Nielsen, M.L.; Olsen, J.V.; Brakebusch, C.; Choudhary, C. Proteomic analyses reveal divergent ubiquitylation site patterns in murine tissues. *Mol. Cell Proteom.* **2012**, *11*, 1578–1585. [CrossRef] [PubMed]

84. Akimov, V.; Barrio-Hernandez, I.; Hansen, S.V.F.; Hallenborg, P.; Pedersen, A.K.; Bekker-Jensen, D.B.; Puglia, M.; Christensen, S.D.K.; Vanselow, J.T.; Nielsen, M.M.; et al. UbiSite approach for comprehensive mapping of lysine and N-terminal ubiquitination sites. *Nat. Struct. Mol. Biol.* **2018**, *25*, 631–640. [CrossRef] [PubMed]

85. Mou, F.; Praskova, M.; Xia, F.; Van Buren, D.; Hock, H.; Avruch, J.; Zhou, D. The Mst1 and Mst2 kinases control activation of rho family GTPases and thymic egress of mature thymocytes. *J. Exp. Med.* **2012**, *209*, 741–759. [CrossRef]

86. Lignitto, L.; Arcella, A.; Sepe, M.; Rinaldi, L.; Delle Donne, R.; Gallo, A.; Stefan, E.; Bachmann, V.A.; Oliva, M.A.; Tiziana Storlazzi, C.; et al. Proteolysis of MOB1 by the ubiquitin ligase praja2 attenuates Hippo signalling and supports glioblastoma growth. *Nat. Commun.* **2013**, *4*, 1822. [CrossRef] [PubMed]

87. Hoa, L.; Kulaberoglu, Y.; Gundogdu, R.; Cook, D.; Mavis, M.; Gomez, M.; Gomez, V.; Hergovich, A. The characterisation of LATS2 kinase regulation in Hippo-YAP signalling. *Cell Signal.* **2016**, *28*, 488–497. [CrossRef] [PubMed]

88. Cook, D.; Hoa, L.Y.; Gomez, V.; Gomez, M.; Hergovich, A. Constitutively active NDR1-PIF kinase functions independent of MST1 and hMOB1 signalling. *Cell Signal.* **2014**, *26*, 1657–1667. [CrossRef]

89. Xiong, S.; Lorenzen, K.; Couzens, A.L.; Templeton, C.M.; Rajendran, D.; Mao, D.Y.L.; Juang, Y.C.; Chiovitti, D.; Kurinov, I.; Guettler, S.; et al. Structural Basis for Auto-Inhibition of the NDR1 Kinase Domain by an Atypically Long Activation Segment. *Structure* **2018**, *26*, 1101–1115. [CrossRef]

90. Vrabioiu, A.M.; Struhl, G. Fat/Dachsous Signaling Promotes Drosophila Wing Growth by Regulating the Conformational State of the NDR Kinase Warts. *Dev. Cell* **2015**, *35*, 737–749. [CrossRef]

91. Vichalkovski, A.; Gresko, E.; Cornils, H.; Hergovich, A.; Schmitz, D.; Hemmings, B.A. NDR kinase is activated by RASSF1A/MST1 in response to Fas receptor stimulation and promotes apoptosis. *Curr. Biol.* **2008**, *18*, 1889–1895. [CrossRef] [PubMed]

92. Couzens, A.L.; Xiong, S.; Knight, J.D.R.; Mao, D.Y.; Guettler, S.; Picaud, S.; Kurinov, I.; Filippakopoulos, P.; Sicheri, F.; Gingras, A.C. MOB1 Mediated Phospho-recognition in the Core Mammalian Hippo Pathway. *Mol. Cell Proteom.* **2017**, *16*, 1098–1110. [CrossRef] [PubMed]

93. Manning, S.A.; Harvey, K.F. Warts Opens Up for Activation. *Dev. Cell* **2015**, *35*, 666–668. [CrossRef] [PubMed]

94. Zhou, D.; Conrad, C.; Xia, F.; Park, J.S.; Payer, B.; Yin, Y.; Lauwers, G.Y.; Thasler, W.; Lee, J.T.; Avruch, J.; et al. Mst1 and Mst2 maintain hepatocyte quiescence and suppress hepatocellular carcinoma development through inactivation of the Yap1 oncogene. *Cancer Cell* **2009**, *16*, 425–438. [CrossRef] [PubMed]

95. Gogl, G.; Schneider, K.D.; Yeh, B.J.; Alam, N.; Nguyen Ba, A.N.; Moses, A.M.; Hetenyi, C.; Remenyi, A.; Weiss, E.L. The Structure of an NDR/LATS Kinase-Mob Complex Reveals a Novel Kinase-Coactivator System and Substrate Docking Mechanism. *PLoS Biol.* **2015**, *13*, e1002146. [CrossRef] [PubMed]

96. Bichsel, S.J.; Tamaskovic, R.; Stegert, M.R.; Hemmings, B.A. Mechanism of activation of NDR (nuclear Dbf2-related) protein kinase by the hMOB1 protein. *J. Biol. Chem.* **2004**, *279*, 35228–35235. [CrossRef] [PubMed]

97. Stegert, M.R.; Tamaskovic, R.; Bichsel, S.J.; Hergovich, A.; Hemmings, B.A. Regulation of NDR2 protein kinase by multi-site phosphorylation and the S100B calcium-binding protein. *J. Biol. Chem.* **2004**, *279*, 23806–23812. [CrossRef] [PubMed]

98. Dutta, S.; Mana-Capelli, S.; Paramasivam, M.; Dasgupta, I.; Cirka, H.; Billiar, K.; McCollum, D. TRIP6 inhibits Hippo signaling in response to tension at adherens junctions. *EMBO Rep.* **2018**, *19*, 337–350. [CrossRef] [PubMed]

99. Han, H.; Qi, R.; Zhou, J.J.; Ta, A.P.; Yang, B.; Nakaoka, H.J.; Seo, G.; Guan, K.L.; Luo, R.; Wang, W. Regulation of the Hippo Pathway by Phosphatidic Acid-Mediated Lipid-Protein Interaction. *Mol. Cell* **2018**, *72*, 328–340. [CrossRef]

100. Xu, W.; Yang, Z.; Xie, C.; Zhu, Y.; Shu, X.; Zhang, Z.; Li, N.; Chai, N.; Zhang, S.; Wu, K.; et al. PTEN lipid phosphatase inactivation links the hippo and PI3K/Akt pathways to induce gastric tumorigenesis. *J. Exp. Clin. Cancer Res.* **2018**, *37*, 198. [CrossRef]

101. Mana-Capelli, S.; McCollum, D. Angiomotins stimulate LATS kinase autophosphorylation and act as scaffolds that promote Hippo signaling. *J. Biol. Chem.* **2018**, *293*, 18230–18241. [CrossRef] [PubMed]

102. Joung, J.; Engreitz, J.M.; Konermann, S.; Abudayyeh, O.O.; Verdine, V.K.; Aguet, F.; Gootenberg, J.S.; Sanjana, N.E.; Wright, J.B.; Fulco, C.P.; et al. Genome-scale activation screen identifies a lncRNA locus regulating a gene neighbourhood. *Nature* **2017**, *548*, 343–346. [CrossRef] [PubMed]

103. Matsumura, S.; Hamasaki, M.; Yamamoto, T.; Ebisuya, M.; Sato, M.; Nishida, E.; Toyoshima, F. ABL1 regulates spindle orientation in adherent cells and mammalian skin. *Nat. Commun.* **2012**, *3*, 626. [CrossRef] [PubMed]

104. Vizeacoumar, F.J.; Arnold, R.; Vizeacoumar, F.S.; Chandrashekhar, M.; Buzina, A.; Young, J.T.; Kwan, J.H.; Sayad, A.; Mero, P.; Lawo, S.; et al. A negative genetic interaction map in isogenic cancer cell lines reveals cancer cell vulnerabilities. *Mol. Syst. Biol.* **2013**, *9*, 696. [CrossRef] [PubMed]

105. Peters, J.M.; Tedeschi, A.; Schmitz, J. The cohesin complex and its roles in chromosome biology. *Genes Dev.* **2008**, *22*, 3089–3114. [CrossRef] [PubMed]

106. Wilmeth, L.J.; Shrestha, S.; Montano, G.; Rashe, J.; Shuster, C.B. Mutual dependence of Mob1 and the chromosomal passenger complex for localization during mitosis. *Mol. Biol. Cell.* **2010**, *21*, 380–392. [CrossRef] [PubMed]

107. Florindo, C.; Perdigao, J.; Fesquet, D.; Schiebel, E.; Pines, J.; Tavares, A.A. Human Mob1 proteins are required for cytokinesis by controlling microtubule stability. *J. Cell Sci.* **2012**, *125*, 3085–3090. [CrossRef] [PubMed]

108. Rines, D.R.; Gomez-Ferreria, M.A.; Zhou, Y.; DeJesus, P.; Grob, S.; Batalov, S.; Labow, M.; Huesken, D.; Mickanin, C.; Hall, J.; et al. Whole genome functional analysis identifies novel components required for mitotic spindle integrity in human cells. *Genome Biol.* **2008**, *9*, R44. [CrossRef]

109. Leng, J.J.; Tan, H.M.; Chen, K.; Shen, W.G.; Tan, J.W. Growth-inhibitory effects of MOB2 on human hepatic carcinoma cell line SMMC-7721. *World J. Gastroenterol.* **2012**, *18*, 7285–7289. [CrossRef]

110. O'Donnell, L.; Panier, S.; Wildenhain, J.; Tkach, J.M.; Al-Hakim, A.; Landry, M.C.; Escribano-Diaz, C.; Szilard, R.K.; Young, J.T.; Munro, M.; et al. The MMS22L-TONSL complex mediates recovery from replication stress and homologous recombination. *Mol. Cell* **2010**, *40*, 619–631. [CrossRef]

111. Paulsen, R.D.; Soni, D.V.; Wollman, R.; Hahn, A.T.; Yee, M.C.; Guan, A.; Hesley, J.A.; Miller, S.C.; Cromwell, E.F.; Solow-Cordero, D.E.; et al. A genome-wide siRNA screen reveals diverse cellular processes and pathways that mediate genome stability. *Mol. Cell* **2009**, *35*, 228–239. [CrossRef] [PubMed]

112. Adamson, B.; Smogorzewska, A.; Sigoillot, F.D.; King, R.W.; Elledge, S.J. A genome-wide homologous recombination screen identifies the RNA-binding protein RBMX as a component of the DNA-damage response. *Nat. Cell Biol.* **2012**, *14*, 318–328. [CrossRef] [PubMed]

113. Cotta-Ramusino, C.; McDonald, E.R., 3rd; Hurov, K.; Sowa, M.E.; Harper, J.W.; Elledge, S.J. A DNA damage response screen identifies RHINO, a 9-1-1 and TopBP1 interacting protein required for ATR signaling. *Science* **2011**, *332*, 1313–1317. [CrossRef] [PubMed]

114. liO'Neill, A.C.; Kyrousi, C.; Einsiedler, M.; Burtscher, I.; Drukker, M.; Markie, D.M.; Kirk, E.P.; Gotz, M.; Robertson, S.P.; Cappello, S. Mob2 Insufficiency Disrupts Neuronal Migration in the Developing Cortex. *Front. Cell Neurosci.* **2018**, *12*, 57. [CrossRef] [PubMed]

115. Cron, K.R.; Zhu, K.; Kushwaha, D.S.; Hsieh, G.; Merzon, D.; Rameseder, J.; Chen, C.C.; D'Andrea, A.D.; Kozono, D. Proteasome inhibitors block DNA repair and radiosensitize non-small cell lung cancer. *PLoS ONE* **2013**, *8*, e73710. [CrossRef]

116. Seoighe, C.; Scally, A. Inference of Candidate Germline Mutator Loci in Humans from Genome-Wide Haplotype Data. *PLoS Genet.* **2017**, *13*, e1006549. [CrossRef] [PubMed]

117. Zhou, Y.; Adolfs, Y.; Pijnappel, W.W.; Fuller, S.J.; Van der Schors, R.C.; Li, K.W.; Sugden, P.H.; Smit, A.B.; Hergovich, A.; Pasterkamp, R.J. MICAL-1 is a negative regulator of MST-NDR kinase signaling and apoptosis. *Mol. Cell Biol.* **2011**, *31*, 3603–3615. [CrossRef]

118. Joffre, C.; Dupont, N.; Hoa, L.; Gomez, V.; Pardo, R.; Goncalves-Pimentel, C.; Achard, P.; Bettoun, A.; Meunier, B.; Bauvy, C.; et al. The Pro-apoptotic STK38 Kinase Is a New Beclin1 Partner Positively Regulating Autophagy. *Curr. Biol.* **2015**, *25*, 2479–2492. [CrossRef]

119. Bettoun, A.; Joffre, C.; Zago, G.; Surdez, D.; Vallerand, D.; Gundogdu, R.; Sharif, A.A.; Gomez, M.; Cascone, I.; Meunier, B.; et al. Mitochondrial clearance by the STK38 kinase supports oncogenic Ras-induced cell transformation. *Oncotarget* **2016**, *7*, 44142–44160. [CrossRef]

120. Bothos, J.; Tuttle, R.L.; Ottey, M.; Luca, F.C.; Halazonetis, T.D. Human LATS1 is a mitotic exit network kinase. *Cancer Res.* **2005**, *65*, 6568–6575. [CrossRef]

121. Rual, J.F.; Venkatesan, K.; Hao, T.; Hirozane-Kishikawa, T.; Dricot, A.; Li, N.; Berriz, G.F.; Gibbons, F.D.; Dreze, M.; Ayivi-Guedehoussou, N.; et al. Towards a proteome-scale map of the human protein-protein interaction network. *Nature* **2005**, *437*, 1173–1178. [CrossRef] [PubMed]

122. Vinayagam, A.; Stelzl, U.; Foulle, R.; Plassmann, S.; Zenkner, M.; Timm, J.; Assmus, H.E.; Andrade-Navarro, M.A.; Wanker, E.E. A directed protein interaction network for investigating intracellular signal transduction. *Sci. Signal.* **2011**, *4*, rs8. [CrossRef] [PubMed]

123. Du, X.; Wang, Q.; Hirohashi, Y.; Greene, M.I. DIPA, which can localize to the centrosome, associates with p78/MCRS1/MSP58 and acts as a repressor of gene transcription. *Exp. Mol. Pathol.* **2006**, *81*, 184–190. [CrossRef] [PubMed]

124. Kim, J.; Lee, J.E.; Heynen-Genel, S.; Suyama, E.; Ono, K.; Lee, K.; Ideker, T.; Aza-Blanc, P.; Gleeson, J.G. Functional genomic screen for modulators of ciliogenesis and cilium length. *Nature* **2010**, *464*, 1048–1051. [CrossRef] [PubMed]

125. Chiba, S.; Amagai, Y.; Homma, Y.; Fukuda, M.; Mizuno, K. NDR2-mediated Rabin8 phosphorylation is crucial for ciliogenesis by switching binding specificity from phosphatidylserine to Sec15. *EMBO J.* **2013**, *32*, 874–885. [CrossRef]

126. Gardiner, K.L.; Downs, L.; Berta-Antalics, A.I.; Santana, E.; Aguirre, G.D.; Genini, S. Photoreceptor proliferation and dysregulation of cell cycle genes in early onset inherited retinal degenerations. *BMC Genom.* **2016**, *17*, 221. [CrossRef] [PubMed]

127. Goldstein, O.; Kukekova, A.V.; Aguirre, G.D.; Acland, G.M. Exonic SINE insertion in STK38L causes canine early retinal degeneration (erd). *Genomics* **2010**, *96*, 362–368. [CrossRef]

128. Leger, H.; Santana, E.; Leu, N.A.; Smith, E.T.; Beltran, W.A.; Aguirre, G.D.; Luca, F.C. Ndr kinases regulate retinal interneuron proliferation and homeostasis. *Sci. Rep.* **2018**, *8*, 12544. [CrossRef]

129. Lobo, G.P.; Fulmer, D.; Guo, L.; Zuo, X.; Dang, Y.; Kim, S.H.; Su, Y.; George, K.; Obert, E.; Fogelgren, B.; et al. The exocyst is required for photoreceptor ciliogenesis and retinal development. *J. Biol. Chem.* **2017**, *292*, 14814–14826. [CrossRef]

130. Tavares, A.; Goncalves, J.; Florindo, C.; Tavares, A.A.; Soares, H. Mob1: Defining cell polarity for proper cell division. *J. Cell Sci.* **2012**, *125*, 516–527. [CrossRef]

131. Bergmann, C.; Guay-Woodford, L.M.; Harris, P.C.; Horie, S.; Peters, D.J.M.; Torres, V.E. Polycystic kidney disease. *Nat. Rev. Dis Primers* **2018**, *4*, 50. [CrossRef] [PubMed]

132. Seixas, C.; Choi, S.Y.; Polgar, N.; Umberger, N.L.; East, M.P.; Zuo, X.; Moreiras, H.; Ghossoub, R.; Benmerah, A.; Kahn, R.A.; et al. Arl13b and the exocyst interact synergistically in ciliogenesis. *Mol. Biol. Cell* **2016**, *27*, 308–320. [CrossRef] [PubMed]

133. Happe, H.; van der Wal, A.M.; Leonhard, W.N.; Kunnen, S.J.; Breuning, M.H.; de Heer, E.; Peters, D.J. Altered Hippo signalling in polycystic kidney disease. *J. Pathol.* **2011**, *224*, 133–142. [CrossRef] [PubMed]

134. Grampa, V.; Delous, M.; Zaidan, M.; Odye, G.; Thomas, S.; Elkhartoufi, N.; Filhol, E.; Niel, O.; Silbermann, F.; Lebreton, C.; et al. Novel NEK8 Mutations Cause Severe Syndromic Renal Cystic Dysplasia through YAP Dysregulation. *PLoS Genet.* **2016**, *12*, e1005894. [CrossRef] [PubMed]

135. Kim, M.; Kim, M.; Lee, M.S.; Kim, C.H.; Lim, D.S. The MST1/2-SAV1 complex of the Hippo pathway promotes ciliogenesis. *Nat. Commun.* **2014**, *5*, 5370. [CrossRef] [PubMed]

136. Frank, V.; Habbig, S.; Bartram, M.P.; Eisenberger, T.; Veenstra-Knol, H.E.; Decker, C.; Boorsma, R.A.; Gobel, H.; Nurnberg, G.; Griessmann, A.; et al. Mutations in NEK8 link multiple organ dysplasia with altered Hippo signalling and increased c-MYC expression. *Hum. Mol. Genet.* **2013**, *22*, 2177–2185. [CrossRef] [PubMed]

137. Habbig, S.; Bartram, M.P.; Muller, R.U.; Schwarz, R.; Andriopoulos, N.; Chen, S.; Sagmuller, J.G.; Hoehne, M.; Burst, V.; Liebau, M.C.; et al. NPHP4, a cilia-associated protein, negatively regulates the Hippo pathway. *J. Cell Biol.* **2011**, *193*, 633–642. [CrossRef] [PubMed]

138. Ma, S.; Guan, K.L. Polycystic kidney disease: A Hippo connection. *Genes Dev.* **2018**, *32*, 737–739. [CrossRef] [PubMed]

139. Cai, J.; Song, X.; Wang, W.; Watnick, T.; Pei, Y.; Qian, F.; Pan, D. A RhoA-YAP-c-Myc signaling axis promotes the development of polycystic kidney disease. *Genes Dev.* **2018**, *32*, 781–793. [CrossRef]

140. Moya, I.M.; Halder, G. Hippo-YAP/TAZ signalling in organ regeneration and regenerative medicine. *Nat. Rev. Mol. Cell Biol.* **2019**, *20*, 211–226. [CrossRef]

141. Shao, D.D.; Xue, W.; Krall, E.B.; Bhutkar, A.; Piccioni, F.; Wang, X.; Schinzel, A.C.; Sood, S.; Rosenbluh, J.; Kim, J.W.; et al. KRAS and YAP1 converge to regulate EMT and tumor survival. *Cell* **2014**, *158*, 171–184. [CrossRef] [PubMed]

142. Luo, J.; Emanuele, M.J.; Li, D.; Creighton, C.J.; Schlabach, M.R.; Westbrook, T.F.; Wong, K.K.; Elledge, S.J. A genome-wide RNAi screen identifies multiple synthetic lethal interactions with the Ras oncogene. *Cell* **2009**, *137*, 835–848. [CrossRef] [PubMed]

143. Grant, T.J.; Mehta, A.K.; Gupta, A.; Sharif, A.A.D.; Arora, K.S.; Deshpande, V.; Ting, D.T.; Bardeesy, N.; Ganem, N.J.; Hergovich, A.; et al. STK38L kinase ablation promotes loss of cell viability in a subset of KRAS-dependent pancreatic cancer cell lines. *Oncotarget* **2017**, *8*, 78556–78572. [CrossRef] [PubMed]

144. Sasaki, H.; Kawano, O.; Endo, K.; Suzuki, E.; Yukiue, H.; Kobayashi, Y.; Yano, M.; Fujii, Y. Human MOB1 expression in non-small-cell lung cancer. *Clin. Lung Cancer* **2007**, *8*, 273–276. [CrossRef] [PubMed]

145. Otsubo, K.; Goto, H.; Nishio, M.; Kawamura, K.; Yanagi, S.; Nishie, W.; Sasaki, T.; Maehama, T.; Nishina, H.; Mimori, K.; et al. MOB1-YAP1/TAZ-NKX2.1 axis controls bronchioalveolar cell differentiation, adhesion and tumour formation. *Oncogene* **2017**, *36*, 4201–4211. [CrossRef] [PubMed]

146. Seve, P.; Mackey, J.R.; Isaac, S.; Tredan, O.; Souquet, P.J.; Perol, M.; Cass, C.; Dumontet, C. cN-II expression predicts survival in patients receiving gemcitabine for advanced non-small cell lung cancer. *Lung Cancer* **2005**, *49*, 363–370. [CrossRef] [PubMed]

147. Chen, M.; Wang, M.; Xu, S.; Guo, X.; Jiang, J. Upregulation of miR-181c contributes to chemoresistance in pancreatic cancer by inactivating the Hippo signaling pathway. *Oncotarget* **2015**, *6*, 44466–44479. [CrossRef]

148. Zhang, M.; Zhao, Y.; Zhang, Y.; Wang, D.; Gu, S.; Feng, W.; Peng, W.; Gong, A.; Xu, M. LncRNA UCA1 promotes migration and invasion in pancreatic cancer cells via the Hippo pathway. *Biochim. Biophys. Acta Mol. Basis. Dis.* **2018**, *1864*, 1770–1782. [CrossRef]

149. Nishio, M.; Sugimachi, K.; Goto, H.; Wang, J.; Morikawa, T.; Miyachi, Y.; Takano, Y.; Hikasa, H.; Itoh, T.; Suzuki, S.O.; et al. Dysregulated YAP1/TAZ and TGF-beta signaling mediate hepatocarcinogenesis in Mob1a/1b-deficient mice. *Proc. Natl. Acad. Sci. USA* **2016**, *113*, E71–80. [CrossRef]

150. Patel, S.H.; Camargo, F.D.; Yimlamai, D. Hippo Signaling in the Liver Regulates Organ Size, Cell Fate, and Carcinogenesis. *Gastroenterology* **2017**, *152*, 533–545. [CrossRef]

151. Wang, Z.X.; Wang, H.Y.; Wu, M.C. Identification and characterization of a novel human hepatocellular carcinoma-associated gene. *Br. J. Cancer* **2001**, *85*, 1162–1167. [CrossRef] [PubMed]

152. Shirvani-Dastgerdi, E.; Schwartz, R.E.; Ploss, A. Hepatocarcinogenesis associated with hepatitis B, delta and C viruses. *Curr. Opin. Virol.* **2016**, *20*, 1–10. [CrossRef] [PubMed]

153. Goodrum, G.; Pelchat, M. Insight into the Contribution and Disruption of Host Processes during HDV Replication. *Viruses* **2018**, *11*. [CrossRef] [PubMed]

154. Chung, H.Y.; Gu, M.; Buehler, E.; MacDonald, M.R.; Rice, C.M. Seed sequence-matched controls reveal limitations of small interfering RNA knockdown in functional and structural studies of hepatitis C virus NS5A-MOBKL1B interaction. *J. Virol.* **2014**, *88*, 11022–11033. [CrossRef] [PubMed]

155. Meistermann, H.; Gao, J.; Golling, S.; Lamerz, J.; Le Pogam, S.; Tzouros, M.; Sankabathula, S.; Gruenbaum, L.; Najera, I.; Langen, H.; et al. A novel immuno-competitive capture mass spectrometry strategy for protein-protein interaction profiling reveals that LATS kinases regulate HCV replication through NS5A phosphorylation. *Mol. Cell Proteom.* **2014**, *13*, 3040–3048. [CrossRef] [PubMed]

156. Shen, J.; Su, J.; Wu, D.; Zhang, F.; Fu, H.; Zhou, H.; Xu, M. Growth Inhibition Accompanied by MOB1 Upregulation in Human Acute Lymphoid Leukemia Cells by 3-Deazaneplanocin, A. *Biochem. Genet.* **2015**, *53*, 268–279. [CrossRef] [PubMed]

157. Galmarini, C.M.; Cros, E.; Thomas, X.; Jordheim, L.; Dumontet, C. The prognostic value of cN-II and cN-III enzymes in adult acute myeloid leukemia. *Haematologica* **2005**, *90*, 1699–1701. [PubMed]

158. Suzuki, K.; Sugawara, T.; Oyake, T.; Uchiyama, T.; Aoki, Y.; Tsukushi, Y.; Onodera, S.; Ito, S.; Murai, K.; Ishida, Y. Clinical significance of high-Km 5′-nucleotidase (cN-II) mRNA expression in high-risk myelodysplastic syndrome. *Leuk. Res.* **2007**, *31*, 1343–1349. [CrossRef] [PubMed]

159. Meyer, J.A.; Wang, J.; Hogan, L.E.; Yang, J.J.; Dandekar, S.; Patel, J.P.; Tang, Z.; Zumbo, P.; Li, S.; Zavadil, J.; et al. Relapse-specific mutations in NT5C2 in childhood acute lymphoblastic leukemia. *Nat. Genet.* **2013**, *45*, 290–294. [CrossRef]

160. Bea, S.; Salaverria, I.; Armengol, L.; Pinyol, M.; Fernandez, V.; Hartmann, E.M.; Jares, P.; Amador, V.; Hernandez, L.; Navarro, A.; et al. Uniparental disomies, homozygous deletions, amplifications, and target genes in mantle cell lymphoma revealed by integrative high-resolution whole-genome profiling. *Blood* **2009**, *113*, 3059–3069. [CrossRef]

161. Hartmann, E.M.; Campo, E.; Wright, G.; Lenz, G.; Salaverria, I.; Jares, P.; Xiao, W.; Braziel, R.M.; Rimsza, L.M.; Chan, W.C.; et al. Pathway discovery in mantle cell lymphoma by integrated analysis of high-resolution gene expression and copy number profiling. *Blood* **2010**, *116*, 953–961. [CrossRef] [PubMed]

162. Kim, S.Y.; Dunn, I.F.; Firestein, R.; Gupta, P.; Wardwell, L.; Repich, K.; Schinzel, A.C.; Wittner, B.; Silver, S.J.; Root, D.E.; et al. CK1epsilon is required for breast cancers dependent on beta-catenin activity. *PLoS ONE* **2010**, *5*, e8979. [CrossRef]

163. Medina-Aguilar, R.; Perez-Plasencia, C.; Marchat, L.A.; Gariglio, P.; Garcia Mena, J.; Rodriguez Cuevas, S.; Ruiz-Garcia, E.; Astudillo-de la Vega, H.; Hernandez Juarez, J.; Flores-Perez, A.; et al. Methylation Landscape of Human Breast Cancer Cells in Response to Dietary Compound Resveratrol. *PLoS ONE* **2016**, *11*, e0157866. [CrossRef] [PubMed]

164. Cheung, H.W.; Cowley, G.S.; Weir, B.A.; Boehm, J.S.; Rusin, S.; Scott, J.A.; East, A.; Ali, L.D.; Lizotte, P.H.; Wong, T.C.; et al. Systematic investigation of genetic vulnerabilities across cancer cell lines reveals lineage-specific dependencies in ovarian cancer. *Proc. Natl. Acad. Sci. USA* **2011**, *108*, 12372–12377. [CrossRef] [PubMed]

165. Kosaka, Y.; Mimori, K.; Tanaka, F.; Inoue, H.; Watanabe, M.; Mori, M. Clinical significance of the loss of MATS1 mRNA expression in colorectal cancer. *Int. J. Oncol.* **2007**, *31*, 333–338. [CrossRef] [PubMed]

166. Bae, J.S.; Jeon, Y.; Kim, S.M.; Jang, J.Y.; Park, M.K.; Kim, I.H.; Hwang, D.S.; Lim, D.S.; Lee, H. Depletion of MOB1A/B causes intestinal epithelial degeneration by suppressing Wnt activity and activating BMP/TGF-beta signaling. *Cell Death Dis.* **2018**, *9*, 1083. [CrossRef] [PubMed]

167. Hong, A.W.; Meng, Z.; Guan, K.L. The Hippo pathway in intestinal regeneration and disease. *Nat. Rev. Gastroenterol. Hepatol.* **2016**, *13*, 324–337. [CrossRef]

168. Schumacher, F.R.; Al Olama, A.A.; Berndt, S.I.; Benlloch, S.; Ahmed, M.; Saunders, E.J.; Dadaev, T.; Leongamornlert, D.; Anokian, E.; Cieza-Borrella, C.; et al. Association analyses of more than 140,000 men identify 63 new prostate cancer susceptibility loci. *Nat. Genet.* **2018**, *50*, 928–936. [CrossRef]

169. Kim, E.A.; Kim, Y.H.; Kang, H.W.; Yoon, H.Y.; Kim, W.T.; Kim, Y.J.; Yun, S.J.; Moon, S.K.; Choi, Y.H.; Kim, I.Y.; et al. Lower Levels of Human MOB3B Are Associated with Prostate Cancer Susceptibility and Aggressive Clinicopathological Characteristics. *J. Korean Med. Sci.* **2015**, *30*, 937–942. [CrossRef]

170. Haldrup, C.; Mundbjerg, K.; Vestergaard, E.M.; Lamy, P.; Wild, P.; Schulz, W.A.; Arsov, C.; Visakorpi, T.; Borre, M.; Hoyer, S.; et al. DNA methylation signatures for prediction of biochemical recurrence after radical prostatectomy of clinically localized prostate cancer. *J. Clin. Oncol.* **2013**, *31*, 3250–3258. [CrossRef]

171. Gerlinger, M.; Santos, C.R.; Spencer-Dene, B.; Martinez, P.; Endesfelder, D.; Burrell, R.A.; Vetter, M.; Jiang, M.; Saunders, R.E.; Kelly, G.; et al. Genome-wide RNA interference analysis of renal carcinoma survival regulators identifies MCT4 as a Warburg effect metabolic target. *J. Pathol.* **2012**, *227*, 146–156. [CrossRef] [PubMed]

172. Cowley, G.S.; Weir, B.A.; Vazquez, F.; Tamayo, P.; Scott, J.A.; Rusin, S.; East-Seletsky, A.; Ali, L.D.; Gerath, W.F.; Pantel, S.E.; et al. Parallel genome-scale loss of function screens in 216 cancer cell lines for the identification of context-specific genetic dependencies. *Sci. Data* **2014**, *1*, 140035. [CrossRef] [PubMed]

173. Behan, F.M.; Iorio, F.; Picco, G.; Goncalves, E.; Beaver, C.M.; Migliardi, G.; Santos, R.; Rao, Y.; Sassi, F.; Pinnelli, M.; et al. Prioritization of cancer therapeutic targets using CRISPR-Cas9 screens. *Nature* **2019**, *568*, 511–516. [CrossRef] [PubMed]

174. Tsherniak, A.; Vazquez, F.; Montgomery, P.G.; Weir, B.A.; Kryukov, G.; Cowley, G.S.; Gill, S.; Harrington, W.F.; Pantel, S.; Krill-Burger, J.M.; et al. Defining a Cancer Dependency Map. *Cell* **2017**, *170*, 564–576. [CrossRef] [PubMed]

175. Kim, Y.R.; Kim, S.Y. Machine learning identifies a core gene set predictive of acquired resistance to EGFR tyrosine kinase inhibitor. *J. Cancer Res. Clin. Oncol.* **2018**, *144*, 1435–1444. [CrossRef]

176. Fang, K.M.; Liu, Y.Y.; Lin, C.H.; Fan, S.S.; Tsai, C.H.; Tzeng, S.F. Mps one binder 2 gene upregulation in the stellation of astrocytes induced by cAMP-dependent pathway. *J. Cell Biochem.* **2012**, *113*, 3019–3028. [CrossRef] [PubMed]

177. Satoh, J.; Kawana, N.; Yamamoto, Y. Pathway Analysis of ChIP-Seq-Based NRF1 Target Genes Suggests a Logical Hypothesis of their Involvement in the Pathogenesis of Neurodegenerative Diseases. *Gene. Regul. Syst. Bio.* **2013**, *7*, 139–152. [CrossRef]

178. Olah, J.; Vincze, O.; Virok, D.; Simon, D.; Bozso, Z.; Tokesi, N.; Horvath, I.; Hlavanda, E.; Kovacs, J.; Magyar, A.; et al. Interactions of pathological hallmark proteins: Tubulin polymerization promoting protein/p25, beta-amyloid, and alpha-synuclein. *J. Biol. Chem.* **2011**, *286*, 34088–34100. [CrossRef]

179. Blondeau, C.; Gaillard, S.; Ternaux, J.P.; Monneron, A.; Baude, A. Expression and distribution of phocein and members of the striatin family in neurones of rat peripheral ganglia. *Histochem. Cell Biol.* **2003**, *119*, 131–138. [CrossRef]

180. Li, D.; Musante, V.; Zhou, W.; Picciotto, M.R.; Nairn, A.C. Striatin-1 is a B subunit of protein phosphatase PP2A that regulates dendritic arborization and spine development in striatal neurons. *J. Biol. Chem.* **2018**, *293*, 11179–11194. [CrossRef]

181. Lee, P.C.; Machner, M.P. The Legionella Effector Kinase LegK7 Hijacks the Host Hippo Pathway to Promote Infection. *Cell Host. Microbe.* **2018**, *24*, 429–438. [CrossRef] [PubMed]

182. Ahola-Olli, A.V.; Wurtz, P.; Havulinna, A.S.; Aalto, K.; Pitkanen, N.; Lehtimaki, T.; Kahonen, M.; Lyytikainen, L.P.; Raitoharju, E.; Seppala, I.; et al. Genome-wide Association Study Identifies 27 Loci Influencing Concentrations of Circulating Cytokines and Growth Factors. *Am. J. Hum. Genet.* **2017**, *100*, 40–50. [CrossRef] [PubMed]

183. Liu, C.; Batliwalla, F.; Li, W.; Lee, A.; Roubenoff, R.; Beckman, E.; Khalili, H.; Damle, A.; Kern, M.; Furie, R.; et al. Genome-wide association scan identifies candidate polymorphisms associated with differential response to anti-TNF treatment in rheumatoid arthritis. *Mol. Med.* **2008**, *14*, 575–581. [CrossRef] [PubMed]

184. Zengini, E.; Hatzikotoulas, K.; Tachmazidou, I.; Steinberg, J.; Hartwig, F.P.; Southam, L.; Hackinger, S.; Boer, C.G.; Styrkarsdottir, U.; Gilly, A.; et al. Genome-wide analyses using UK Biobank data provide insights into the genetic architecture of osteoarthritis. *Nat. Genet.* **2018**, *50*, 549–558. [CrossRef]

185. Almoguera, B.; Vazquez, L.; Mentch, F.; Connolly, J.; Pacheco, J.A.; Sundaresan, A.S.; Peissig, P.L.; Linneman, J.G.; McCarty, C.A.; Crosslin, D.; et al. Identification of Four Novel Loci in Asthma in European American and African American Populations. *Am. J. Respir Crit Care Med.* **2017**, *195*, 456–463. [CrossRef] [PubMed]

186. Warner, N.; Burberry, A.; Pliakas, M.; McDonald, C.; Nunez, G. A genome-wide small interfering RNA (siRNA) screen reveals nuclear factor-kappaB (NF-kappaB)-independent regulators of NOD2-induced interleukin-8 (IL-8) secretion. *J. Biol. Chem.* **2014**, *289*, 28213–28224. [CrossRef] [PubMed]

187. Nishio, M.; Miyachi, Y.; Otani, J.; Tane, S.; Omori, H.; Ueda, F.; Togashi, H.; Sasaki, T.; Mak, T.W.; Nakao, K.; et al. Hippo pathway controls cell adhesion and context-dependent cell competition to influence skin engraftment efficiency. *FASEB J.* **2019**. [CrossRef] [PubMed]

188. Goto, H.; Nishio, M.; To, Y.; Oishi, T.; Miyachi, Y.; Maehama, T.; Nishina, H.; Akiyama, H.; Mak, T.W.; Makii, Y.; et al. Loss of Mob1a/b in mice results in chondrodysplasia due to YAP1/TAZ-TEAD-dependent repression of SOX9. *Development* **2018**, *145*. [CrossRef]

189. de Miguel-Yanes, J.M.; Shrader, P.; Pencina, M.J.; Fox, C.S.; Manning, A.K.; Grant, R.W.; Dupuis, J.; Florez, J.C.; D'Agostino, R.B.; Cupples, L.A.; et al. Genetic risk reclassification for type 2 diabetes by age below or above 50 years using 40 type 2 diabetes risk single nucleotide polymorphisms. *Diabetes Care* **2011**, *34*, 121–125. [CrossRef]

190. Hanson, R.L.; Guo, T.; Muller, Y.L.; Fleming, J.; Knowler, W.C.; Kobes, S.; Bogardus, C.; Baier, L.J. Strong parent-of-origin effects in the association of KCNQ1 variants with type 2 diabetes in American Indians. *Diabetes* **2013**, *62*, 2984–2991. [CrossRef]

191. Cui, X.; Guo, Z.; Song, L.; Wang, Y.; Cheng, Y. NCP1/AtMOB1A Plays Key Roles in Auxin-Mediated Arabidopsis Development. *PLoS Genet.* **2016**, *12*, e1005923. [CrossRef] [PubMed]

192. Xiong, J.; Cui, X.; Yuan, X.; Yu, X.; Sun, J.; Gong, Q. The Hippo/STE20 homolog SIK1 interacts with MOB1 to regulate cell proliferation and cell expansion in Arabidopsis. *J. Exp. Bot.* **2016**, *67*, 1461–1475. [CrossRef] [PubMed]

cells

MDPI

Review

Regulation of TEAD Transcription Factors in Cancer Biology

Hyunbin D. Huh [1], Dong Hyeon Kim [1], Han-Sol Jeong [2],* and Hyun Woo Park [1],*

[1] Department of Biochemistry, College of Life Science and Biotechnology, Yonsei University, Seoul 03722, Korea; huhhb621@yonsei.ac.kr (H.D.H.); dhk1107@yonsei.ac.kr (D.H.K.)
[2] Division of Applied Medicine, School of Korean Medicine, Pusan National University, Yangsan, Gyeongnam 50612, Korea
* Correspondence: jhsol33@pusan.ac.kr (H.-S.J.); hwp003@yonsei.ac.kr (H.W.P.); Tel.: +82-51-510-8461 (H.-S.J.); +82-2-2123-2698 (H.W.P.)

Received: 20 April 2019; Accepted: 11 June 2019; Published: 17 June 2019

Abstract: Transcriptional enhanced associate domain (TEAD) transcription factors play important roles during development, cell proliferation, regeneration, and tissue homeostasis. TEAD integrates with and coordinates various signal transduction pathways including Hippo, Wnt, transforming growth factor beta (TGFβ), and epidermal growth factor receptor (EGFR) pathways. TEAD deregulation affects well-established cancer genes such as KRAS, BRAF, LKB1, NF2, and MYC, and its transcriptional output plays an important role in tumor progression, metastasis, cancer metabolism, immunity, and drug resistance. To date, TEADs have been recognized to be key transcription factors of the Hippo pathway. Therefore, most studies are focused on the Hippo kinases and YAP/TAZ, whereas the Hippo-dependent and Hippo-independent regulators and regulations governing TEAD only emerged recently. Deregulation of the TEAD transcriptional output plays important roles in tumor progression and serves as a prognostic biomarker due to high correlation with clinicopathological parameters in human malignancies. In addition, discovering the molecular mechanisms of TEAD, such as post-translational modifications and nucleocytoplasmic shuttling, represents an important means of modulating TEAD transcriptional activity. Collectively, this review highlights the role of TEAD in multistep-tumorigenesis by interacting with upstream oncogenic signaling pathways and controlling downstream target genes, which provides unprecedented insight and rationale into developing TEAD-targeted anticancer therapeutics.

Keywords: TEAD; Hippo pathway; cancer; stem cell

1. Introduction

The TEAD family of transcription factors are the final nuclear effectors of the Hippo pathway, which regulate cell growth, proliferation, and tissue homeostasis via their transcriptional target genes. Since their initial discovery three decades ago, TEADs have been best studied in the context of the Hippo-YAP/TAZ signaling pathway and tumorigenesis. To date, studies on TEAD activity are limited to serving as the functional readout of the Hippo-YAP/TAZ pathway. However, recent evidence suggests that nucleocytoplasmic shuttling, post-translational modifications, and crosstalk between oncogenic signaling pathways are important determinants of TEAD activity both in vitro and in vivo. Importantly, since the Hippo pathway components are hardly druggable, TEADs have emerged as critical drug candidates to treat human diseases including cancer, cardiovascular diseases, and neurodegenerative disorders. The current review underscores and reinterprets the oncogenic role of TEADs in tumorigenesis in past reports from a TEAD point of view in order to provide unprecedented insight and rationale into developing TEAD-targeted anticancer therapeutics.

2. The TEAD Family of Transcription Factors

All four TEADs are transcription factors that are evolutionarily conserved and broadly expressed in most tissues of the human body [1,2]. Each family member has multiple names TEAD1 (TEF-1/NTEF), TEAD2 (TEF-4/ETF), TEAD3 (TEF-5/ETFR-1), and TEAD4 (TEF-3/ETFR-2/FR-19). Despite the high homology and expression pattern shared between TEAD1-4, animal model experiments indicate that each TEAD has tissue-specific roles, such as cardiogenesis, neural development, and trophectoderm lineage determination, during embryonic development [3–8]. One of their major roles in cell biology is to regulate cell proliferation and contact inhibition [9,10]. TEADs also share highly similar domain architectures (Figure 1). TEAD N-terminus share a highly conserved 68-amino acid TEA/ATTS DNA-binding domain, which binds to the MCAT element (5'-CATTCCA/T-3') originally defined as the GT-IIC motif (5'-ACATTCCAC-3') of the simian virus 40 (SV40) enhancer [11–13]. Based on these sequences, a synthetic TEAD luciferase reporter 8xGTIIC-luciferase plasmid, which contains eight GT-IIC motifs, is being widely used to measure YAP/TAZ and TEAD activity [14]. Unlike several other transcription factors, TEAD is known to be mostly bound to the DNA since the majority of TEADs are found in the chromatin fraction [15]. However, they hardly exhibit any transcriptional activity by themselves [16]. Hence, TEAD activity mainly relies on the C-terminus, in which all TEADs share their transactivation domain in order to recruit transcriptional coactivators YAP/TAZ [17–20], corepressors VGLL1-4 [21–26], chromatin remodeling factors NuRD [27], and the Mediator [28]. Although TEAD has been recognized as the final effector of the Hippo-YAP/TAZ pathway, TEAD also interacts with other signaling transduction pathway transcription factors including TCF, SMAD, OCT4, AP-1, and MRTF [29–33] (Figure 2a). TEAD-driven transcriptional targets include well-established genes that are involved in cell growth, proliferation, and tissue homeostasis. In addition to classical TEAD target genes such as *CTGF* and *CYR61* [18], recent studies also identify *WNT5A/B* [34], *DKK1* [34,35], *TGFB2* [36], *BMP4* [37], *AREG* [38], *EGFR* [39], *PD-L1* [40–43], *MYC* [44,45], *LATS2* [46], amino acid transporters *SLC38A1/SLC7A5* [47,48], and glucose transporter *GLUT3* [49] as direct TEAD target genes (Figure 2a). These signaling inputs, protein-protein interactions, and target genes further expand the roles of TEAD to directly control Wnt, TGFβ, RTK, mTOR, and Hippo signaling in the context of tumorigenesis, cancer immunity, stem cell pluripotency, metabolism, and development.

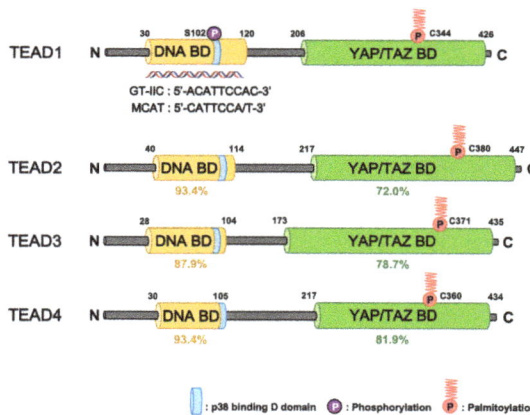

Figure 1. Domain architecture of human TEADs. The N-terminal DNA binding domain (DNA-BD) and C-terminal YAP/TAZ binding domain (YAP/TAZ-BD) of TEAD1-4 harbor high similarity across four different paralogs. The percent (%) represents the identity for each domain of TEADs compared to that of TEAD1 [50]. TEAD post-translation modifications include palmitoylation and PKA-, PKC-mediated phosphorylation that occur in the YAP/TAZ-BD and DNA-BD, respectively. Palmitoylation is required for proper TEAD functions. TEAD cytoplasmic translocation occurs through protein-protein interaction with p38 MAPK that binds the p38-binding motif within the DNA-BD of all TEADs.

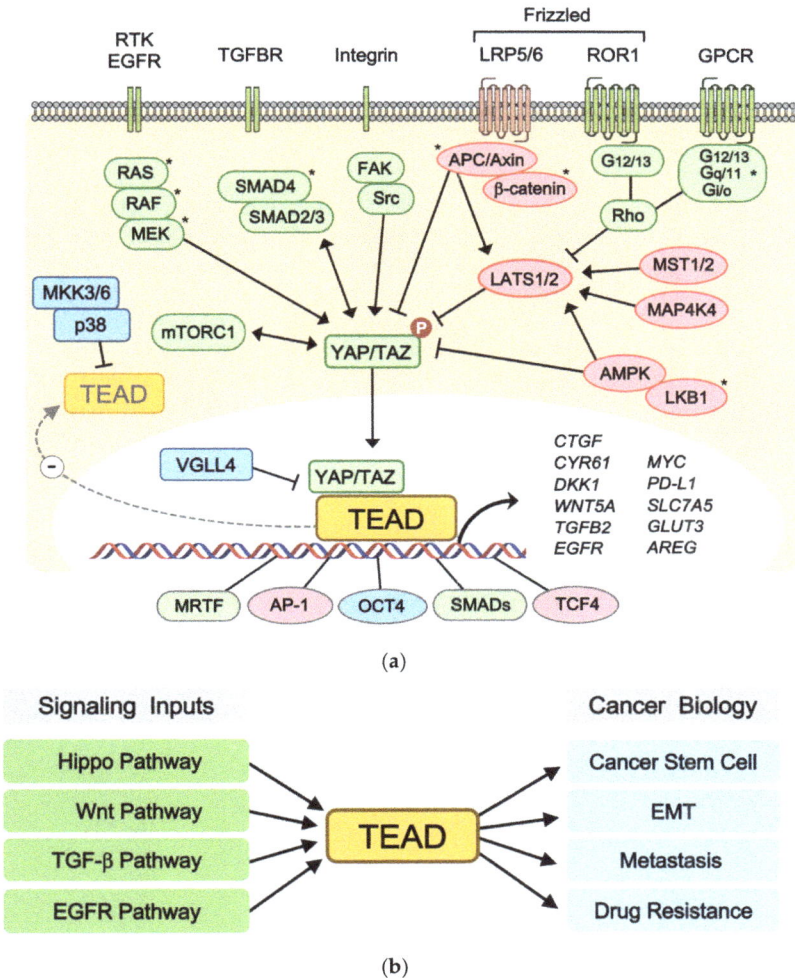

Figure 2. The regulatory mechanisms of TEAD in cancer biology. (**a**) Upstream signaling and downstream transcriptional outputs of TEAD. Various oncogenic signal transduction pathways, such as EGFR signaling, TGFβ signaling, Wnt signaling, GPCR signaling, and cancer genes (*), such as KRAS, BRAF, LKB1, APC, GNAQ/11 regulate TEAD activity through multiple signaling mechanisms. The TEAD transcriptional outputs have critical functions in tumorigenesis, stem cell maintenance, cancer immunology, metabolism as well as formation of signaling feedback loops. (**b**) Role of TEAD in multiple stages of tumorigenesis. TEAD activation via various oncogenic pathways play critical roles in cancer biology including EMT, metastasis, drug resistance, and cancer stem cells.

3. Signaling Inputs and Transcriptional Outputs of TEAD

3.1. Hippo Pathway

Since TEADs exhibit minimal transcriptional activity by themselves, they require coactivators to induce target gene expression [16]. The most well-established cofactors that activate TEAD-mediated transcription are YAP and its paralog TAZ, which are transcriptional coactivators of the Hippo pathway that play major roles in organ size control, cell proliferation, tumorigenesis, and stem cell self-renewal [51–54] (Figure 2a). The N-terminus of YAP/TAZ interact with the C-terminal transactivation domain of TEAD to form a YAP/TAZ-TEAD complex that constitutes the nuclear transcriptional module of the Hippo pathway [55]. On the other hand, the cytosolic kinase modules of the Hippo pathway, which consists of MST1/2, MAP4K4, and LATS1/2, phosphorylate YAP/TAZ at multiple sites. This promotes YAP/TAZ cytoplasmic retention, ubiquitination, and protein degradation [56]. Cytoplasmic YAP/TAZ are degraded by both the ubiquitin-proteasome system and autophagy [57–60], which renders TEAD transcriptionally inactive.

To date, numerous studies and ChIP-seq analyses highlight YAP/TAZ to be the major TEAD coactivators. In MDA-MB-231 breast cancer cells that harbor genetic inactivation of the Hippo pathway (*NF2* null), approximately 80% of TEAD4-bound promoters and enhancer regions were co-occupied with YAP/TAZ, while the TEAD consensus sequence was present in 75% of DNA-bound YAP/TAZ peaks [32]. In MCF10A mammary gland epithelial cells, YAP and TEAD1 co-occupied 80% of the promoters [18]. Furthermore, in glioblastoma cells, 86% of all YAP peak regions contained at least one TEAD binding site [61]. Although YAP/TAZ can interact with different transcription factors such as RUNX, p73, KLF4, TBX5, SMAD, and others, TEADs are the predominant factors that facilitate YAP/TAZ recruitment to the chromatin. In mouse studies, dominant-negative TEAD2 was found to be sufficient in suppressing YAP overexpression-, or NF2 inactivating mutation-induced hepatomegaly and tumorigenesis, which indicates that TEAD mostly attributes to YAP-induced tumorigenesis [62]. Although oncogenic driver mutations have not been reported in TEADs, numerous studies demonstrate their pro-tumorigenic roles due to their crosstalk with other cancer genes, which is discussed in later sections.

Furthermore, studies have shown that TEAD interaction is required for YAP/TAZ nuclear retention. Osmotic stress-activated p38 MAPK induces TEAD cytoplasmic translocation independent of the Hippo-YAP/TAZ pathway. In this context, YAP/TAZ fails to accumulate in the nucleus regardless of its phosphorylation status [63]. In TEAD knockout cells, YAP/TAZ remained in the cytoplasm even after treating YAP/TAZ activating stimuli such as LPA and serum [63], and, similarly, TAZ mutants defective in terms of TEAD interaction showed impaired nuclear accumulation [64]. In addition, forced expression of wild-type TEAD2, but not the YAP/TAZ binding deficient-TEAD2 mutant, induced nuclear localization of YAP/TAZ, which was required for tumorigenesis [65]. These results suggest a possible mechanism by which the TEAD interaction is required for YAP/TAZ accumulation and activation in the nucleus.

Moreover, the TEAD transcriptional output also regulates Hippo-YAP/TAZ signaling via target gene expression. TEAD target genes include important membrane transporters and secreted ligands that act through autocrine or paracrine signaling. The *AREG*, *WNT5A/B*, and *TGFB2* ligands activate YAP/TAZ via EGFR [66,67], Frizzled/ROR [34,68,69], and TGFBR [70–72], respectively, while *SLC38A1/SLC7A5* and *GLUT3* activate YAP/TAZ via mTORC1 [58] and glycolytic enzymes [49,73] respectively. Therefore, these TEAD-dependent positive feedback loops function as a nexus that coordinates important biological pathways and deregulation may lead to human malignancies. TEADs also induce *LATS2*, *NF2*, and *AMOTL2*, which are major components of the Hippo pathway, to form a negative feedback loop [46,74]. Thus, these TEAD-mediated feedback loops provide an efficient mechanism by which the robustness and homeostasis of Hippo-YAP/TAZ regulation is established.

3.2. Wnt Pathway

The canonical Wnt/β-catenin pathway and alternative Wnt pathway are one of the best characterized upstream signal transduction pathways that regulate TEAD. Crosstalk between the Wnt and Hippo pathways converge on TEAD via destruction complex-dependent and -independent mechanisms, which have been studied in the context of tumorigenesis, stem cell biology, and development. In these contexts, TEAD activity is indispensable for Wnt-induced biological responses (Figure 2a).

Canonical Wnt/β-catenin pathway regulates TEAD and YAP/TAZ via both Hippo-dependent and -independent mechanisms. Using immunoprecipitation assay, YAP/TAZ were shown to directly interact with the β-catenin destruction complex consisting of Axin1, APC, and GSK3β [75,76]. Within the destruction complex, YAP bridges β-TrCP E3 ligase to degrade β-catenin. Another study shows that APC binds and activates the Hippo pathway components, LATS1 and Sav1, which, in turn, inhibits YAP/TAZ [77]. In both cases, Wnt stimulation or loss-of-function mutations in APC trigger YAP/TAZ nuclear translocation and TEAD-mediated transcription. Furthermore, YAP/TAZ mediates >50% of the Wnt target genes induced by APC deletion in transformed mammary epithelial cells [75]. TEAD4 also directly interacts with TCF4, and the TEAD4-TCF4 complex directly links Wnt/β-catenin and Hippo/YAP signaling at the transcription factor level [29]. Thus, TEAD transcriptional output underlies the pathogenesis of APC mutation-induced colorectal tumorigenesis and Wnt-induced crypt regeneration.

The alternative Wnt pathway, which is independent of β-catenin and the destruction complex, also plays an important role in tumorigenesis, differentiation, development, and Wnt/β-catenin signaling inhibition. The alternative Wnt ligand Wnt5A/B is a potent TEAD activator that signal via the Frizzled/ROR1-G$\alpha_{12/13}$-Rho GTPases-Hippo-YAP/TAZ pathway in the context of cancer progression and mesenchymal stem cell differentiation [34]. Subsequent studies further demonstrate YAP/TAZ and TEAD activation via the alternative Wnt pathway in the context of breast cancer progression, chemotherapy resistance, stem cell maintenance, and macrophage polarization [68,69,78,79].

Moreover, the TEAD transcriptional output regulates the Wnt pathway. Major TEAD target genes that inhibit Wnt/β-catenin pathway are *DKK1* and *WNT5A/B*. These are well-established secreted inhibitors of the canonical Wnt/β-catenin pathway [34,35]. Wnt3a stimulation induces *WNT5A/B* gene expression via TEAD, which suppresses Wnt/β-catenin-induced mesenchymal stem cell differentiation [34]. In addition, YAP represses *WNT3* gene expression to inhibit the Wnt/β-catenin pathway possibly via TEAD [80]. Although YAP has often been shown to inhibit Wnt-induced biological responses in cancer and stem cells [35,81–83], the precise involvement of the TEAD transcriptional output requires further investigation. Collectively, TEADs are activated via the upstream Wnt pathway, while the TEAD transcriptional output concomitantly inhibits Wnt signaling, thus forming a negative feedback loop.

3.3. TGFβ Pathway

The TGFβ pathway regulates multiple biological processes including embryonic development, stem cell differentiation, immune regulation, wound healing, and inflammation. The crosstalk between the TGFβ and Hippo pathway centers on Smad and TEAD transcription factors, respectively. TGFβ stimulation triggers TEAD-mediated biological responses in the context of cell fate determination, tumorigenesis, and fibrosis, which are either dependent or independent of Smads (Figure 2a). TGFβ increases the expression level and activity of TEAD, and vice versa, TEAD can also directly trigger TGFβ signaling. TGFβ induces TAZ expression via a Smad3-independent, p38-mediated, and MRTF-mediated mechanism [84]. TGFβ also induces TEAD2 expression during epithelial-to-mesenchymal transition (EMT) [65]. Thus, TGFβ-induced TEAD target gene expression promotes EMT in mammary gland epithelial cells and malignant tumor phenotypes. Notably, the TGFBII ligand itself is a direct target gene of TEAD that evokes a positive feedback regulation [36,85].

Upon TGFβ stimulation, Smad2 and Smad3 form complexes with Smad4 and accumulate in the nucleus. It is important to note that TGFβ-induced Smad nuclear translocation is dependent on YAP/TAZ. For example, lung and breast cancers lacking RASSF1A display hyperactive TGFβ signaling and tumor invasion via the YAP-dependent nuclear localization of Smad2 [70]. In human embryonic stem cells, TAZ is also required for TGFβ-induced nuclear translocation of the Smad2/3-4 complex, and TAZ-dependent Smad2/3 nuclear translocation is required to maintain self-renewal markers [71]. YAP/TAZ also regulates the localization of the Smad complex in response to cell density-mediated formation of polarity complexes [86]. However, the direct involvement of TEAD in YAP/TAZ-dependent Smad localization requires further investigations.

It is still unknown how TGFβ signaling switches from enforcing pluripotency to promoting mesendodermal differentiation during development. Smad2/3 forms an enhancer complex with TEADs and OCT4, which suppress the gene expression of differentiation markers and modulates the levels of core pluripotency genes. This maintains the pluripotency of embryonic stem cells [30]. An independent study showed that TEAD binds the negative elongation factor and blocks Wnt3a/β-catenin and Activin/Smad2/3-induced mesendodermal differentiation [87]. TEAD also mediates TGFβ-induced tumorigenesis. The majority of malignant mesotheliomas harbor genetic inactivation of the Hippo pathway components, which collaborate with the TGFβ pathway by forming the YAP-TEAD4-Smad3-p300 complex on the CTGF promoter to induce gene expression and tumor growth [88]. Furthermore, a multitude of genes harbor both Smad-binding and TEAD-binding elements in their promoters. In breast cancer cells, TGFβ-YAP/TAZ-TEAD signaling is crucial in driving late-stage metastatic phenotypes via Smad2/3-induced *NERG1* and *UCA1* transcription [72]. Accumulating evidence highlights the importance of the TEAD transcriptional output acting as a bona fide effector of the TGFβ pathway during development and tumorigenesis.

4. Molecular Mechanisms Controlling TEAD Activity

4.1. Regulation of TEAD via Subcellular Localization

Transcription factors have been shown to form distinct protein complexes both on and off the chromatin [15]. Important signaling pathways function by ultimately regulating the activity and subcellular localization of transcription factors. For example, the final effectors of the TGFβ, Wnt, NFκB, and Hippo signaling pathways are Smads, TCF/LEF, p65, and TEAD transcription factors, respectively, and their chromatin association is tightly controlled via complex upstream signals. Unlike many other transcription factors, a majority of TEAD proteins reside on the chromatin [15], which are known to be passively regulated and dependent on cofactors such as YAP/TAZ and VGLL4. However, mechanistic insight that governs dynamic TEAD subcellular localization was recently elucidated.

There is a wide range of cellular stress, such as energy starvation, oxidative stress, cytotoxic agent, which induce Hippo kinase activation, YAP/TAZ phosphorylation, cytoplasmic translocation, protein degradation, and thus inhibition of TEAD transcriptional outputs. However, these stimuli did not alter TEAD subcellular localization or expression levels. Interestingly, certain environmental stresses, such as osmotic stress, high cell density, and cell suspension, promoted TEAD cytoplasmic translocation [89]. Although controversial, the cytoplasmic localization of TEAD4 has been observed in the inner cell mass of mouse embryonic stem cells [90,91]. Cytoplasmic TEADs were also detected in lung, spleen, and kidney tissues, but not in renal cell carcinoma tissues [63]. A splicing isoform of TEAD4, which lacks the N-terminal DNA binding domain, is found in the cytoplasm acting as a dominant negative isoform that inhibits YAP activity [92]. Because TEADs are the major effectors that dictate the transcriptional output of the Hippo-YAP/TAZ pathway, physiologic and pathologic conditions affecting TEAD localization significantly impact the functional output of the Hippo pathway.

The molecular mechanism of stress-induced cytoplasmic translocation of TEAD involves the p38 MAPK pathway, which is independent of Hippo pathway components [89]. Therefore, the crosstalk between p38 MAPK and the Hippo pathway impinges on TEAD (Figure 2a). The p38-binding motif is located near the N-terminus nuclear localization signal (NLS) of TEADs through which TEAD-p38 forms a complex via direct protein-protein interaction (Figure 1). Although p38 does not phosphorylate TEAD directly, its kinase activity is required for TEAD-p38 complex formation and subsequent cytoplasmic translocation. Cellular responses to environmental stresses are mediated by distinct gene expression during the acute and adaptation phases [93]. During the acute phase of osmotic stress, YAP-TEAD is activated via NLK-mediated YAP S128 phosphorylation [94,95]. However, during the adaptation phase to stress, p38 binds and inhibits TEAD by inducing its cytoplasmic translocation, which indicates that the TEAD transcriptional output is required for cell survival upon the acute stress response, but is indispensable during the adaptation phase.

More importantly, the nuclear absence of TEAD impairs YAP/TAZ nuclear accumulation [63–65]. For example, in TEAD 1/2/4 KO cells, LPA-stimulated YAP/TAZ fails to accumulate in the nucleus even after it is completely dephosphorylated [63]. Therefore, stimuli that evoke TEAD cytoplasmic translocation overrides YAP/TAZ activating signals. The mechanism of stress- and p38-induced TEAD cytoplasmic retention is intact in various cancer cells. Notably, YAP-driven cancer cells, such as the GNAQ/11 mutant uveal melanoma cells and Hippo mutant mesothelioma cells, were specifically sensitive to TEAD cytoplasmic translocation when compared to YAP-independent cancer cells [63]. These findings suggest that signal transductions and/or chemical compounds that modulate TEAD subcellular localization potentially predominate the biological function of the Hippo-YAP/TAZ pathway.

4.2. Regulation of TEAD via Post-Translational Modifications

The post-translational modifications in TEAD have recently gained significant interest after its important roles in human pathophysiology came to light. To date, TEAD phosphorylation and palmitoylation have been shown to regulate its function. TEAD is phosphorylated by protein kinase A (PKA) and protein kinase C (PKC), which have been shown to inhibit TEAD by disrupting its DNA-binding [96,97] (Figure 1). However, the precise mechanism and context of TEAD phosphorylation requires further investigation. Recently, TEAD palmitoylation emerged as an important post-translational regulation mechanism [98–100]. Protein palmitoylation is important for protein trafficking and membrane localization [101]. S-palmitoylation of TEAD occurs by attaching a fatty acid (palmitate) to conserved cysteine residues in the TEAD C-terminus, which are within the YAP-binding domain (YBD) that share > 70% sequence identity among TEAD1-4 (Figure 1). The palmitoyl group is buried deep inside a hydrophobic pocket of TEAD as revealed by structural studies. All four TEAD paralogs are found palmitoylated in mammalian cells via an autopalmitoylation process that could be removed via a certain depalmitoylating enzyme such as APT2 [99,102]. Although the molecular mechanisms of YAP/TAZ and TEAD regulation via TEAD-YBD palmitoylation have been proposed, the functional role of YBD palmitoylation still remains unclear. Noland et al. and Mesrouze et al. showed that TEAD palmitoylation did not alter protein localization and YAP/TAZ binding. However, it was required for proper TEAD folding and protein stability [99,100]. Impaired TEAD2 palmitoylation decreased TEAD protein stability and resulted in the loss of protein abundance. On the other hand, Chan et al. demonstrated that TEAD palmitoylation did not affect its protein stability. However, it was required for the YAP/TAZ interaction. Palmitoylation-deficient TEAD mutants were not able to bind YAP/TAZ. Therefore, TEAD transcriptional activity was impaired, which indicates that TEAD palmitoylation plays important roles in regulating its binding to the transcriptional coactivators. Palmitoylation-deficient mutants have been reported to be properly folded and could still bind VGLL4, which suggests that the loss of YAP/TAZ binding is not due to TEAD misfolding [98]. The palmitoylation-deficient TEAD1 mutant also impaired TAZ-mediated muscle differentiation and YAP-mediated tissue overgrowth in Drosophila.

Thus, the precise role of TEAD palmitoylation requires in-depth investigation since the palmitate-binding hydrophobic pocket located in the YBD is likely to be an important site for therapeutic intervention. The first indication that the hydrophobic pocket could be a therapeutic target was obtained from a high-throughput screening that attempt to identify ligands that stabilized the TEAD-YBD. The results indicated that NSAIDs, such as flufenamic acid (FA) and niflumic acid (NA), were small-molecular inhibitors that bound the central TEAD hydrophobic pocket at its palmitoylation site [103]. Although the affinity of these drugs to TEAD requires optimization, FA- and NA-treatments decreased TEAD transcriptional activity and TEAD-induced cell migration and proliferation without disrupting TEAD-YAP interaction. Whether FA and NA compete against TEAD palmitoylation requires further investigation.

5. Roles of TEAD in Cancer Biology

5.1. TEAD Expression in Human Cancers

Numerous studies suggest the importance of TEAD in the development of human cancers. TEAD overexpression and hyperactivity has been implicated in multiple stages of cancer progression (Figure 2b). Although TEAD may be downregulated in some breast cancers and renal or bladder tumors [104], high TEAD expression levels have been correlated with poor clinical outcome, which serves as a prognostic marker in various solid tumors, such as prostate cancers [105], colorectal cancers [106,107], gastric cancers [108,109], breast cancers [110,111], germ cell tumors [112,113], head and neck squamous cell carcinomas [114], renal cell carcinomas [115], and medulloblastomas [116]. On the other hand, a loss-of-function mutation in TEAD1 (Y421H) was shown to cause Sveinsson's chorioretinal atrophy, which is a genetic disorder that results in degeneration by disrupting TEAD1-YAP interaction [117,118]. In accordance with these studies, meta-analysis studies revealed that both total and nuclear YAP and TAZ expression are intimately associated with adverse overall survival (OS) and disease-free survival (DFS) in numerous cancers, which suggests the prognostic role of TEADs and YAP/TAZ expression in patients with various malignancies [119,120]. Notably, TEAD was shown to be a critical drug target in YAP-driven tumorigenesis. In hepatocellular carcinoma (HCC), dominant-negative TEAD reversed YAP-induced hepatomegaly and tumorigenesis in vivo. Moreover, VGLL4-mimicking peptide (which binds the YBD) and verteporfin (a small YAP-binding chemical) were demonstrated to harbor therapeutic effects against YAP-induced tumorigenesis by interrupting TEAD-YAP interaction [62,121,122].

5.2. Role of TEAD in EMT

TEADs have emerged as important drivers of cancer development, tumor growth, EMT, metastasis, and drug resistance (Figure 2b). EMT is a natural developmental process that is phenocopied by cancer cells of epithelial origin. This process is crucial for cancer cells since it promotes cell migration, invasion, and anoikis resistance. Therefore, EMT emerged as a critical regulator of the cancer stem cell phenotype and a prerequisite for metastasis [123]. TEADs are critical mediators of EMT and metastasis during cancer progression. Numerous studies indicate that the TEAD transcriptional output induced by YAP/TAZ activation drives cell transformation by inducing EMT [17,18,124–126]. In these studies, TEAD activation disrupted cell-cell junctions, promoted mesenchymal gene expression, and increased cell migration and invasion. Aberrant TEAD activity was shown to promote mammary carcinoma and melanoma metastasis in a manner that was highly dependent on the YAP-interaction domain, which suggests that the interaction between TEAD and YAP is essential for EMT and metastasis [126]. TEAD also contributes to EMT by upregulating Slug and ZEB1 [127,128], which are major EMT transcription factors that promote cell migration and invasion by inhibiting epithelial markers and upregulating mesenchymal markers. TEAD triggers transcriptional induction of ZEB1 that drives metastatic squamous cell carcinoma. ZEB1 and TEAD also forms a complex to promote cancer stem cell traits and predict poor survival, therapy resistance, and increased metastatic risk in breast cancer [129]. In line with ZEB1, TEAD directly transcribes ZEB2 and represses DNp63 to regulate cell fate and lineage conversion in lung cancer progression [130]. Moreover, TEAD mediated non-small-cell lung carcinoma (NSCLC) aggressiveness by inducing Slug transcription and EMT [128]. Collectively, accumulating evidence indicates that TEAD functions as a critical EMT transcription factor in tumorigenesis. It will be interesting to answer key questions regarding the role of TEAD in EMT during development and MET (mesenchymal-to-epithelial transition), which occurs during metastatic colonization.

5.3. Role of TEAD in Metastasis

The majority of cancer-associated deaths occur due to metastasis, which is the dissemination of cancer cells from the primary tumor site to secondary organs. In order to metastasize, cancer cells must avoid anoikis (detachment-induced apoptosis) and must enter and exit from the blood vessels (intravasation and extravasation, respectively), complete metastatic colonization, and acquire drug resistance. These are all hallmarks of cancer stem cells. Upon interaction with platelets, metastatic cancer cells induce TEAD activation via the RhoA-MYPT1-PP1-YAP pathway. The platelet-activated TEAD transcriptional program in detached cancer cells induced anoikis-resistance and promoted cell survival and metastasis [131]. Similarly, YAP promotes anoikis-resistance upon cell detachment in cancer cells [132]. Cancer cells exposed to shear stress or disturbed flow also activates TEAD via ROCK-LIMK-cofilin signaling, which promotes cancer cell motility and metastasis [133]. Furthermore, the activation of TEAD target gene, CTGF, mediates the metastatic colonization of breast cancer through leukemia inhibitory factor receptor (LIFR) suppression [134]. In colorectal cancer, RARγ promoted TEAD activation through the Hippo pathway, which induces EMT, invasion, and metastasis [135]. However, increased TEAD expression and its nuclear localization in colorectal cancer cells promote EMT and metastasis via a Hippo-independent mechanism [107]. Moreover, TEAD activation induces ROR1-HER3-mediated osteoclast differentiation and bone metastasis of cancer cells via the Hippo-YAP pathway [136]. MRTF-activated TEAD also promotes breast cancer cell metastasis to the lung [33]. In breast cancer and melanoma cells, SRC tyrosine kinase-induced cell-ECM adhesion activates TEAD to promote tumor growth and enhance metastasis [137]. In addition, TGFβ-induced TEAD transcriptional activity is required to promote metastatic phenotypes in breast cancer cells [72]. *ARHGAP29* is a TEAD target gene that increases cell invasion and metastasis by regulating actin dynamics in cancer cells [138]. TEADs are also involved in metastatic seeding via disseminated cancer cells. The spreading of circulating tumor cells induced TEAD activation via L1CAM-ILK-YAP signaling, which is critical for metastatic colonization [139]. Collectively, these studies suggest that TEAD activation enhances metastatic tumor formation and raises the possibility that TEAD inhibition can prevent the survival and outgrowth of disseminated tumor cells.

6. TEADs as Mediators of Cancer Genes

Driver mutations in cancer-associated genes alter downstream signaling and transcription patterns, which are critical in tumor progression. Deregulation of TEAD transcriptional output have been demonstrated to mediate the pathology of critical oncogenes and tumor suppressor genes including NF2, BRAF, KRAS, MYC, PTEN, LKB1, and PKA (Figure 2a). This section will discuss the molecular mechanisms of TEAD regulation in the context of cancer gene-induced tumor development, drug resistance, and metastasis.

6.1. Hippo Pathway and TEADs

The Hippo pathway functions as a tumor suppressor pathway, whose activity is deregulated in many cancers [140]. However, mutations directly linked to alterations in the Hippo-YAP/TAZ pathway is uncommon. Inhibition of the core Hippo pathway components via point mutations and epigenetic alterations are found in subsets of human cancers including mutations in NF2 [141–143], MST1/2 [144,145], SAV1 [146], MOB1A/B [147], LATS1 [148–152], and LATS2 [88,153–155]. In addition, approximately 80% of uveal melanomas harbor activating mutations in GNAQ or GNA11, which are associated with the inhibition of the Hippo pathway and activation of TEAD [156,157]. Moreover, gain-of-function mutations were recently found in YAP and TAZ. Hyperactivating mutations in YAP were identified in melanoma and lung cancer patients [158–160]. Furthermore, chromosomal translocations that generated fusion proteins containing YAP (YAP-TFE3) and TAZ (WWTR1-CAMTA1) are oncogenic drivers in epithelioid hemangioendothelioma [161–163]. Similarly, in poromas and poro-carcinomas, fusion proteins containing YAP (YAP-MAML2, YAP-NUTM1) and TAZ (WWTR1-NUTM1) induced tumorigenesis via TEAD hyperactivation [164]. Several studies demonstrated that TEAD hyperactivation drives tumorigenesis driven via Hippo-YAP/TAZ pathway mutations, which suggests TEADs are therapeutic targets [141,143,157,158,162]. However, further investigations are required to link TEAD activity to the pathophysiology of each cancer gene within the Hippo pathway.

6.2. EGFR-RAS-RAF-MAPK Pathway and TEADs

TEADs are mediators of the EGFR-RAS-RAF-MAPK pathway, which is one of the most deregulated molecular pathways in human cancers. TEADs play important roles in tumor progression and drug resistance downstream of the hyperactivating mutations on EGFR, KRAS, or BRAF (Figure 2a). In NSCLC patients, TEAD activity correlates with the EGFR mutation status. EGFR mutant lung cancer tissues and cell lines show upregulated YAP expression, followed by increased TEAD activity [165]. Although EGFR acquires drug resistance to TKI via the T790M mutation, TEAD inhibition effectively reduces the viability of TKI-resistant lung adenocarcinoma cells [166,167]. TEAD also contributes to the immune escape of NSCLC cells by directly transcribing *PD-L1*, which, in turn, causes CD8$^+$ T cell exhaustion [43,168]. Furthermore, TEAD increases EGFR expression by directly binding to the *EGFR* promotor, which induces tumorigenesis and drug resistance in esophageal cancer [39]. The KRAS oncogene, which is downstream of EGFR, is one of the most frequently mutated proteins in human carcinoma. Activating mutations in KRAS are particularly prominent in pancreatic ductal adenocarcinoma (PDAC). Unlike EGFR and BRAF, KRAS has, thus far, not been considered to be a viable therapeutic target, which renders downstream effectors critical in the treatment of PDAC. Activation of TEAD-induced target genes including *COX2* and *MMP7*, which fueled KRAS^{G12D} driven-PDAC progression both in vivo and in vitro, suggests Celebrex (COX2 inhibitor) and marimastat (a clinical MMP inhibitor) to be possible therapeutic agents for PDAC treatment [169]. Moreover, numerous TEAD target genes have been associated with unfavorable prognosis for PDAC [170]. TEAD2 was shown to act cooperatively with the E2F transcription factor to promote a cell-cycle gene expression program, which enabled the bypass of oncogenic KRAS addiction in PDAC and evoked KRAS-independent tumor relapse [171]. Approximately half of metastatic melanoma patients harbor the BRAF^{V600} mutation, with the most common being BRAF^{V600E} [172]. Although vemurafenib (PLX4032) and dabrafenib were developed to treat BRAF^{V600}-mutant metastatic melanoma, a majority of patients ultimately became resistant [173]. Multiple studies have demonstrated that TEAD activity contributes toward BRAF inhibitor-resistance in melanoma cells. TEAD activity was increased in drug resistant-melanoma cells due to actin cytoskeletal remodeling that induced cancer stemness and invasion [174,175]. BRAF inhibitor-resistant cells can evade the immune system via TEAD activation. TEAD-mediated direct transcription of *PD-L1* was responsible for PD-1-dependent CD8$^+$ T cell exhaustion, which allowed BRAF inhibitor-resistant cells to escape the immune responses [41,42]. Moreover, TEAD induced cytokines *IL-6* and *CSF1-3* in KRAS mutant PDAC cells to recruit myeloid-derived suppressor cells (MDSCs), which formed an immunosuppressive tumor microenvironment [176]. Collectively, the TEAD transcriptional output played critical functions in the pathogenesis of the EGFR-RAS-RAF-MAPK pathway and mutation-induced tumor progression.

6.3. LKB1-AMPK Signaling, Energy Stress, and TEADs

LKB1 (STK11) is a well-characterized tumor suppressor that governs diverse cellular processes, including cell growth, polarity, and metabolism. Approximately 15% to 30% of NSCLC patients harbor LKB1 inactivating mutations. An RNAi-based kinome screen identified LKB1-MARK signaling as a potent inhibitor of TEAD transcription activity by inactivating YAP, which suggests TEAD to be a therapeutic target in the treatment of LKB1-mutant human malignancies [177]. Another study showed that LKB1 was required by Dishevelled (DVL) to inhibit TEAD activity, which facilitated the co-activation of the Wnt/β-catenin and Hippo/YAP pathways [178]. Furthermore, the loss of LKB1 in lung adenocarcinoma cells activated TEAD, which directly transcribed *ZEB2* and repressed DNp63 to regulate cell fate and lineage conversion in lung cancer progression [130]. Similarly, the oncogenic protein Survivin, which is a TEAD target gene, promoted malignant progression of LKB1-deficient lung adenocarcinoma cells [179]. Downstream signaling of LKB1 led to the activation of AMPK, which is a key cellular energy sensor that functions as a tumor suppressor. Similar to LKB1, AMPK activation also inhibited TEAD activity. Upon energy stress, AMPK regulated TEAD by inhibiting YAP via both Hippo-dependent and Hippo-independent mechanisms, which resulted in impaired tumor growth [49,180,181]. Furthermore, the glucose transporter *GLUT3* is a direct target gene of TEAD. Studies have shown that the glycolytic enzyme PFK1 interacts with TEAD to increase glucose metabolism and tumor growth [49,73]. Further studies on the role of TEAD as a downstream mediator of LKB1-AMPK signaling and energy stress in growth control and cancer metabolism will provide important means of treating LKB1 mutant cancer cells.

6.4. MYC and TEADs

The MYC oncogene is deregulated in >50% of human cancers and its hyperactivation is associated with a poor prognosis and unfavorable patient survival [182]. TEADs have emerged as transcription factors that induce *MYC* gene expression (Figure 2a). Transcriptome analysis of a Hippo-deficient gastric cancer model showed that TEAD activation directly upregulates *MYC* and its target genes, which, in turn, induced tumor progression [183]. Similarly, in oral squamous cell carcinoma (OSCC), TEAD induced *MYC* and *BCL-2* gene expression, which accelerated OSCC tumorigenesis [184]. In *Drosophila*, TEAD has also been shown to transcribe *MYC*, which induces tissue growth and increased organ size [44]. Furthermore, an in vitro cell competition assay shows that direct TEAD-induced *MYC* expression is required to become super-competitors, which indicates that TEAD and MYC cooperatively control cell proliferation and cell competition [185]. Despite several studies demonstrating MYC to be a TEAD target gene, MYC is also shown to act in parallel with or upstream of TEAD to modulate cellular function. MYC and TEAD coordinate the gene expression required for cell proliferation upon activation of mitogenic signals. TEADs were found to be constitutively bound to the promoters of the MYC target genes. The subsequent binding of MYC to these sites recruited YAP to bind TEAD, which induced MYC/TEAD target gene expression. This cooperativity between MYC and TEAD induced liver growth and tumorigenesis in mouse models [186]. Notably, the TEAD4-MYCN positive feedback loop was identified to drive high-risk neuroblastoma associated with MYCN amplification [45]. Apart from these reports, TEAD and MYC also showed an anti-correlation in breast cancer patients that enabled the stratification of breast cancer subtypes [187]. Another study also shows that MYC inhibits TEAD transcriptional activity in MYC-driven breast cancer by suppressing the binding between TEAD and YAP/TAZ via AMPK-induced phosphorylation [188]. Similarly, a transcriptional repressor TRPS1 and MYC are commonly co-amplified in breast cancers and TRPS1 inhibits YAP-dependent TEAD activity by direct binding [189]. Collectively, TEAD and MYC form a regulatory feedback mechanism that is important for proper growth control. Thus, further research on the TEAD-MYC circuitry may yield new insights into the treatment of diverse cancer subtypes.

7. Conclusions

In this review, we underscore the molecular mechanisms and functions of TEAD, which is involved in the multistep process of tumor progression, including tumor development, EMT, drug resistance, and metastasis (Figure 2b). Although oncogenic driver mutations in TEAD are yet to be identified, the TEAD transcriptional output was found to be responsible for human malignancies via the crosstalk between various oncogenic signaling pathways and cancer genes such as the Hippo pathway, EGFR-RAS-RAF-MEK pathway, LKB1, GNAQ/11, and MYC (Figure 2a). Despite the emerging role of TEADs as critical effectors in cancer biology, most research regarding TEAD regulation focuses on its major coactivators, YAP and TAZ. YAP/TAZ-independent regulatory mechanisms that govern TEAD activity, which include post-translation modifications, subcellular localizations, and upstream activating and inhibiting stimuli, have been hardly elucidated. Notably, recent studies revealed unprecedented regulatory mechanisms of TEAD such as nucleocytoplasmic shuttling and palmitoylation. These provide the first mechanistic insight in manipulating and targeting TEAD subcellular localization and post-translational modifications to be attractive therapeutic interventions to treat human malignancies. Although flufenamates, originally developed as a COX inhibitor, have been demonstrated to bind and inhibit TEAD functions, additional drugs that specifically bind TEAD with high selectivity and efficacy is yet to be developed. Since TEAD functions as the signaling nexus for critical pathways in tumor progression, future progress in elucidating regulatory mechanisms of TEAD and developing therapeutic interventions will initiate an exciting new arena for basic science as well as pharmaceutics.

Funding: This work was supported by a grant from the Korea Health Technology R&D Project through the Korea Health Industry Development Institute (KHIDI), funded by the Ministry of Health & Welfare, Republic of Korea (HI17C1560), and the National Research Foundation of Korea (NRF) grant funded by the Korean government (NRF-2018R1C1B6004301) and (NRF-2017R1A4A1015328), and Yonsei University Future leading Research Initiative of 2017 (2017-22-0071) to H.W.P., and the National Research Foundation of Korea (NRF) grant funded by the Korean government (NRF-2014R1A5A2009936) and (NRF-2015R1A2A2A01004240) to H.-S.J., and H.D.H. and D.H.K were supported by the Brain Korea (BK21) PLUS Program.

Acknowledgments: We apologize to those colleagues whose work could not be cited because of space limitations. The authors thank all the members of the TCR laboratory for insightful discussions and critical comments pertaining to this review article.

Conflicts of Interest: The authors declare no conflict of interest.

References

1. Yasunami, M.; Suzuki, K.; Ohkubo, H. A novel family of TEA domain-containing transcription factors with distinct spatiotemporal expression patterns. *Biochem. Biophys. Res. Commun.* **1996**, *228*, 365–370. [CrossRef] [PubMed]

2. Lin, K.C.; Park, H.W.; Guan, K.L. Regulation of the Hippo Pathway Transcription Factor TEAD. *Trends Biochem. Sci.* **2017**, *42*, 862–872. [CrossRef] [PubMed]

3. Chen, Z.; Friedrich, G.A.; Soriano, P. Transcriptional Enhancer Factor-1 Disruption by a Retroviral Gene Trap Leads to Heart-Defects and Embryonic Lethality in Mice. *Gene Dev.* **1994**, *8*, 2293–2301. [CrossRef] [PubMed]

4. Kaneko, K.J.; Kohn, M.J.; Liu, C.; DePamphilis, M.L. Transcription factor TEAD2 is involved in neural tube closure. *Genesis* **2007**, *45*, 577–587. [CrossRef] [PubMed]

5. Sawada, A.; Kiyonari, H.; Ukita, K.; Nishioka, N.; Imuta, Y.; Sasaki, H. Redundant roles of Tead1 and Tead2 in notochord development and the regulation of cell proliferation and survival. *Mol. Cell Biol.* **2008**, *28*, 3177–3189. [CrossRef] [PubMed]

6. Yagi, R.; Kohn, M.J.; Karavanova, I.; Kaneko, K.J.; Vullhorst, D.; DePamphilis, M.L.; Buonanno, A. Transcription factor TEAD4 specifies the trophectoderm lineage at the beginning of mammalian development. *Development* **2007**, *134*, 3827–3836. [CrossRef]

7. Nishioka, N.; Yamamoto, S.; Kiyonari, H.; Sato, H.; Sawada, A.; Ota, M.; Nakao, K.; Sasaki, H. Tead4 is required for specification of trophectoderm in pre-implantation mouse embryos. *Mech. Dev.* **2008**, *125*, 270–283. [CrossRef]

8. Nishioka, N.; Inoue, K.; Adachi, K.; Kiyonari, H.; Ota, M.; Ralston, A.; Yabuta, N.; Hirahara, S.; Stephenson, R.O.; Ogonuki, N.; et al. The Hippo signaling pathway components Lats and Yap pattern Tead4 activity to distinguish mouse trophectoderm from inner cell mass. *Dev. Cell* **2009**, *16*, 398–410. [CrossRef]

9. Ota, M.; Sasaki, H. Mammalian Tead proteins regulate cell proliferation and contact inhibition as transcriptional mediators of Hippo signaling. *Development* **2008**, *135*, 4059–4069. [CrossRef]

10. Zhao, B.; Wei, X.; Li, W.; Udan, R.S.; Yang, Q.; Kim, J.; Xie, J.; Ikenoue, T.; Yu, J.; Li, L.; et al. Inactivation of YAP oncoprotein by the Hippo pathway is involved in cell contact inhibition and tissue growth control. *Gene Dev.* **2007**, *21*, 2747–2761. [CrossRef]

11. Jacquemin, P.; Hwang, J.J.; Martial, J.A.; Dolle, P.; Davidson, I. A novel family of developmentally regulated mammalian transcription factors containing the TEA/ATTS DNA binding domain. *J. Biol. Chem.* **1996**, *271*, 21775–21785. [CrossRef] [PubMed]

12. Jiang, S.W.; Desai, D.; Khan, S.; Eberhardt, N.L. Cooperative binding of TEF-1 to repeated GGAATG-related consensus elements with restricted spatial separation and orientation. *DNA Cell. Biol.* **2000**, *19*, 507–514. [CrossRef] [PubMed]

13. Anbanandam, A.; Albarado, D.C.; Nguyen, C.T.; Halder, G.; Gao, X.L.; Veeraraghavan, S. Insights into transcription enhancer factor 1 (TEF-1) activity from the solution structure of the TEA domain. *Proc. Natl. Acad. Sci. USA* **2006**, *103*, 17225–17230. [CrossRef] [PubMed]

14. Dupont, S.; Morsut, L.; Aragona, M.; Enzo, E.; Giulitti, S.; Cordenonsi, M.; Zanconato, F.; Le Digabel, J.; Forcato, M.; Bicciato, S.; et al. Role of YAP/TAZ in mechanotransduction. *Nature* **2011**, *474*, 179–183. [CrossRef] [PubMed]

15. Li, X.; Wang, W.Q.; Wang, J.D.; Malovannaya, A.; Xi, Y.X.; Li, W.; Guerra, R.; Hawke, D.H.; Qin, J.; Chen, J.J. Proteomic analyses reveal distinct chromatin-associated and soluble transcription factor complexes. *Mol. Syst. Biol.* **2015**, *11*. [CrossRef] [PubMed]

16. Xiao, J.H.; Davidson, I.; Matthes, H.; Garnier, J.M.; Chambon, P. Cloning, Expression, and Transcriptional Properties of the Human Enhancer Factor Tef-1. *Cell* **1991**, *65*, 551–568. [CrossRef]

17. Zhang, H.; Liu, C.Y.; Zha, Z.Y.; Zhao, B.; Yao, J.; Zhao, S.M.; Xiong, Y.; Lei, Q.Y.; Guan, K.L. TEAD Transcription Factors Mediate the Function of TAZ in Cell Growth and Epithelial-Mesenchymal Transition. *J. Biol. Chem.* **2009**, *284*, 13355–13362. [CrossRef]

18. Zhao, B.; Ye, X.; Yu, J.D.; Li, L.; Li, W.Q.; Li, S.M.; Yu, J.J.; Lin, J.D.; Wang, C.Y.; Chinnaiyan, A.M.; et al. TEAD mediates YAP-dependent gene induction and growth control. *Gene Dev.* **2008**, *22*, 1962–1971. [CrossRef]

19. Vassilev, A.; Kaneko, K.J.; Shu, H.J.; Zhao, Y.M.; DePamphilis, M.L. TEAD/TEF transcription factors utilize the activation domain of YAP65, a Src/Yes-associated protein localized in the cytoplasm. *Gene Dev.* **2001**, *15*, 1229–1241. [CrossRef]

20. Mahoney, W.M.; Hong, J.H.; Yaffe, M.B.; Farrance, K.G. The transcriptional co-activator TAZ interacts differentially with transcriptional enhancer factor-1 (TEF-1) family members. *Biochem. J.* **2005**, *388*, 217–225. [CrossRef]

21. Chen, H.H.; Maeda, T.; Mullett, S.J.; Stewart, A.F.R. Transcription cofactor Vgl-2 is required for skeletal muscle differentiation. *Genesis* **2004**, *39*, 273–279. [CrossRef] [PubMed]

22. Koontz, L.M.; Liu-Chittenden, Y.; Yin, F.; Zheng, Y.G.; Yu, J.Z.; Huang, B.; Chen, Q.; Wu, S.; Pan, D.J. The Hippo Effector Yorkie Controls Normal Tissue Growth by Antagonizing Scalloped-Mediated Default Repression. *Dev. Cell* **2013**, *25*, 388–401. [CrossRef] [PubMed]

23. Chen, H.H.; Mullett, S.J.; Stewart, A.F. Vgl-4, a novel member of the vestigial-like family of transcription cofactors, regulates alpha1-adrenergic activation of gene expression in cardiac myocytes. *J. Biol. Chem.* **2004**, *279*, 30800–30806. [CrossRef] [PubMed]

24. Pobbati, A.V.; Chan, S.W.; Lee, I.; Song, H.W.; Hong, W.J. Structural and Functional Similarity between the Vgll1-TEAD and the YAP-TEAD Complexes. *Structure* **2012**, *20*, 1135–1140. [CrossRef] [PubMed]

25. Honda, M.; Hidaka, K.; Fukada, S.; Sugawa, R.; Shirai, M.; Ikawa, M.; Morisaki, T. Vestigial-like 2 contributes to normal muscle fiber type distribution in mice. *Sci. Rep.* **2017**, *7*. [CrossRef] [PubMed]

26. Figeac, N.; Mohamed, A.D.; Sun, C.; Schonfelder, M.; Matallanas, D.; Garcia-Munoz, A.; Missiaglia, E.; Collie-Duguid, E.; De Mello, V.; Pobbati, A.V.; et al. Vgll3 operates via Tead1, Tead3 and Tead4 to influence myogenesis in skeletal muscle. *J. Cell Sci.* **2019**. [CrossRef]

27. Kim, M.; Kim, T.; Johnson, R.L.; Lim, D.S. Transcriptional Co-repressor Function of the Hippo Pathway Transducers YAP and TAZ. *Cell Rep.* **2015**, *11*, 270–282. [CrossRef]

28. Galli, G.G.; Carrara, M.; Yuan, W.C.; Valdes-Quezada, C.; Gurung, B.; Pepe-Mooney, B.; Zhang, T.H.; Geeven, G.; Gray, N.S.; de Laat, W.; et al. YAP Drives Growth by Controlling Transcriptional Pause Release from Dynamic Enhancers. *Mol. Cell* **2015**, *60*, 328–337. [CrossRef]

29. Jiao, S.; Li, C.C.; Hao, Q.; Miao, H.F.; Zhang, L.; Li, L.; Zhou, Z.C. VGLL4 targets a TCF4-TEAD4 complex to coregulate Wnt and Hippo signalling in colorectal cancer. *Nat. Commun.* **2017**, *8*. [CrossRef]

30. Beyer, T.A.; Weiss, A.; Khomchuk, Y.; Huang, K.; Ogunjimi, A.A.; Varelas, X.; Wrana, J.L. Switch enhancers interpret TGF-beta and Hippo signaling to control cell fate in human embryonic stem cells. *Cell Rep.* **2013**, *5*, 1611–1624. [CrossRef]

31. Liu, X.; Li, H.; Rajurkar, M.; Li, Q.; Cotton, J.L.; Ou, J.; Zhu, L.J.; Goel, H.L.; Mercurio, A.M.; Park, J.S.; et al. Tead and AP1 Coordinate Transcription and Motility. *Cell Rep.* **2016**, *14*, 1169–1180. [CrossRef] [PubMed]

32. Zanconato, F.; Forcato, M.; Battilana, G.; Azzolin, L.; Quaranta, E.; Bodega, B.; Rosato, A.; Bicciato, S.; Cordenonsi, M.; Piccolo, S. Genome-wide association between YAP/TAZ/TEAD and AP-1 at enhancers drives oncogenic growth. *Nat. Cell Biol.* **2015**, *17*, 1218. [CrossRef] [PubMed]

33. Kim, T.; Hwang, D.; Lee, D.; Kim, J.H.; Kim, S.Y.; Lim, D.S. MRTF potentiates TEAD-YAP transcriptional activity causing metastasis. *EMBO J.* **2017**, *36*, 520–535. [CrossRef] [PubMed]

34. Park, H.W.; Kim, Y.C.; Yu, B.; Moroishi, T.; Mo, J.S.; Plouffe, S.W.; Meng, Z.P.; Lin, K.C.; Yu, F.X.; Alexander, C.M.; et al. Alternative Wnt Signaling Activates YAP/TAZ. *Cell* **2015**, *162*, 780–794. [CrossRef]

35. Seo, E.; Basu-Roy, U.; Gunaratne, P.H.; Coarfa, C.; Lim, D.S.; Basilico, C.; Mansukhani, A. SOX2 Regulates YAP1 to Maintain Stemness and Determine Cell Fate in the Osteo-Adipo Lineage. *Cell Rep.* **2013**, *3*, 2075–2087. [CrossRef]

36. Lee, D.H.; Park, J.O.; Kim, T.S.; Kim, S.K.; Kim, T.H.; Kim, M.C.; Park, G.S.; Kim, J.H.; Kuninaka, S.; Olson, E.N.; et al. LATS-YAP/TAZ controls lineage specification by regulating TGF beta signaling and Hnf4 alpha expression during liver development. *Nat. Commun.* **2016**, *7*. [CrossRef]

37. Lai, D.; Yang, X.L. BMP4 is a novel transcriptional target and mediator of mammary cell migration downstream of the Hippo pathway component TAZ. *Cell Signal.* **2013**, *25*, 1720–1728. [CrossRef]

38. Zhang, J.M.; Ji, J.Y.; Yu, M.; Overholtzer, M.; Smolen, G.A.; Wang, R.; Brugge, J.S.; Dyson, N.J.; Haber, D.A. YAP-dependent induction of amphiregulin identifies a non-cell-autonomous component of the Hippo pathway. *Nat. Cell Biol.* **2009**, *11*, 1444–1450. [CrossRef]

39. Song, S.M.; Honjo, S.; Jin, J.K.; Chang, S.S.; Scott, A.W.; Chen, Q.R.; Kalhor, N.; Correa, A.M.; Hofstetter, W.L.; Albarracin, C.T.; et al. The Hippo Coactivator YAP1 Mediates EGFR Overexpression and Confers Chemoresistance in Esophageal Cancer. *Clin. Cancer Res.* **2015**, *21*, 2580–2590. [CrossRef]

40. Feng, J.; Yang, H.; Zhang, Y.; Wei, H.; Zhu, Z.; Zhu, B.; Yang, M.; Cao, W.; Wang, L.; Wu, Z. Tumor cell-derived lactate induces TAZ-dependent upregulation of PD-L1 through GPR81 in human lung cancer cells. *Oncogene* **2017**, *36*, 5829–5839. [CrossRef]

41. Kim, M.H.; Kim, C.G.; Kim, S.K.; Shin, S.J.; Choe, E.A.; Park, S.H.; Shin, E.C.; Kim, J. YAP-Induced PD-L1 Expression Drives Immune Evasion in BRAFi-Resistant Melanoma. *Cancer Immunol. Res.* **2018**, *6*, 255–266. [CrossRef] [PubMed]

42. Van Rensburg, H.J.J.; Azad, T.; Ling, M.; Hao, Y.W.; Snetsinger, B.; Khanal, P.; Minassian, L.M.; Graham, C.H.; Rauh, M.J.; Yang, X.L. The Hippo Pathway Component TAZ Promotes Immune Evasion in Human Cancer through PD-L1. *Cancer Res.* **2018**, *78*, 1457–1470. [CrossRef] [PubMed]

43. Miao, J.B.; Hsu, P.C.; Yang, Y.L.; Xu, Z.D.; Dai, Y.Y.; Wang, Y.C.; Chan, G.; Huang, Z.; Hu, B.; Li, H.; et al. YAP regulates PD-L1 expression in human NSCLC cells. *Oncotarget* **2017**, *8*, 114576–114587. [CrossRef] [PubMed]

44. Neto-Silva, R.M.; de Beco, S.; Johnston, L.A. Evidence for a Growth-Stabilizing Regulatory Feedback Mechanism between Myc and Yorkie, the Drosophila Homolog of Yap. *Dev. Cell* **2010**, *19*, 507–520. [CrossRef] [PubMed]

45. Rajbhandari, P.; Lopez, G.; Capdevila, C.; Salvatori, B.; Yu, J.Y.; Rodriguez-Barrueco, R.; Martinez, D.; Yarmarkovich, M.; Weichert-Leahey, N.; Abraham, B.J.; et al. Cross-Cohort Analysis Identifies a TEAD4-MYCN Positive Feedback Loop as the Core Regulatory Element of High-Risk Neuroblastoma. *Cancer Discov.* **2018**, *8*, 582–599. [CrossRef] [PubMed]

46. Moroishi, T.; Park, H.W.; Qin, B.D.; Chen, Q.; Meng, Z.P.; Plouffe, S.W.; Taniguchi, K.; Yu, F.X.; Karin, M.; Pan, D.J.; et al. A YAP/TAZ-induced feedback mechanism regulates Hippo pathway homeostasis. *Gene Dev.* **2015**, *29*, 1271–1284. [CrossRef] [PubMed]

47. Park, Y.Y.; Sohn, B.H.; Johnson, R.L.; Kang, M.H.; Kim, S.B.; Shim, J.J.; Mangala, L.S.; Kim, J.H.; Yoo, J.E.; Rodriguez-Aguayo, C.; et al. Yes-associated protein 1 and transcriptional coactivator with PDZ-binding motif activate the mammalian target of rapamycin complex 1 pathway by regulating amino acid transporters in hepatocellular carcinoma. *Hepatology* **2016**, *63*, 159–172. [CrossRef] [PubMed]

48. Hansen, C.G.; Ng, Y.L.D.; Lam, W.L.M.; Plouffe, S.W.; Guan, K.L. The Hippo pathway effectors YAP and TAZ promote cell growth by modulating amino acid signaling to mTORC1. *Cell Res.* **2015**, *25*, 1299–1313. [CrossRef]

49. Wang, W.; Xiao, Z.D.; Li, X.; Aziz, K.E.; Gan, B.; Johnson, R.L.; Chen, J. AMPK modulates Hippo pathway activity to regulate energy homeostasis. *Nat. Cell Biol.* **2015**, *17*, 490–499. [CrossRef]

50. Holden, J.K.; Cunningham, C.N. Targeting the Hippo Pathway and Cancer through the TEAD Family of Transcription Factors. *Cancers* **2018**, *10*. [CrossRef]

51. Park, J.H.; Shin, J.E.; Park, H.W. The Role of Hippo Pathway in Cancer Stem Cell Biology. *Mol. Cells* **2018**, *41*, 83–92. [CrossRef] [PubMed]

52. Moya, I.M.; Halder, G. Hippo-YAP/TAZ signalling in organ regeneration and regenerative medicine. *Nat. Rev. Mol. Cell Biol.* **2018**. [CrossRef] [PubMed]

53. Moon, S.; Yeon Park, S.; Woo Park, H. Regulation of the Hippo pathway in cancer biology. *Cell Mol. Life Sci.* **2018**, *75*, 2303–2319. [CrossRef] [PubMed]

54. Kim, C.L.; Choi, S.H.; Mo, J.S. Role of the Hippo Pathway in Fibrosis and Cancer. *Cells* **2019**, *8*. [CrossRef]

55. Meng, Z.P.; Moroishi, T.; Guan, K.L. Mechanisms of Hippo pathway regulation. *Gene Dev.* **2016**, *30*, 1–17. [CrossRef] [PubMed]

56. Yu, F.X.; Guan, K.L. The Hippo pathway: regulators and regulations. *Gene Dev.* **2013**, *27*, 355–371. [CrossRef] [PubMed]

57. Pavel, M.; Renna, M.; Park, S.J.; Menzies, F.M.; Ricketts, T.; Fullgrabe, J.; Ashkenazi, A.; Frake, R.A.; Lombarte, A.C.; Bento, C.F.; et al. Contact inhibition controls cell survival and proliferation via YAP/TAZ-autophagy axis. *Nat. Commun.* **2018**, *9*. [CrossRef] [PubMed]

58. Liang, N.; Zhang, C.; Dill, P.; Panasyuk, G.; Pion, D.; Koka, V.; Gallazzini, M.; Olson, E.N.; Lam, H.; Henske, E.P.; et al. Regulation of YAP by mTOR and autophagy reveals a therapeutic target of tuberous sclerosis complex. *J. Exp. Med.* **2014**, *211*, 2249–2263. [CrossRef]

59. Liu, C.Y.; Zha, Z.Y.; Zhou, X.; Zhang, H.; Huang, W.; Zhao, D.; Li, T.T.; Chan, S.W.; Lim, C.J.; Hong, W.J.; et al. The Hippo Tumor Pathway Promotes TAZ Degradation by Phosphorylating a Phosphodegron and Recruiting the SCF beta-TrCP E3 Ligase. *J. Biol. Chem.* **2010**, *285*, 37159–37169. [CrossRef]

60. Zhao, B.; Li, L.; Tumaneng, K.; Wang, C.Y.; Guan, K.L. A coordinated phosphorylation by Lats and CK1 regulates YAP stability through SCF beta-TRCP. *Gene Dev.* **2010**, *24*, 72–85. [CrossRef]

61. Stein, C.; Bardet, A.F.; Roma, G.; Bergling, S.; Clay, I.; Ruchti, A.; Agarinis, C.; Schmelzle, T.; Bouwmeester, T.; Schubeler, D.; et al. YAP1 Exerts Its Transcriptional Control via TEAD-Mediated Activation of Enhancers. *PLoS Genet.* **2015**, *11*. [CrossRef] [PubMed]

62. Liu-Chittenden, Y.; Huang, B.; Shim, J.S.; Chen, Q.; Lee, S.J.; Anders, R.A.; Liu, J.O.; Pan, D.J. Genetic and pharmacological disruption of the TEAD-YAP complex suppresses the oncogenic activity of YAP. *Gene Dev.* **2012**, *26*, 1300–1305. [CrossRef] [PubMed]

63. Lin, K.C.; Moroishi, T.; Meng, Z.P.; Jeong, H.S.; Plouffe, S.W.; Sekido, Y.; Han, J.H.; Park, H.W.; Guan, K.L. Regulation of Hippo pathway transcription factor TEAD by p38 MAPK-induced cytoplasmic translocation. *Nat. Cell Biol.* **2017**, *19*, 996. [CrossRef] [PubMed]

64. Chan, S.W.; Lim, C.J.; Loo, L.S.; Chong, Y.F.; Huang, C.X.; Hong, W.J. TEADs Mediate Nuclear Retention of TAZ to Promote Oncogenic Transformation. *J. Biol. Chem.* **2009**, *284*, 14347–14358. [CrossRef] [PubMed]

65. Diepenbruck, M.; Waldmeier, L.; Ivanek, R.; Berninger, P.; Arnold, P.; van Nimwegen, E.; Christofori, G. Tead2 expression levels control the subcellular distribution of Yap and Taz, zyxin expression and epithelial-mesenchymal transition. *J. Cell Sci.* **2014**, *127*, 1523–1536. [CrossRef] [PubMed]

66. He, C.B.; Mao, D.G.; Hua, G.H.; Lv, X.M.; Chen, X.C.; Angeletti, P.C.; Dong, J.X.; Remmenga, S.W.; Rodabaugh, K.J.; Zhou, J.; et al. The Hippo/YAP pathway interacts with EGFR signaling and HPV oncoproteins to regulate cervical cancer progression. *EMBO Mol. Med.* **2015**, *7*, 1426–1449. [CrossRef]

67. Fan, R.; Kim, N.G.; Gumbiner, B.M. Regulation of Hippo pathway by mitogenic growth factors via phosphoinositide 3-kinase and phosphoinositide-dependent kinase-1. *Proc. Natl. Acad. Sci. USA* **2013**, *110*, 2569–2574. [CrossRef]

68. Samanta, S.; Guru, S.; Elaimy, A.L.; Amante, J.J.; Ou, J.H.; Yu, J.; Zhu, L.H.J.; Mercurio, A.M. IMP3 Stabilization of WNT5B mRNA Facilitates TAZ Activation in Breast Cancer. *Cell Rep.* **2018**, *23*, 2559–2567. [CrossRef]

69. Zhang, S.P.; Zhang, H.; Ghia, E.M.; Huang, J.J.; Wu, L.F.; Zhang, J.C.; Lam, S.; Lei, Y.; He, J.S.; Cui, B.; et al. Inhibition of chemotherapy resistant breast cancer stem cells by a ROR1 specific antibody. *Proc. Natl. Acad. Sci. USA* **2019**, *116*, 1370–1377. [CrossRef]

70. Pefani, D.E.; Pankova, D.; Abraham, A.G.; Grawenda, A.M.; Vlahov, N.; Scrace, S.; E, O.N. TGF-beta Targets the Hippo Pathway Scaffold RASSF1A to Facilitate YAP/SMAD2 Nuclear Translocation. *Mol. Cell* **2016**, *63*, 156–166. [CrossRef]

71. Varelas, X.; Sakuma, R.; Samavarchi-Tehrani, P.; Peerani, R.; Rao, B.M.; Dembowy, J.; Yaffe, M.B.; Zandstra, P.W.; Wrana, J.L. TAZ controls Smad nucleocytoplasmic shuttling and regulates human embryonic stem-cell self-renewal. *Nat. Cell Biol.* **2008**, *10*, 837–848. [CrossRef]

72. Hiemer, S.E.; Szymaniak, A.D.; Varelas, X. The transcriptional regulators TAZ and YAP direct transforming growth factor beta-induced tumorigenic phenotypes in breast cancer cells. *J. Biol. Chem.* **2014**, *289*, 13461–13474. [CrossRef]

73. Enzo, E.; Santinon, G.; Pocaterra, A.; Aragona, M.; Bresolin, S.; Forcato, M.; Grifoni, D.; Pession, A.; Zanconato, F.; Guzzo, G.; et al. Aerobic glycolysis tunes YAP/TAZ transcriptional activity. *EMBO J.* **2015**, *34*, 1349–1370. [CrossRef] [PubMed]

74. Dai, X.M.; Liu, H.; Shen, S.Y.; Guo, X.C.; Yan, H.; Ji, X.Y.; Li, L.; Huang, J.; Feng, X.H.; Zhao, B. YAP activates the Hippo pathway in a negative feedback loop. *Cell Res.* **2015**, *25*, 1175–1178. [CrossRef] [PubMed]

75. Azzolin, L.; Zanconato, F.; Bresolin, S.; Forcato, M.; Basso, G.; Bicciato, S.; Cordenonsi, M.; Piccolo, S. Role of TAZ as Mediator of Wnt Signaling. *Cell* **2012**, *151*, 1443–1456. [CrossRef] [PubMed]

76. Azzolin, L.; Panciera, T.; Soligo, S.; Enzo, E.; Bicciato, S.; Dupont, S.; Bresolin, S.; Frasson, C.; Basso, G.; Guzzardo, V.; et al. YAP/TAZ Incorporation in the beta-Catenin Destruction Complex Orchestrates the Wnt Response. *Cell* **2014**, *158*, 157–170. [CrossRef]

77. Cai, J.; Maitra, A.; Anders, R.A.; Taketo, M.M.; Pan, D.J. beta-Catenin destruction complex-independent regulation of Hippo-YAP signaling by APC in intestinal tumorigenesis. *Gene Dev.* **2015**, *29*, 1493–1506. [CrossRef] [PubMed]

78. Feng, Y.; Liang, Y.; Zhu, X.; Wang, M.; Gui, Y.; Lu, Q.; Gu, M.; Xue, X.; Sun, X.; He, W.; et al. The signaling protein Wnt5a promotes TGFbeta1-mediated macrophage polarization and kidney fibrosis by inducing the transcriptional regulators Yap/Taz. *J. Biol. Chem.* **2018**, *293*, 19290–19302. [CrossRef]

79. Wang, C.; Han, X.; Zhou, Z.; Uyunbilig, B.; Huang, X.; Li, R.; Li, X. Wnt3a Activates the WNT-YAP/TAZ Pathway to Sustain CDX2 Expression in Bovine Trophoblast Stem Cells. *DNA Cell Biol.* **2019**. [CrossRef]

80. Estaras, C.; Hsu, H.T.; Huang, L.; Jones, K.A. YAP repression of the WNT3 gene controls hESC differentiation along the cardiac mesoderm lineage. *Genes Dev.* **2017**, *31*, 2250–2263. [CrossRef]

81. Gregorieff, A.; Liu, Y.; Inanlou, M.R.; Khomchuk, Y.; Wrana, J.L. Yap-dependent reprogramming of Lgr5(+) stem cells drives intestinal regeneration and cancer. *Nature* **2015**, *526*, 715–718. [CrossRef] [PubMed]

82. Barry, E.R.; Morikawa, T.; Butler, B.L.; Shrestha, K.; de la Rosa, R.; Yan, K.S.; Fuchs, C.S.; Magness, S.T.; Smits, R.; Ogino, S.; et al. Restriction of intestinal stem cell expansion and the regenerative response by YAP. *Nature* **2013**, *493*, 106. [CrossRef] [PubMed]

83. Qin, H.; Hejna, M.; Liu, Y.X.; Percharde, M.; Wossidlo, M.; Blouin, L.; Durruthy-Durruthy, J.; Wong, P.; Qi, Z.X.; Yu, J.W.; et al. YAP Induces Human Naive Pluripotency. *Cell Rep.* **2016**, *14*, 2301–2312. [CrossRef] [PubMed]

84. Miranda, M.Z.; Bialik, J.F.; Speight, P.; Dan, Q.; Yeung, T.; Szaszi, K.; Pedersen, S.F.; Kapus, A. TGF-beta1 regulates the expression and transcriptional activity of TAZ protein via a Smad3-independent, myocardin-related transcription factor-mediated mechanism. *J. Biol. Chem.* **2017**, *292*, 14902–14920. [CrossRef] [PubMed]

85. Kimura, T.E.; Duggirala, A.; Smith, M.C.; White, S.; Sala-Newby, G.B.; Newby, A.C.; Bond, M. The Hippo pathway mediates inhibition of vascular smooth muscle cell proliferation by cAMP. *J. Mol. Cell Cardiol.* **2016**, *90*, 1–10. [CrossRef]

86. Varelas, X.; Samavarchi-Tehrani, P.; Narimatsu, M.; Weiss, A.; Cockburn, K.; Larsen, B.G.; Rossant, J.; Wrana, J.L. The Crumbs complex couples cell density sensing to Hippo-dependent control of the TGF-beta-SMAD pathway. *Dev. Cell* **2010**, *19*, 831–844. [CrossRef]

87. Estaras, C.; Benner, C.; Jones, K.A. SMADs and YAP compete to control elongation of beta-catenin:LEF-1-recruited RNAPII during hESC differentiation. *Mol. Cell* **2015**, *58*, 780–793. [CrossRef]

88. Fujii, M.; Toyoda, T.; Nakanishi, H.; Yatabe, Y.; Sato, A.; Matsudaira, Y.; Ito, H.; Murakami, H.; Kondo, Y.; Kondo, E.; et al. TGF-beta synergizes with defects in the Hippo pathway to stimulate human malignant mesothelioma growth. *J. Exp. Med.* **2012**, *209*, 479–494. [CrossRef]

89. Lin, K.C.; Moroishi, T.; Meng, Z.P.; Jeong, H.S.; Plouffe, S.W.; Sekido, Y.; Han, J.H.; Park, H.W.; Guan, K.L. Regulation of Hippo pathway transcription factor TEAD by p38 MAPK-induced cytoplasmic translocation (vol 19, pg 996, 2017). *Nat. Cell Biol.* **2018**, *20*, 1098. [CrossRef]

90. Home, P.; Saha, B.; Ray, S.; Dutta, D.; Gunewardena, S.; Yoo, B.; Pal, A.; Vivian, J.L.; Larson, M.; Petroff, M.; et al. Altered subcellular localization of transcription factor TEAD4 regulates first mammalian cell lineage commitment. *Proc. Natl. Acad. Sci. USA* **2012**, *109*, 7362–7367. [CrossRef]

91. Hirate, Y.; Cockburn, K.; Rossant, J.; Sasaki, H. Tead4 is constitutively nuclear, while nuclear vs. cytoplasmic Yap distribution is regulated in preimplantation mouse embryos. *Proc. Natl. Acad. Sci. USA* **2012**, *109*, E3389–E3390. [CrossRef] [PubMed]

92. Qi, Y.F.; Yu, J.; Han, W.; Fan, X.J.; Qian, H.L.; Wei, H.H.; Tsai, Y.H.S.; Zhao, J.Y.; Zhang, W.J.; Liu, Q.T.; et al. A splicing isoform of TEAD4 attenuates the Hippo-YAP signalling to inhibit tumour proliferation. *Nat. Commun.* **2016**, *7*. [CrossRef] [PubMed]

93. De Nadal, E.; Ammerer, G.; Posas, F. Controlling gene expression in response to stress. *Nat. Rev. Genet.* **2011**, *12*, 833–845. [CrossRef] [PubMed]

94. Hong, A.W.; Meng, Z.P.; Yuan, H.X.; Plouffe, S.W.; Moon, S.; Kim, W.; Jho, E.H.; Guan, K.L. Osmotic stress-induced phosphorylation by NLK at Ser128 activates YAP. *EMBO Rep.* **2017**, *18*, 72–86. [CrossRef] [PubMed]

95. Moon, S.; Kim, W.; Kim, S.; Kim, Y.; Song, Y.; Bilousov, O.; Kim, J.; Lee, T.; Cha, B.; Kim, M.; et al. Phosphorylation by NLK inhibits YAP-14-3-3-interactions and induces its nuclear localization. *EMBO Rep.* **2017**, *18*, 61–71. [CrossRef]

96. Gupta, M.P.; Kogut, P.; Gupta, M. Protein kinase-A dependent phosphorylation of transcription enhancer factor-1 represses its DNA-binding activity but enhances its gene activation ability. *Nucleic Acids Res.* **2000**, *28*, 3168–3177. [CrossRef]

97. Jiang, S.W.; Dong, M.Q.; Trujillo, M.A.; Miller, L.J.; Eberhardt, N.L. DNA binding of TEA/ATTS domain factors is regulated by protein kinase C phosphorylation in human choriocarcinoma cells. *J. Biol. Chem.* **2001**, *276*, 23464–23470. [CrossRef]

98. Chan, P.; Han, X.; Zheng, B.; DeRan, M.; Yu, J.; Jarugumilli, G.K.; Deng, H.; Pan, D.; Luo, X.; Wu, X. Autopalmitoylation of TEAD proteins regulates transcriptional output of the Hippo pathway. *Nat. Chem. Biol.* **2016**, *12*, 282–289. [CrossRef]

99. Noland, C.L.; Gierke, S.; Schnier, P.D.; Murray, J.; Sandoval, W.N.; Sagolla, M.; Dey, A.; Hannoush, R.N.; Fairbrother, W.J.; Cunningham, C.N. Palmitoylation of TEAD Transcription Factors Is Required for Their Stability and Function in Hippo Pathway Signaling. *Structure* **2016**, *24*, 179–186. [CrossRef]

100. Mesrouze, Y.; Meyerhofer, M.; Bokhovchuk, F.; Fontana, P.; Zimmermann, C.; Martin, T.; Delaunay, C.; Izaac, A.; Kallen, J.; Schmelze, T.; et al. Effect of the acylation of TEAD4 on its interaction with co-activators YAP and TAZ. *Protein Sci.* **2017**, *26*, 2399–2409. [CrossRef]

101. Resh, M.D. Palmitoylation of ligands, receptors, and intracellular signaling molecules. *Sci. STKE* **2006**, *2006*, re14. [CrossRef] [PubMed]

102. Kim, N.G.; Gumbiner, B.M. Cell contact and Nf2/Merlin-dependent regulation of TEAD palmitoylation and activity. *Proc. Natl. Acad. Sci. USA* **2019**, *116*, 9877–9882. [CrossRef] [PubMed]

103. Pobbati, A.V.; Han, X.; Hung, A.W.; Weiguang, S.; Huda, N.; Chen, G.Y.; Kang, C.B.; Chia, C.S.B.; Luo, X.L.; Hong, W.J.; et al. Targeting the Central Pocket in Human Transcription Factor TEAD as a Potential Cancer Therapeutic Strategy. *Structure* **2015**, *23*, 2076–2086. [CrossRef] [PubMed]

104. Pobbati, A.V.; Hong, W. Emerging roles of TEAD transcription factors and its coactivators in cancers. *Cancer Biol. Ther.* **2013**, *14*, 390–398. [CrossRef] [PubMed]

105. Knight, J.F.; Shepherd, C.J.; Rizzo, S.; Brewer, D.; Jhavar, S.; Dodson, A.R.; Cooper, C.S.; Eeles, R.; Falconer, A.; Kovacs, G.; et al. TEAD1 and c-Cbl are novel prostate basal cell markers that correlate with poor clinical outcome in prostate cancer. *Br. J. Cancer* **2008**, *99*, 1849–1858. [CrossRef]

106. Liang, K.; Zhou, G.X.; Zhang, Q.; Li, J.; Zhang, C.P. Expression of Hippo Pathway in Colorectal Cancer. *Saudi J. Gastroentero* **2014**, *20*, 188–194. [CrossRef]

107. Liu, Y.; Wang, G.; Yang, Y.; Mei, Z.; Liang, Z.; Cui, A.; Wu, T.; Liu, C.Y.; Cui, L. Increased TEAD4 expression and nuclear localization in colorectal cancer promote epithelial-mesenchymal transition and metastasis in a YAP-independent manner. *Oncogene* **2016**, *35*, 2789–2800. [CrossRef]

108. Zhou, Y.; Huang, T.; Zhang, J.; Wong, C.C.; Zhang, B.; Dong, Y.; Wu, F.; Tong, J.H.M.; Wu, W.K.K.; Cheng, A.S.L.; et al. TEAD1/4 exerts oncogenic role and is negatively regulated by miR-4269 in gastric tumorigenesis. *Oncogene* **2017**, *36*, 6518–6530. [CrossRef]

109. Zhou, G.X.; Li, X.Y.; Zhang, Q.; Zhao, K.; Zhang, C.P.; Xue, C.H.; Yang, K.; Tian, Z.B. Effects of the Hippo Signaling Pathway in Human Gastric Cancer. *Asian Pac. J. Cancer Prev.* **2013**, *14*, 5199–5205. [CrossRef]

110. Han, W.; Jung, E.M.; Cho, J.; Lee, J.W.; Hwang, K.T.; Yang, S.J.; Kang, J.J.; Bae, J.Y.; Jeon, Y.K.; Park, I.A.; et al. DNA copy number alterations and expression of relevant genes in triple-negative breast cancer. *Gene Chromosome Cancer* **2008**, *47*, 490–499. [CrossRef]

111. Wang, C.Y.; Nie, Z.; Zhou, Z.M.; Zhang, H.L.; Liu, R.; Wu, J.; Qin, J.Y.; Ma, Y.; Chen, L.; Li, S.M.; et al. The interplay between TEAD4 and KLF5 promotes breast cancer partially through inhibiting the transcription of p27(Kip1). *Oncotarget* **2015**, *6*, 17685–17697. [CrossRef] [PubMed]

112. Korkola, J.E.; Houldsworth, J.; Chadalavada, R.S.V.; Olshen, A.B.; Dobrzynski, D.; Reuter, V.E.; Bosl, G.J.; Chaganti, R.S.K. Down-regulation of stem cell genes, including those in a 200-kb gene cluster at 12p13.31, is associated with in vivo differentiation of human male germ cell tumors. *Cancer Res.* **2006**, *66*, 820–827. [CrossRef]

113. Skotheim, R.I.; Autio, R.; Lind, G.E.; Kraggerud, S.M.; Andrews, P.W.; Monni, O.; Kallioniemi, O.; Lothe, R.A. Novel genomic aberrations in testicular germ cell tumors by array-CGH, and associated gene expression changes. *Cell Oncol.* **2006**, *28*, 315–326. [PubMed]

114. Zhang, W.; Li, J.; Wu, Y.P.; Ge, H.; Song, Y.; Wang, D.M.; Yuan, H.; Jiang, H.B.; Wang, Y.L.; Cheng, J. TEAD4 overexpression promotes epithelial-mesenchymal transition and associates with aggressiveness and adverse prognosis in head neck squamous cell carcinoma. *Cancer Cell Int.* **2018**, *18*. [CrossRef]

115. Schutte, U.; Bisht, S.; Heukamp, L.C.; Kebschull, M.; Florin, A.; Haarmann, J.; Hoffmann, P.; Bendas, G.; Buettner, R.; Brossart, P.; et al. Hippo Signaling Mediates Proliferation, Invasiveness, and Metastatic Potential of Clear Cell Renal Cell Carcinoma. *Transl. Oncol.* **2014**, *7*, 309–321. [CrossRef] [PubMed]

116. Fernandez, A.; Northcott, P.A.; Dalton, J.; Fraga, C.; Ellison, D.; Angers, S.; Taylor, M.D.; Kenney, A.M. YAP1 is amplified and up-regulated in hedgehog-associated medulloblastomas and mediates Sonic hedgehog-driven neural precursor proliferation. *Gene Dev.* **2009**, *23*, 2729–2741. [CrossRef]

117. Bokhovchuk, F.; Mesrouze, Y.; Izaac, A.; Meyerhofer, M.; Zimmermann, C.; Fontana, P.; Schmelzle, T.; Erdmann, D.; Furet, P.; Kallen, J.; et al. Molecular and structural characterization of a TEAD mutation at the origin of Sveinsson's chorioretinal atrophy. *FEBS J.* **2019**. [CrossRef]

118. Fossdal, R.; Jonasson, F.; Kristjansdottir, G.T.; Kong, A.; Stefansson, H.; Gosh, S.; Gulcher, J.R.; Stefansson, K. A novel TEAD1 mutation is the causative allele in Sveinsson's chorioretinal atrophy (helicoid peripapillary chorioretinal degeneration). *Hum. Mol. Genet.* **2004**, *13*, 975–981. [CrossRef]

119. Sun, Z.Q.; Xu, R.W.; Li, X.Y.; Ren, W.G.; Ou, C.L.; Wang, Q.S.; Zhang, H.; Zhang, X.M.; Ma, J.; Wang, H.J.; et al. Prognostic Value of Yes-Associated Protein 1 (YAP1) in Various Cancers: A Meta-Analysis. *PLoS ONE* **2015**, *10*. [CrossRef]

120. Feng, J.T.; Ren, P.W.; Gou, J.H.; Li, Z.Y. Prognostic significance of TAZ expression in various cancers: A meta-analysis. *Oncotargets Ther.* **2016**, *9*, 5235–5244. [CrossRef]

121. Jiao, S.; Wang, H.Z.; Shi, Z.B.; Dong, A.M.; Zhang, W.J.; Song, X.M.; He, F.; Wang, Y.C.; Zhang, Z.Z.; Wang, W.J.; et al. A Peptide Mimicking VGLL4 Function Acts as a YAP Antagonist Therapy against Gastric Cancer. *Cancer Cell* **2014**, *25*, 166–180. [CrossRef] [PubMed]

122. Park, H.W.; Guan, K.L. Regulation of the Hippo pathway and implications for anticancer drug development. *Trends Pharm. Sci.* **2013**, *34*, 581–589. [CrossRef] [PubMed]

123. Shibue, T.; Weinberg, R.A. EMT, CSCs, and drug resistance: the mechanistic link and clinical implications. *Nat. Rev. Clin. Oncol.* **2017**, *14*, 611–629. [CrossRef] [PubMed]

124. Overholtzer, M.; Zhang, J.; Smolen, G.A.; Muir, B.; Li, W.; Sgroi, D.C.; Deng, C.X.; Brugge, J.S.; Haber, D.A. Transforming properties of YAP, a candidate oncogene on the chromosome 11q22 amplicon. *Proc. Natl. Acad. Sci. USA* **2006**, *103*, 12405–12410. [CrossRef] [PubMed]

125. Lei, Q.Y.; Zhang, H.; Zhao, B.; Zha, Z.Y.; Bai, F.; Pei, X.H.; Zhao, S.; Xiong, Y.; Guan, K.L. TAZ promotes cell proliferation and epithelial-mesenchymal transition and is inhibited by the hippo pathway. *Mol. Cell Biol.* **2008**, *28*, 2426–2436. [CrossRef] [PubMed]

126. Lamar, J.M.; Stern, P.; Liu, H.; Schindler, J.W.; Jiang, Z.G.; Hynes, R.O. The Hippo pathway target, YAP, promotes metastasis through its TEAD-interaction domain. *Proc. Natl. Acad. Sci. USA* **2012**, *109*, E2441–E2450. [CrossRef]

127. Vincent-Mistiaen, Z.; Elbediwy, A.; Vanyai, H.; Cotton, J.; Stamp, G.; Nye, E.; Spencer-Dene, B.; Thomas, G.J.; Mao, J.H.; Thompson, B. YAP drives cutaneous squamous cell carcinoma formation and progression. *eLife* **2018**, *7*. [CrossRef] [PubMed]

128. Yu, M.X.; Chen, Y.Z.; Li, X.L.; Yang, R.; Zhang, L.J.; Huangfu, L.T.; Zheng, N.; Zhao, X.G.; Lv, L.F.; Hong, Y.Z.; et al. YAP1 contributes to NSCLC invasion and migration by promoting Slug transcription via the transcription co-factor TEAD. *Cell Death Dis.* **2018**, *9*. [CrossRef]

129. Lehmann, W.; Mossmann, D.; Kleemann, J.; Mock, K.; Meisinger, C.; Brummer, T.; Herr, R.; Brabletz, S.; Stemmler, M.P.; Brabletz, T. ZEB1 turns into a transcriptional activator by interacting with YAP1 in aggressive cancer types. *Nat. Commun.* **2016**, *7*. [CrossRef]

130. Gao, Y.J.; Zhang, W.J.; Han, X.K.; Li, F.M.; Wang, X.J.; Wang, R.; Fang, Z.Y.; Tong, X.Y.; Yao, S.; Li, F.; et al. YAP inhibits squamous transdifferentiation of Lkb1-deficient lung adenocarcinoma through ZEB2-dependent DNp63 repression (vol 5, 4629, 2014). *Nat. Commun.* **2015**, *6*. [CrossRef]

131. Haemmerle, M.; Taylor, M.L.; Gutschner, T.; Pradeep, S.; Cho, M.S.; Sheng, J.T.; Lyons, Y.M.; Nagaraja, A.S.; Dood, R.L.; Wen, Y.F.; et al. Platelets reduce anoikis and promote metastasis by activating YAP1 signaling. *Nat. Commun.* **2017**, *8*. [CrossRef] [PubMed]

132. Zhao, B.; Li, L.; Wang, L.; Wang, C.Y.; Yu, J.D.; Guan, K.L. Cell detachment activates the Hippo pathway via cytoskeleton reorganization to induce anoikis. *Gene Dev.* **2012**, *26*, 54–68. [CrossRef] [PubMed]

133. Lee, H.J.; Diaz, M.F.; Price, K.M.; Ozuna, J.A.; Zhang, S.L.; Sevick-Muraca, E.M.; Hagan, J.P.; Wenzel, P.L. Fluid shear stress activates YAP1 to promote cancer cell motility. *Nat. Commun.* **2017**, *8*. [CrossRef] [PubMed]

134. Chen, D.H.; Sun, Y.T.; Wei, Y.K.; Zhang, P.J.; Rezaeian, A.H.; Teruya-Feldstein, J.; Gupta, S.; Liang, H.; Lin, H.K.; Hung, M.C.; et al. LIFR is a breast cancer metastasis suppressor upstream of the Hippo-YAP pathway and a prognostic marker. *Nat. Med.* **2012**, *18*, 1511–1517. [CrossRef] [PubMed]

135. Guo, P.D.; Lu, X.X.; Gan, W.J.; Li, X.M.; He, X.S.; Zhang, S.; Ji, Q.H.; Zhou, F.; Cao, Y.; Wang, J.R.; et al. RAR gamma Downregulation Contributes to Colorectal Tumorigenesis and Metastasis by Derepressing the Hippo-Yap Pathway. *Cancer Res.* **2016**, *76*, 3813–3825. [CrossRef] [PubMed]

136. Li, C.; Wang, S.; Xing, Z.; Lin, A.; Liang, K.; Song, J.; Hu, Q.; Yao, J.; Chen, Z.; Park, P.K.; et al. A ROR1-HER3-lncRNA signalling axis modulates the Hippo-YAP pathway to regulate bone metastasis. *Nat. Cell Biol.* **2017**, *19*, 106–119. [CrossRef] [PubMed]

137. Lamar, J.M.; Xiao, Y.; Norton, E.; Jiang, Z.G.; Gerhard, G.M.; Kooner, S.; Warren, J.S.A.; Hynes, R.O. SRC tyrosine kinase activates the YAP/TAZ axis and thereby drives tumor growth and metastasis. *J. Biol. Chem.* **2019**, *294*, 2302–2317. [CrossRef] [PubMed]

138. Qiao, Y.T.; Chen, J.X.; Lim, Y.B.; Finch-Edmondson, M.L.; Seshachalam, V.P.; Qin, L.; Jiang, T.; Low, B.C.; Singh, H.; Lim, C.T.; et al. YAP Regulates Actin Dynamics through ARHGAP29 and Promotes Metastasis. *Cell Rep.* **2017**, *19*, 1495–1502. [CrossRef] [PubMed]

139. Er, E.E.; Valiente, M.; Ganesh, K.; Zou, Y.L.; Agrawal, S.; Hu, J.; Griscom, B.; Rosenblum, M.; Boire, A.; Brogi, E.; et al. Pericyte-like spreading by disseminated cancer cells activates YAP and MRTF for metastatic colonization (vol 20, pg 966, 2018). *Nat. Cell Biol.* **2019**, *21*, 408. [CrossRef]

140. Lin, K.C.; Park, H.W.; Guan, K.L. Deregulation and Therapeutic Potential of the Hippo Pathway in Cancer. *Annu. Rev. Cancer Biol.* **2018**, *2*, 59–79. [CrossRef]

141. Sourbier, C.; Liao, P.J.; Ricketts, C.J.; Wei, D.; Yang, Y.; Baranes, S.M.; Gibbs, B.K.; Ohanjanian, L.; Spencer Krane, L.; Scroggins, B.T.; et al. Targeting loss of the Hippo signaling pathway in NF2-deficient papillary kidney cancers. *Oncotarget* **2018**, *9*, 10723–10733. [CrossRef] [PubMed]

142. Rouleau, G.A.; Merel, P.; Lutchman, M.; Sanson, M.; Zucman, J.; Marineau, C.; Hoang-Xuan, K.; Demczuk, S.; Desmaze, C.; Plougastel, B.; et al. Alteration in a new gene encoding a putative membrane-organizing protein causes neuro-fibromatosis type 2. *Nature* **1993**, *363*, 515–521. [CrossRef] [PubMed]

143. Woodard, G.A.; Yang, Y.L.; You, L.; Jablons, D.M. Drug development against the hippo pathway in mesothelioma. *Transl. Lung Cancer Res.* **2017**, *6*, 335–342. [CrossRef] [PubMed]

144. Abdollahpour, H.; Appaswamy, G.; Kotlarz, D.; Diestelhorst, J.; Beier, R.; Schaffer, A.A.; Gertz, E.M.; Schambach, A.; Kreipe, H.H.; Pfeifer, D.; et al. The phenotype of human STK4 deficiency. *Blood* **2012**, *119*, 3450–3457. [CrossRef] [PubMed]

145. Seidel, C.; Schagdarsurengin, U.; Blumke, K.; Wurl, P.; Pfeifer, G.P.; Hauptmann, S.; Taubert, H.; Dammann, R. Frequent hypermethylation of MST1 and MST2 in soft tissue sarcoma. *Mol. Carcinog.* **2007**, *46*, 865–871. [CrossRef]

146. Tapon, N.; Harvey, K.F.; Bell, D.W.; Wahrer, D.C.R.; Schiripo, T.A.; Haber, D.A.; Hariharan, I.K. salvador promotes both cell cycle exit and apoptosis in Drosophila and is mutated in human cancer cell lines. *Cell* **2002**, *110*, 467–478. [CrossRef]

147. Lai, Z.C.; Wei, X.M.; Shimizu, T.; Ramos, E.; Rohrbaugh, M.; Nikolaidis, N.; Ho, L.L.; Li, Y. Control of cell proliferation and apoptosis by Mob as tumor suppressor, Mats. *Cell* **2005**, *120*, 675–685. [CrossRef]

148. Rutherford, S.; Yu, Y.T.; Rumpel, C.A.; Frierson, H.F.; Moskaluk, C.A. Chromosome 6 deletion and candidate tumor suppressor genes in adenoid cystic carcinoma. *Cancer Lett.* **2006**, *236*, 309–317. [CrossRef]

149. Saadeldin, M.K.; Shawer, H.; Mostafa, A.; Kassem, N.M.; Amleh, A.; Siam, R. New genetic variants of LATS1 detected in urinary bladder and colon cancer. *Front. Genet.* **2015**, *5*. [CrossRef]

150. Oh, J.E.; Ohta, T.; Satomi, K.; Foll, M.; Durand, G.; McKay, J.; Le Calvez-Kelm, F.; Mittelbronn, M.; Brokinkel, B.; Paulus, W.; et al. Alterations in the NF2/LATS1/LATS2/YAP Pathway in Schwannomas. *J. Neuropath. Exp. Neur.* **2015**, *74*, 952–959. [CrossRef]

151. Bonilla, X.; Parmentier, L.; King, B.; Bezrukov, F.; Kaya, G.; Zoete, V.; Seplyarskiy, V.B.; Sharpe, H.J.; McKee, T.; Letourneau, A.; et al. Genomic analysis identifies new drivers and progression pathways in skin basal cell carcinoma. *Nat. Genet.* **2016**, *48*, 398. [CrossRef] [PubMed]

152. Jiang, Z.; Li, X.; Hu, J.; Zhou, W.; Jiang, Y.; Li, G.; Lu, D. Promoter hypermethylation-mediated down-regulation of LATS1 and LATS2 in human astrocytoma. *Neurosci. Res.* **2006**, *56*, 450–458. [CrossRef] [PubMed]

153. Kuijjer, M.L.; Rydbeck, H.; Kresse, S.H.; Buddingh, E.P.; Roelofs, H.; Burger, H.; Myklebost, O.; Hogendoorn, P.C.W.; Meza-Zepeda, L.A.; Cleton-Jansen, A.M. Identification of osteosarcoma driver genes by integrative analysis of copy number and gene expression data. *Cancer Res.* **2012**, *72*. [CrossRef]

154. Miyanaga, A.; Masuda, M.; Tsuta, K.; Kawasaki, K.; Nakamura, Y.; Sakuma, T.; Asamura, H.; Gemma, A.; Yamada, T. Hippo Pathway Gene Mutations in Malignant Mesothelioma Revealed by RNA and Targeted Exon Sequencing. *J. Thorac. Oncol.* **2015**, *10*, 844–851. [CrossRef] [PubMed]

155. Takahashi, Y.; Miyoshi, Y.; Takahata, C.; Irahara, N.; Taguchi, T.; Tamaki, Y.; Noguchi, S. Down-regulation of LATS1 and LATS2 mRNA expression by promoter hypermethylation and its association with biologically aggressive phenotype in human breast cancers. *Clin. Cancer Res.* **2005**, *11*, 1380–1385. [CrossRef] [PubMed]

156. Van Raamsdonk, C.D.; Griewank, K.G.; Crosby, M.B.; Garrido, M.C.; Vemula, S.; Wiesner, T.; Obenauf, A.C.; Wackernagel, W.; Green, G.; Bouvier, N.; et al. Mutations in GNA11 in uveal melanoma. *N. Engl. J. Med.* **2010**, *363*, 2191–2199. [CrossRef] [PubMed]

157. Yu, F.X.; Luo, J.; Mo, J.S.; Liu, G.; Kim, Y.C.; Meng, Z.; Zhao, L.; Peyman, G.; Ouyang, H.; Jiang, W.; et al. Mutant Gq/11 promote uveal melanoma tumorigenesis by activating YAP. *Cancer Cell* **2014**, *25*, 822–830. [CrossRef] [PubMed]

158. Zhang, X.; Tang, J.Z.; Vergara, I.A.; Zhang, Y.; Szeto, P.; Yang, L.; Mintoff, C.; Colebatch, A.; McIntosh, L.; Mitchell, K.A.; et al. Somatic hypermutation of the YAP oncogene in a human cutaneous melanoma. *Mol. Cancer Res.* **2019**. [CrossRef]

159. Menzel, M.; Meckbach, D.; Weide, B.; Toussaint, N.C.; Schilbach, K.; Noor, S.; Eigentler, T.; Ikenberg, K.; Busch, C.; Quintanilla-Martinez, L.; et al. In melanoma, Hippo signaling is affected by copy number alterations and YAP1 overexpression impairs patient survival. *Pigm. Cell Melanoma R* **2014**, *27*. [CrossRef] [PubMed]

160. Chen, H.Y.; Yu, S.L.; Ho, B.C.; Su, K.Y.; Hsu, Y.C.; Chang, C.S.; Li, Y.C.; Yang, S.Y.; Hsu, P.Y.; Ho, H.; et al. R331W Missense Mutation of Oncogene YAP1 Is a Germline Risk Allele for Lung Adenocarcinoma With Medical Actionability. *J. Clin. Oncol.* **2015**, *33*, 2303–2310. [CrossRef]

161. Tanas, M.R.; Sboner, A.; Oliveira, A.M.; Erickson-Johnson, M.R.; Hespelt, J.; Hanwright, P.J.; Flanagan, J.; Luo, Y.; Fenwick, K.; Natrajan, R.; et al. Identification of a disease-defining gene fusion in epithelioid hemangioendothelioma. *Sci. Transl. Med.* **2011**, *3*, 98ra82. [CrossRef] [PubMed]

162. Errani, C.; Zhang, L.; Sung, Y.S.; Hajdu, M.; Singer, S.; Maki, R.G.; Healey, J.H.; Antonescu, C.R. A Novel WWTR1-CAMTA1 Gene Fusion Is a Consistent Abnormality in Epithelioid Hemangioendothelioma of Different Anatomic Sites. *Gene Chromosome Cancer* **2011**, *50*, 644–653. [CrossRef] [PubMed]

163. Antonescu, C.R.; Le Loarer, F.; Mosquera, J.M.; Sboner, A.; Zhang, L.; Chen, C.L.; Chen, H.W.; Pathan, N.; Krausz, T.; Dickson, B.C.; et al. Novel YAP1-TFE3 fusion defines a distinct subset of epithelioid hemangioendothelioma. *Gene Chromosome Cancer* **2013**, *52*, 775–784. [CrossRef] [PubMed]

164. Sekine, S.; Kiyono, T.; Ryo, E.; Ogawa, R.; Wakai, S.; Ichikawa, H.; Suzuki, K.; Arai, S.; Tsuta, K.; Ishida, M.; et al. Recurrent YAP1-MAML2 and YAP1-NUTM1 fusions in poroma and porocarcinoma. *J. Clin. Investig.* **2019**, *130*. [CrossRef] [PubMed]

165. Lee, T.F.; Tseng, Y.C.; Chang, W.C.; Chen, Y.C.; Kao, Y.R.; Chou, T.Y.; Ho, C.C.; Wu, C.W. YAP1 is essential for tumor growth and is a potential therapeutic target for EGFR-dependent lung adenocarcinomas. *Oncotarget* **2017**, *8*, 89539–89551. [CrossRef] [PubMed]

166. Lee, T.F.; Tseng, Y.C.; Nguyen, P.A.; Li, Y.C.; Ho, C.C.; Wu, C.W. Enhanced YAP expression leads to EGFR TKI resistance in lung adenocarcinomas. *Sci. Rep.* **2018**, *8*. [CrossRef] [PubMed]

167. Xu, W.; Wei, Y.Y.; Wu, S.S.; Wang, Y.; Wang, Z.; Sun, Y.; Cheng, S.Y.; Wu, J.Q. Up-regulation of the Hippo pathway effector TAZ renders lung adenocarcinoma cells harboring EGFR-T790M mutation resistant to gefitinib. *Cell Biosci.* **2015**, *5*. [CrossRef]

168. Yamauchi, T.; Moroishi, T. Hippo Pathway in Mammalian Adaptive Immune System. *Cells* **2019**, *8*. [CrossRef]

169. Zhang, W.Y.; Nandakumar, N.; Shi, Y.H.; Manzano, M.; Smith, A.; Graham, G.; Gupta, S.; Vietsch, E.E.; Laughlin, S.Z.; Wadhwa, M.; et al. Downstream of Mutant KRAS, the Transcription Regulator YAP Is Essential for Neoplastic Progression to Pancreatic Ductal Adenocarcinoma. *Sci. Signal.* **2014**, *7*. [CrossRef]

170. Rozengurt, E.; Sinnett-Smith, J.; Eibl, G. Yes-associated protein (YAP) in pancreatic cancer: At the epicenter of a targetable signaling network associated with patient survival. *Signal. Transduct. Tar.* **2018**, *3*. [CrossRef]

171. Kapoor, A.; Yao, W.T.; Ying, H.Q.; Hua, S.J.; Liewen, A.; Wang, Q.Y.; Zhong, Y.; Wu, C.J.; Sadanandam, A.; Hu, B.L.; et al. Yap1 Activation Enables Bypass of Oncogenic Kras Addiction in Pancreatic Cancer. *Cell* **2014**, *158*, 185–197. [CrossRef] [PubMed]

172. Davies, H.; Bignell, G.R.; Cox, C.; Stephens, P.; Edkins, S.; Clegg, S.; Teague, J.; Woffendin, H.; Garnett, M.J.; Bottomley, W.; et al. Mutations of the BRAF gene in human cancer. *Nature* **2002**, *417*, 949–954. [CrossRef] [PubMed]

173. Hauschild, A.; Grob, J.J.; Demidov, L.V.; Jouary, T.; Gutzmer, R.; Millward, M.; Rutkowski, P.; Blank, C.U.; Miller, W.H.; Kaempgen, E.; et al. Dabrafenib in BRAF-mutated metastatic melanoma: A multicentre, open-label, phase 3 randomised controlled trial. *Lancet* **2012**, *380*, 358–365. [CrossRef]

174. Fisher, M.L.; Grun, D.; Adhikary, G.; Xu, W.; Eckert, R.L. Inhibition of YAP function overcomes BRAF inhibitor resistance in melanoma cancer stem cells. *Oncotarget* **2017**, *8*, 110257–110272. [CrossRef] [PubMed]

175. Kim, M.H.; Kim, J.; Hong, H.; Lee, S.H.; Lee, J.K.; Jung, E.; Kim, J. Actin remodeling confers BRAF inhibitor resistance to melanoma cells through YAP/TAZ activation. *EMBO J.* **2016**, *35*, 462–478. [CrossRef] [PubMed]

176. Murakami, S.; Shahbazian, D.; Surana, R.; Zhang, W.; Chen, H.; Graham, G.T.; White, S.M.; Weiner, L.M.; Yi, C. Yes-associated protein mediates immune reprogramming in pancreatic ductal adenocarcinoma. *Oncogene* **2017**, *36*, 1232–1244. [CrossRef] [PubMed]

177. Mohseni, M.; Sun, J.; Lau, A.; Curtis, S.; Goldsmith, J.; Fox, V.L.; Wei, C.; Frazier, M.; Samson, O.; Wong, K.K.; et al. A genetic screen identifies an LKB1-MARK signalling axis controlling the Hippo-YAP pathway. *Nat. Cell Biol.* **2014**, *16*, 108–117. [CrossRef]

178. Lee, Y.; Kim, N.H.; Cho, E.S.; Yang, J.H.; Cha, Y.H.; Kang, H.E.; Yun, J.S.; Cho, S.B.; Lee, S.H.; Paclikova, P.; et al. Dishevelled has a YAP nuclear export function in a tumor suppressor context-dependent manner. *Nat. Commun.* **2018**, *9*. [CrossRef]

179. Zhang, W.J.; Gao, Y.J.; Li, F.M.; Tong, X.Y.; Ren, Y.; Han, X.K.; Yao, S.; Long, F.; Yang, Z.Z.; Fan, H.Y.; et al. YAP Promotes Malignant Progression of Lkb1-Deficient Lung Adenocarcinoma through Downstream Regulation of Survivin. *Cancer Res.* **2015**, *75*, 4450–4457. [CrossRef]

180. Mo, J.S.; Meng, Z.P.; Kim, Y.C.; Park, H.W.; Hansen, C.G.; Kim, S.; Lim, D.S.; Guan, K.L. Cellular energy stress induces AMPK-mediated regulation of YAP and the Hippo pathway. *Nat. Cell Biol.* **2015**, *17*, 500. [CrossRef]

181. DeRan, M.; Yang, J.; Shen, C.H.; Peters, E.C.; Fitamant, J.; Chan, P.; Hsieh, M.; Zhu, S.; Asara, J.M.; Zheng, B.; et al. Energy stress regulates hippo-YAP signaling involving AMPK-mediated regulation of angiomotin-like 1 protein. *Cell Rep.* **2014**, *9*, 495–503. [CrossRef] [PubMed]

182. Chen, H.; Liu, H.D.; Qing, G.L. Targeting oncogenic Myc as a strategy for cancer treatment. *Signal. Transduct. Tar.* **2018**, *3*. [CrossRef] [PubMed]

183. Choi, W.; Kim, J.; Park, J.; Lee, D.H.; Hwang, D.; Kim, J.H.; Ashktorab, H.; Smoot, D.; Kim, S.Y.; Choi, C.; et al. YAP/TAZ Initiates Gastric Tumorigenesis via Upregulation of MYC. *Cancer Res.* **2018**, *78*, 3306–3320. [CrossRef] [PubMed]

184. Chen, X.Y.; Gu, W.T.; Wang, Q.; Fu, X.C.; Wang, Y.; Xu, X.; Wen, Y. C-MYC and BCL-2 mediate YAP-regulated tumorigenesis in OSCC. *Oncotarget* **2018**, *9*, 668–679. [CrossRef] [PubMed]

185. Mamada, H.; Sato, T.; Ota, M.; Sasaki, H. Cell competition in mouse NIH3T3 embryonic fibroblasts is controlled by the activity of Tead family proteins and Myc. *J. Cell Sci.* **2015**, *128*, 790–803. [CrossRef] [PubMed]

186. Croci, O.; De Fazio, S.; Biagioni, F.; Donato, E.; Caganova, M.; Curti, L.; Doni, M.; Sberna, S.; Aldeghi, D.; Biancotto, C.; et al. Transcriptional integration of mitogenic and mechanical signals by Myc and YAP. *Gene Dev.* **2017**, *31*, 2017–2022. [CrossRef]

187. Elster, D.; Jaenicke, L.A.; Eilers, M.; von Eyss, B. TEAD activity is restrained by MYC and stratifies human breast cancer subtypes. *Cell Cycle* **2016**, *15*, 2551–2556. [CrossRef]

188. Von Eyss, B.; Jaenicke, L.A.; Kortlever, R.M.; Royla, N.; Wiese, K.E.; Letschert, S.; McDuffus, L.A.; Sauer, M.; Rosenwald, A.; Evan, G.I.; et al. A MYC-Driven Change in Mitochondrial Dynamics Limits YAP/TAZ Function in Mammary Epithelial Cells and Breast Cancer. *Cancer Cell* **2015**, *28*, 743–757. [CrossRef]

189. Elster, D.; Tollot, M.; Schlegelmilch, K.; Ori, A.; Rosenwald, A.; Sahai, E.; von Eyss, B. TRPS1 shapes YAP/TEAD-dependent transcription in breast cancer cells. *Nat. Commun.* **2018**, *9*, 3115. [CrossRef]

Review

The Roles of Hippo Signaling Transducers Yap and Taz in Chromatin Remodeling

Ryan E. Hillmer and Brian A. Link *

Department of Cell Biology, Neurobiology and Anatomy, Medical College of Wisconsin,
8701 Watertown Plank Road, Milwaukee, WI 53226, USA; rhillmer@mcw.edu
* Correspondence: blink@mcw.edu; Tel.: +(414)-955-8072

Received: 19 April 2019; Accepted: 19 May 2019; Published: 24 May 2019

Abstract: Hippo signaling controls cellular processes that ultimately impact organogenesis and homeostasis. Consequently, disease states including cancer can emerge when signaling is deregulated. The major pathway transducers Yap and Taz require cofactors to impart transcriptional control over target genes. Research into Yap/Taz-mediated epigenetic modifications has revealed their association with chromatin-remodeling complex proteins as a means of altering chromatin structure, therefore affecting accessibility and activity of target genes. Specifically, Yap/Taz have been found to associate with factors of the GAGA, Ncoa6, Mediator, Switch/sucrose nonfermentable (SWI/SNF), and Nucleosome Remodeling and Deacetylase (NuRD) chromatin-remodeling complexes to alter the accessibility of target genes. This review highlights the different mechanisms by which Yap/Taz collaborate with other factors to modify DNA packing at specific loci to either activate or repress target gene transcription.

Keywords: chromatin; epigenetic; transcription; Hippo pathway

---

1. Introduction

Recruitment and/or activation of transcription factors and proteins capable of altering target gene transcriptional activity is a hallmark of many signaling pathways. Alterations in transcriptional activity can be achieved via recruitment of proteins capable of remodeling the chromatin structure through the modification of nucleosome positioning and histone proteins at regulatory regions of target genes [1–3]. Transcriptional activity can also be modulated via direct or indirect recruitment of transcriptional machinery to target gene loci [4].

The Hippo signaling pathway governs tissue growth and homeostasis through control over cell proliferation, differentiation, fate, metabolism, and apoptosis [5,6]. Regulation of these cellular processes is ultimately achieved though pathway-mediated localization of the downstream mammalian effectors Yes-associated protein (Yap) and transcriptional activator with PDZ binding motif (Taz/WWTR1), or Yorkie (Yki) in *Drosophila* [7,8]. The pathway itself is comprised of a core kinase cascade involving the subsequent phosphorylation of mammalian STE20-like protein 1/2 (Mst1/2) and large tumor suppressor 1/2 (Lats1/2) kinases. Core kinase activity is bolstered by interaction with Salvador-homolog 1 (Sav1) and MOB kinase activator 1 (Mob1). Mst1/2 interaction with Sav1 reinforces its phosphorylation of Lats1/2, whose phosphorylation activity is enhanced through Mst1/2 phosphorylation of Mob1 [9–11]. Sequential phosphorylation events ultimately lead to the phosphorylation of Lats1/2 to induce its interaction with and phosphorylation of downstream pathway effectors Yap and Taz [12,13]. Yap/Taz phosphorylation prevents their nuclear localization and results in cytoplasmic sequestration via binding to the 14-3-3 adaptor protein [14]. Furthermore, targeted degradation of Yap/Taz can be achieved through subsequent phosphorylation by casein kinase 1 [15,16]. Overall activity of the Hippo signaling kinase cascade serves to prevent the transcriptional activity of downstream effectors Yap and Taz.

When signaling is not active, Yap/Taz can enter the nucleus and bind to DNA through interaction with cofactors to impart effects on transcription. Canonically, Yap/Taz binding to TEA-domain (TEAD) family members has been shown to induce transcription of target genes [17]. However, direct interaction of Yap/Taz with other DNA-bound cofactors including p73, Tbx5, SMADs, and RUNX1/2 has also been demonstrated [18–21].

Mechanistic studies into how Hippo signaling effectors Yap and Taz influence target gene transcriptional activity has revealed the significance of imparting chromatin alterations at target loci. For example, recent chromatin conformation and transcript expression experiments performed on cardiomyocytes overexpressing YAP suggest a function for YAP in modulating chromatin accessibility [22]. The chromatin landscape of cardiomyocytes expressing a constitutively active form of YAP was found to be in a more accessible conformation at TEAD binding motifs within the genome. Genomic regions characterized by decreased chromatin accessibility with YAP overexpression were also apparent [22]. It is possible, however, that these chromatin changes simply reflect a block in differentiation, as these cells maintain a proliferative, fetal-like state. Transient overexpression of Yap following cellular differentiation will be insightful to discriminate these possibilities. However, bona fide interactions with chromatin-remodeling complexes have been established with the transcriptional output factors of Hippo signaling. Yki/Yap/Taz recruitment of and interaction with chromatin remodelers of the SWI/SNF complex, GAGA factor, Mediator complex, Ncoa6, and NuRD complexes have all been documented as means for Yki/Yap/Taz mediated alterations of target gene transcriptional activity [23–33] and are reviewed in the following sections (Table 1).

Table 1. Documented Yki/YAP/TAZ interactions with chromatin-modifying proteins.

Interacting Chromatin Modifying Protein or Complex	Conclusion	Reference(s)
SWI/SNF complex (direct)	Brahma–Yki/Sd interact in the nucleus.	[29,30,32]
	Brahma–Yki/Sd bind to Yki targets.	[30,32]
	Knockdown of Brahma inhibits Yki-mediated tissue overgrowth.	[32]
	Knockdown of Brahma exacerbates Yki-mediated tissue overgrowth.	[33]
	BAP knockdown increases Yki target expression.	[33]
	TAZ/BRM interact via PPXY-WW domains to increase TAZ target expression.	[24]
SWI/SNF complex (indirect)	*ACTL6A-p63* inhibits *KIBRA* expression to increase YAP activity.	[31]
	ARID1A sequesters YAP/TAZ from binding to TEAD to decrease YAP activity.	[34]
GAGA factor (direct)	GAF–Yki/dE2f1 bind to Yki targets, increasing their expression and overall cell proliferation.	[28]
	GAF ChIP-seq peaks overlap with Yki ChIP-seq peaks.	[30]
	Yki–GAF interactions occur in a WW domain-independent manner.	[30]
Mediator complex (direct)	Mediator–Yki interact in the nucleus and increase Yki target transcription.	[30]
Histone methyltransferase complex (direct)	Ncoa6–Yki/Sd interact via PPXY-WW domains at Yki targets to drive transcription.	[25,26]
	NCOA6–YAP interact and increase YAP target gene transcription.	[26]
NuRD complex (direct)	YAP/TAZ/TEAD–NuRD interact within the TSO complex to buffer/inhibit expression of pluripotency/ME specification genes.	[23]
	YAP/TAZ/TEAD–NuRD interact to epigenetically repress target gene activity to promote proliferation.	[27]

Note: For full names of gene symbols, see Abbreviations list at the end of this article.

2. Interactions of Yki/Yap/Taz with the SWI/SNF Family of ATP-Dependent Chromatin-Remodeling Complexes

The switch/sucrose nonfermentable (SWI/SNF) complex is an ATP-dependent chromatin-remodeling complex first described in yeast and named for the effects of its subunits on altered gene expression related to mating type switching (SWI) and sucrose fermentation (SNF) [35,36]. *Drosophila* SWI/SNF complexes include the Brahma-associated protein complex (BAP) and the Polybromo-containing BAP complex (PBAP), with the Brahma ATPase being a common component of both complexes [37]. Brahma

(Brm), Brahma-related gene 1 (Brg1), and associated factors (BAFs) comprise the SWI/SNF complex in vertebrates [38]. Mechanistically, SWI/SNF complex activity is thought to be achieved through chromatin binding and histone positioning mediated by actin-related proteins (Arps), and subsequent DNA-dependent ATPase activity at acetylated histone tails [39,40]. Functionally, SWI/SNF has mainly been implicated in activating gene transcription, although instances of SWI/SNF involvement in gene repression have also been documented [41,42]. Mechanistically, SWI/SNF chromatin-remodeling complexes function to modify nucleosome organization in an ATPase-dependent manner, altering the accessibility of transcription factors to genomic loci. As such, SWI/SNF complexes play important gene regulatory roles in multiple contexts [43–46]. In relation to Yki/Yap/Taz transcriptional functionality, the Brahma subunit of the SWI/SNF complex has been documented as an important cofactor for transcriptional regulation.

2.1. Brahma–Yki Interactions Documented in Drosophila

Analysis of SWI/SNF functionality in *Drosophila* midgut intestinal stem cell (ISC) proliferation and regeneration revealed a requirement of the Brahma subunit in proper regulation of these cellular processes. Known involvement of Yki activity in governing ISC proliferation led to investigations of Brahma–Yki interactions in driving the midgut ISC proliferative capacity [47,48]. Coimmunoprecipitation products of Yki or Scalloped (Sd, the fly homolog of TEAD) subjected to mass spectrometry analyses revealed interactions with multiple BAP complex components, thus suggesting a role of this complex in Yki-Sd-mediated transcriptional activity governing ISC proliferation [29]. Of note, when cotransfected with the Hippo kinase in a cell culture assay, Brahma protein levels were decreased. This reduction in protein levels was found to be induced by Hippo mediated cleavage of Brahma, negatively affecting overall complex stability. Mechanistically, Hippo kinase activity was found to stimulate caspase proteolysis, resulting in cleavage of Brahma. Coincidently, phospho-mediated caspase activation is a known function of the Hippo pathway in *Drosophila* [49]. Furthermore, a cleavage-resistant Brahma mutant was found to promote ISC proliferation [29]. Ultimately, these results suggest that Yki-Sd form a complex with Brahma in the nucleus and that Brahma protein stability is mediated by Hippo kinase activity (Figure 1A). These findings provide support for a regulatory role of Hippo signaling in control of Brahma protein stability and chromatin alterations imparted by BAP SWI/SNF complex recruitment at Yki targets; and subsequent Brahma regulation by the Hippo kinase.

Other investigations into modulation of the Hippo effector Yki by the SWI/SNF component Brahma in *Drosophila* implicated Brahma–Yki interactions in inducing the transcription of *crumbs*. Crumbs is a large cell junction-associated transmembrane protein known to negatively regulate the Hippo signaling cascade. Knockdown of *brahma* resulted in wing growth reduction and a small eye phenotype, placing Brahma as an integral factor for cell proliferation and overall tissue growth regulation [32]. To genetically test Brahma–Hippo pathway interactions, phenotypes resulting from loss-of-function mutations of expanded or *hippo* coupled with *brahma* knockdown were characterized. Mutations of *expanded* and *hippo* resulted in increased Yki activity and were characterized by tissue overgrowth. This effect was inhibited by *brahma* knockdown. Tissue overgrowth was also observed with direct overexpression of *yki*, and likewise was inhibited by *brahma* knockdown. In addition, Yki–Brahma activity depended on Sd interaction, a similar finding to previous reports [29]. Furthermore, Yki and Brahma were localized to the *crumbs* promoter as shown by chromatin immunoprecipitation (ChIP), therefore implicating Brahma in a feed-forward loop of Crumbs-mediated Yki activation in governing tissue growth (Figure 1B) [32]. Together, these experiments defined Brahma as critical for Yki–Sd function in regards to tissue growth regulation [32].

Tissue growth regulation by Brahma–Yki interactions provides the potential for this protein complex to cause cancer and affect tumor growth. Indeed, dysregulation of both SWI/SNF complexes and Hippo pathway activity can result in cancer phenotypes [50,51]. This tumor-promoting activity of Yki is, in certain circumstances, dependent on the activity of the SWI/SNF BAP complex [33].

In the wing imaginal disc, *yki* overexpression coupled with knockdown of *brm* (or other BAP-specific components) was shown to result in wing disc overgrowth, which was exacerbated as compared to *yki* overexpression alone. This demonstrates that BAP can limit *yki*-driven tissue overgrowth. Hyperproliferative wing discs were characteristic of larvae overexpressing Yki or depleted of BAP subunits. Additional malignant features included defects in cell polarity and induction of secreted matrix metalloproteinase 1, which promotes basement membrane degradation [33,52]. Furthermore, depletion of BAP led to ectopic expression of Yki target genes, along with ectopic expression of wing disc growth factors *decapentaplegic* (*dpp*) and Wingless (Wg) [53]. Ectopic expression of *dpp* and Wg was found to augment the tumor-forming phenotype observed in *yki* overexpression/*brm* knockdown wing discs [33]. This research therefore describes a role of the BAP complex as a tumor suppressor in tissues with gain of Yki activity (Figure 1C) and links BAP-mediated chromatin remodeling to cancer phenotypes resulting from dysregulated Hippo signaling.

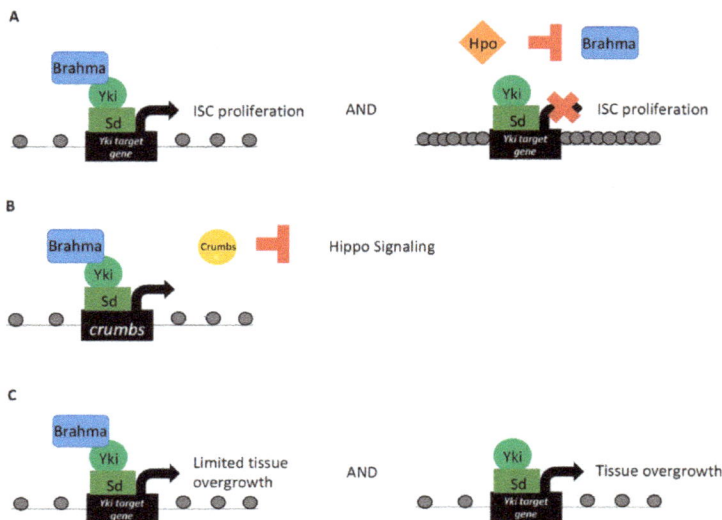

Figure 1. Yorkie interactions with the SWI/SNF complex. (**A**) Brahma and Yki–Sd interact to drive *Drosophila* midgut intestinal stem cell (ISC) proliferation. Notably, cotransfection with Hippo results in caspase-mediated proteolysis of Brahma and loss of complex stability. (**B**) Brahma has also been shown to interact with Yki–Sd at Yki target genes to promote tissue overgrowth in the wing disc and eye. This interaction occurred at the *crumbs* locus, providing evidence for a feed-forward loop of Crumbs-mediated Yki activation in regulation of tissue growth in *Drosophila*. (**C**) In the *Drosophila* wing imaginal disc, *yki* overexpression drives tissue overgrowth. This overgrowth phenotype was exacerbated by *brahma* knockdown, providing evidence for a regulatory role of Brahma in limiting Yki-mediated tissue overgrowth. Spacing of grey dots, representing nucleosomes, represents chromatin compaction.

2.2. BAF-YAP/TAZ Interactions Documented in Mammalian Cells

In addition to SWI/SNF–Hippo signaling interactions in *Drosophila*, cooperation is evident in mammalian cells as well. Investigation into human mammary epithelial cell (MEC) lineage switching between basal and luminal cell fates has implicated TAZ as important in controlling this process. Specifically, TAZ depletion in basal cells, where it is localized to the nucleus, results in lineage switching towards a luminal cell fate. Conversely, ectopic nuclear TAZ expression in luminal cells switches their identity to a more basal cell type [24]. To determine how TAZ induces this cell fate switch in MECs, coimmunoprecipitation/mass spectrometry studies were conducted. These experiments revealed interactions between TAZ and the BAF–SWI/SNF complex catalytic subunits BRG1 and

BRM. The interaction was dependent on the PPXY motif of BRG1/BRM and the WW domain of TAZ. Of these particular SWI/SNF catalytic subunits, only BRM depletion resulted in a decrease in TAZ target expression and ChIP experiments revealed an enrichment of BRM at genomic regions containing TEAD binding motifs. Furthermore, the TAZ–BRM interaction was necessary for the repression of MECs towards a luminal differentiation fate (Figure 2A), thus providing evidence of TAZ–BRM interactions in mammalian cell lines and importance of this association in governing lineage specification [24].

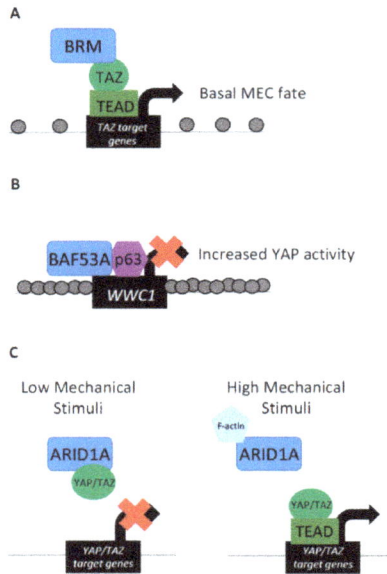

Figure 2. YAP/TAZ interactions with the SWI/SNF complex. (**A**) BRM interaction with TAZ–TEAD at TAZ target loci drives transcription of genes specifying a basal cell fate in mammary epithelial cells (MECs). (**B**) The SWI/SNF subunit Brahma-associated factor 53a (*BAF53A*) interacts with *p63* to inhibit the expression of *WWC1/KIBRA*. This results in the overactivation of YAP and is a hallmark of head and neck squamous cell carcinoma (HNSCC). (**C**) Direct interaction of the SWI/SNF subunit ARID1A with YAP/TAZ under conditions of low mechanical stimuli sequester YAP/TAZ from activating target gene expression. Under conditions of high mechanical stimuli, ARID1A itself is sequestered by nuclear F-actin, allowing YAP/TAZ to bind TEAD at target loci and induce transcription. Of note, interactions of YAP/TAZ with ARID1A does not alter chromatin accessibility at YAP/TAZ target loci. Spacing of grey dots, representing nucleosomes, represents chromatin compaction.

A role for YAP/TAZ working in conjunction with chromatin-remodeling factors may be a more generalized feature of multiple cancers. Dysregulation of the SWI/SNF complex in concert with increased Yki/YAP/TAZ activity has been documented to promote tumor formation in *Drosophila* and mammals [33]. In human cells, transcript levels of the BAF–SWI/SNF subunit *Brahma-associated factor 53a (BAF53A)* has been found to be amplified in head and neck squamous cell carcinoma (HNSCC) along with *p63*. *BAF53A* and *p63*, a DNA-binding transcription factor, form a complex to activate the expression of target genes that promote proliferation and prevent differentiation in HSNCC [31]. These effects on target gene transcription are mediated by *BAF53A-p63* repression of *WWC1* (also known as *KIBRA*), which is a regulator of the Hippo–YAP signaling pathway [54]. Decreased *WWC1* expression was achieved through a *BAF53A-p63*-mediated reduction in chromatin accessibility upstream of the *WWC1* transcription start site. Reduced levels of WWC1 subsequently resulted in increased YAP activity (Figure 2B). Furthermore, the finding that the SWI/SNF subunit complexed with *p63* to ultimately activate YAP-mediated transcription in HSCC was correlated with poor survival of HSCC patients [31]. These results lend additional support to the involvement of SWI/SNF complex dysregulation in tumor progression and demonstrate another role of the SWI/SNF chromatin-remodeling complex in mediating Hippo pathway effector activity.

In contrast to the above-mentioned report of BAF–SWI/SNF in activating TAZ target transcription, this complex has also been shown to directly inhibit YAP and TAZ in cultured mammalian cells. In MCF10AT and HEK293T cells, representing breast and kidney epithelia, YAP/TAZ forms complexes with BAF–SWI/SNF through the ARID1A subunit. Interestingly, regulation of ARID1A–YAP/TAZ is promoted through mechanotransduction, where the interactions of these proteins form under situations in which cells are exposed to low levels of mechanical stimuli [34]. Under situations in which cells are exposed to high levels of mechanical stimuli, ARID1A is sequestered by nuclear F-actin and canonical YAP/TAZ–TEAD complexes form (Figure 2C). In this manner, YAP/TAZ transcriptional activity can be inhibited not only by Hippo signaling activity, but also by interactions with the SWI/SNF complex under differing levels of mechanical stimulation [34]. In this example, BAF–SWI/SNF components also maintain regulatory roles with YAP/TAZ independent of chromatin accessibility regulation.

3. Interactions of *Drosophila* Yki with the Chromatin Protein GAGA Factor and the Mediator Complex

The chromatin protein GAGA factor (GAF) associates with transcriptional machinery components and regulatory protein complexes to alter chromatin structure. GAF is encoded by the *trithorax-like* (*trl*) gene in *Drosophila* and was initially thought to act as a transcriptional activator [55]. GAF has been shown to recruit chromatin-remodeling complexes to induce and maintain nucleosome-free chromatin regions [56]; including PBAP, NURF, and FACT complexes [57–60]. In addition to involvement in transcriptional activation, GAF has been implicated in transcriptional silencing via interaction with Polycomb group proteins [61–63]. Furthermore, in addition to associating with numerous chromatin remodelers, GAF has also been documented to recruit RNA polymerase II [64]. The Mediator complex also associates with transcriptional machinery components, linking the preinitiation complex to RNA polymerase II to recruit transcriptional machinery [4]. The Mediator complex has also been shown to act with p300 to activate chromatin templates, inducing chromatin remodeling and preinitiation complex formation [65,66].

Although not characterized by intrinsic chromatin-remodeling capabilities itself, GAF acts as a hub for the recruitment of protein complexes with chromatin-remodeling capacities. GAF has been documented to recruit chromatin-remodeling complexes capable of altering chromatin accessibility in a bidirectional manner. In this way, it can either complex with remodelers that induce a nucleosome-free, open chromatin region, or recruit repressor complexes to inhibit chromatin accessibility [57–63]. In *Drosophila*, GAF has been shown to act as a pertinent cofactor for the Yki/Sd–dE2f1 transcriptional program driving cell proliferation (Figure 3A) [28]. Knockdown of GAF protein in larval wing imaginal discs resulted in reduced cell proliferation. Furthermore, GAF is found at Yki/Sd–dE2f1 target gene promoters, which exhibit reduced expression when GAF is knocked down. The wing discs of larvae in which GAF was knocked down were characterized by slowed proliferation and were smaller in size when compared to control wing discs [28]. These studies in *Drosophila* defined GAF as required for Yki to induce target transcription and ultimately affect cell proliferation in wing progenitor cells.

Investigation into the chromatin binding of Yki and its interaction with chromatin-remodeling complexes has been studied by ChIP-DNA sequencing (ChIP-seq) characterization [30]. DNA-binding motif analysis of Yki-bound genomic sites revealed an enrichment of *GAGA* sequences. GAF and Yki ChIP-seq peaks were also found to overlap. Furthermore, GAF and Yki proteins bound each other in a manner independent of the Yki WW domain. GAF was also noted as required for the expression of genes that were cobound by Yki and GAF. Consistent with this result, GAF depletion resulted in the reduction of Yki target genes, even under conditions of Yki overexpression [30].

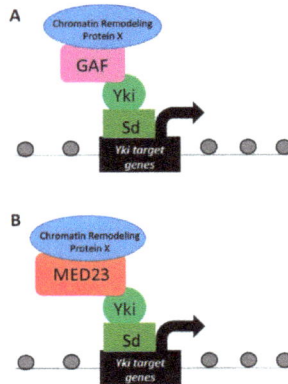

Figure 3. Yorkie interactions with GAF and the Mediator complex. Yki–Sd can interact with the chromatin protein GAF (**A**) or the Mediator complex subunit MED23 (**B**) to drive the transcription of Yki target genes that govern tissue proliferation. Although GAF and the Mediator complex do not contain intrinsic chromatin-remodeling capabilities, they are thought to recruit chromatin-remodeling proteins, yet to be identified, capable of modifying DNA organization.

Similar to GAF, the Mediator complex has also been shown to alter DNA packaging via recruitment of chromatin-remodeling complexes [65,66]. The Mediator complex subunit Med23 was shown as a nuclear Yki-associated protein (Figure 3B) [30]. Furthermore, the expression of direct transcriptional targets of Yki was found to be dependent on the Mediator complex, such that downregulation of Mediator decreased Yki target gene expression under conditions of activated Yki [30]. Taken together, these studies provide evidence for Yki recruitment of, and dependency on, chromatin GAF and Mediator complexes in governing Yki-mediated transcriptional regulation over target gene expression. Despite the requirement for the GAF and Mediator proteins in Yki transcriptional output efficiency, the detailed mechanisms by which these complexes modulate chromatin structure at Yki target loci remains elusive and requires further investigation. In addition, studies directed at whether Hippo proteins recruit and precisely regulate factors of the GAF and Mediator complexes are also warranted.

4. Yki/Yap and the Histone Methyltransferase Complex

Posttranslational modifications to histones, including methylation, aid in regulation of transcriptional activity. Histone H3 methylation in particular promotes alterations to chromatin structure that affect transcriptional activity [1]. In this manner, differential methylation of particular lysine residues on histone H3 at different genomic regions promotes transcriptional specificity [67]. The methylation status of histone H3 is regulated by a distinct complex of proteins associated with the Set1 (COMPASS) family of H3K4 histone methyltransferase (HMT) complexes [3,68]. COMPASS H3K4 HMTs include global H3K4 HMT Set1, along with Trithorax (Trx) and Trithorax-related (Trr) histone methyltransferase complexes [68,69]. Trx is primarily dedicated to the regulation of homeotic genes, while Trr is implicated in steroid hormone signaling and H3K4 monomethylation [70]. Of particular importance, the nuclear receptor coactivator 6 (Ncoa6) subunit of the Trr histone methyltransferase complex has been documented to interact with *Drosophila* Yki and mammalian YAP to methylate H3K4 and activate transcription [25,26].

Ncoa6, a subunit of the Trithorax-related H3K4 methyltransferase complex, was found to be a positive regulator of the Yki–Sd-driven Hippo-responsive transcriptional reporter in *Drosophila* S2R+ cells [71]. Ncoa6 was also found to harbor PPXY motifs, required for its physical interaction with Yki WW domains [25]. Overexpression of Ncoa6 enhanced Yki–Sd-mediated transcriptional reporter activity and bound the Hippo-responsive DNA elements of the Yki target gene *diap1*. Consistent with a functional role, knockdown of Ncoa6 resulted in decreased expression of *diap1* in adult wings. Furthermore, increased eye size induced by Yki overexpression was ameliorated by knockdown of Ncoa6 [25]. These phenotypes place Ncoa6 as an important player in the regulation of Hippo pathway-mediated target gene expression and subsequent tissue growth. Of note, Ncoa6–Sd was incapable of inducing target transcription in the absence of Yki, suggesting that a complete Ncoa6–Yki–Sd complex is required for Hippo pathway target gene transcription. Furthermore, the H3K4 methylation status of Yki target genes was deleteriously affected by knockdown of either Yki, Ncoa6, or Trr. Therefore, in wing and eye imaginal disc differentiation, Yki activates transcription through the recruitment of the Ncoa6 histone methyltransferase complex (Figure 4A) [25].

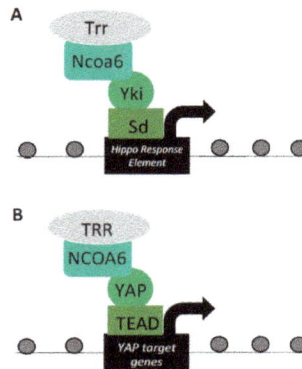

Figure 4. Yorkie/YAP interactions with the histone methyltransferase (HMT) complex. (**A**) Ncoa6, a component of the Trithorax-related (Trr) H3K4 methyltransferase complex, binds to Yki–Sd and is capable of activating a Hippo response element (HRE) reporter in *Drosophila*. Ncoa6 interaction with Yki–Sd drives the expression of Yki target genes, inducing tissue growth and H3K4 methylation at these loci. (**B**) Human NCOA6 interaction with YAP–TEAD drives the expression of YAP target genes, providing evidence for evolutionary conservation between this Yki/YAP–HMT interaction.

A similar study investigating Yki recruitment of a histone methyl transferase to induce target gene transcription also found that ChIP-seq peaks for Yki overlapped with peaks for the H3K4me3 histone modification. Furthermore, targeting Yki to a novel chromosomal locus induced H3K4me3 modifications in a WW domain-dependent manner [26]. Yki was similarly found to bind to Ncoa6 in cultured *Drosophila* S2 cells and Ncoa6 and Trr were both shown to interact with Yki. Binding of Trr on genomic DNA overlaps with Yki-bound regions, suggesting co-occupancy at Yki target genes. As reported in a similar study, Ncoa6 recruitment was sufficient to drive Yki transcriptional activity [25,26] and was required for Yki-mediated eye-overgrowth phenotypes. Concerning the phylogenetic extension of this Ncoa6–Yki interaction, human YAP and NCOA6 were found to bind in a PPXY and WW domain-dependent manner (Figure 4B). Reduction of NCOA6 decreased transcriptional activation of YAP targets [26]. This observed NCOA6–YAP association in mammalian cells supports an evolutionarily conserved mechanism by which YAP recruits histone methyltransferases to modulate chromatin structure and regulate target gene transcription.

5. Interactions of YAP/TAZ/TEAD with the NuRD Complex

While the cases documented above describe how Hippo signaling components collaborate with chromatin-remodeling factors to increase gene expression, alterations of chromatin structure to decrease underlying gene activity are also critical for proper genome regulation. One such method of decreasing chromatin accessibility and transcriptional activity is through recruitment of the repressive nucleosome-remodeling and deacetylase (NuRD) complex [72]. The NuRD complex is distinct in that it couples both ATP-dependent chromatin remodeling with histone deacetylase activity as a means of repressing transcriptional activity [73]. In this manner, NuRD induces the compaction of nucleosomes around bound regulatory regions to restrict genomic accessibility. The main ATP-dependent chromatin-remodeling NuRD components are chromodomain helicase DNA-binding proteins 3/4 (CHD3/4), while histone deacetylase 1/2 (HDAC1/2) activity is responsible for the removal of methyl groups from lysine residues at bound genomic loci [72,73]. Relevant to Hippo signaling, human YAP/TAZ/TEAD have been shown to interact with the NuRD complex as a means of repressing target gene activity [23,27]. Although the NuRD complex can activate target transcription in some circumstances, documented instances of interactions with YAP/TAZ/TEAD are all repressive in nature [74].

A regulatory complex comprised of Hippo effectors YAP/TAZ–TEAD, TGF-β effectors SMAD2/3, and the pluripotency regulator OCT4, termed the TEAD–SMAD–OCT4 (TSO) complex, has been implicated in governing the switch of human embryonic stem cells (hESCs) to either maintain their pluripotency status or specify their fate towards a mesendoderm (ME) cell type. Cell fate is regulated in part through differential interactions of the NuRD repressor complex to first inhibit expression of loci involved in ME specification until fate induction is triggered, at which point the TSO complex is remodeled to allow for gene activation [23]. The interaction of TSO and NuRD was ameliorated by YAP/TAZ knockdown, suggesting that YAP/TAZ are the main proteins in TSO responsible for the functional assembly of NuRD. Upon addition of ME fate-driving factors to the culture media, NuRD was replaced by FOXH1 to facilitate activation of the ME target loci [23]. This study was the first to document a role for YAP/TAZ–TEAD in NuRD complex activity to repress target transcription.

YAP/TAZ–TEAD collaboration with the NuRD complex to repress target gene transcription has also been verified in recent studies. A screen for genes directly regulated by *YAP* overexpression revealed transcripts that both increased and decreased under these conditions. This represents another example where YAP/TAZ, which have canonically been thought of as transcriptional activators, are capable of repressing target gene expression. This transcriptional corepressive role of YAP/TAZ was found to depend on TEAD binding, and YAP/TAZ–TEAD were found to directly bind repressed target loci [27]. The observed repression of YAP/TAZ targets was shown to result from alterations in chromatin at repressed targets, as indicated by effects on histone acetylation and factor occupancy at target promoter regions. This alteration in DNA organization of loci repressed by YAP/TAZ was mediated by recruitment of the NuRD complex, which bound repressed YAP/TAZ targets in a TEAD-dependent manner (Figure 5). Of note, the genes repressed by YAP/TAZ–TEAD–NuRD included those that encode proteins that promote senescence or drive apoptosis [27]. These results document the capacity of YAP/TAZ as transcriptional corepressors and define a mechanism in which YAP/TAZ–TEAD recruit and/or activate the NuRD complex at regulatory regions of target genes to repress their expression and allow for cell proliferation and survival.

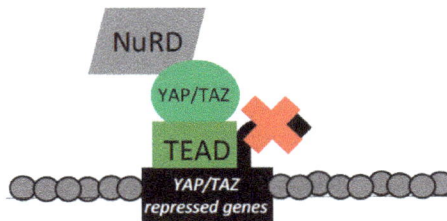

Figure 5. YAP/TAZ/TEAD interactions with the NuRD complex. YAP/TAZ–TEAD bind targets and recruit the NuRD complex to repress target expression. This repression is mediated through dual ATP-dependent chromatin remodeling and histone deacetylase (HDAC)-mediated histone deacetylase functions of the NuRD complex to ultimately reduce chromatin accessibility. YAP/TAZ targets repressed by NuRD recruitment included genes that drive apoptosis and promote senescence.

6. Summary and Future Directions

The Hippo signaling pathway has been documented to control tissue growth and homeostasis through its regulation of downstream effectors Yki/YAP and TAZ [5,7]. Canonically, Yki/YAP/TAZ have been described as transcriptional coactivators, and investigations into the mechanisms behind this coactivator role have revealed the importance of direct partnership with chromatin-modifying protein complexes. Specifically, Yki/YAP/TAZ have been shown to interact with chromatin-remodeling complexes of the SWI/SNF family, GAGA factor, Mediator complex, and multiple histone methyltransferases to aid in their activation of target gene transcription through modification of DNA packing and organization [24–26,28–34]. More recently, a transcriptional repressive role of YAP/TAZ/TEAD has also been described, which is mediated through recruitment of the NuRD complex [23,27]. Therefore, effects of Yki/YAP/TAZ on target gene activity are context-specific and highly dependent on the interaction with protein complexes capable of remodeling nucleosome positioning in an ATP-dependent manner and imparting posttranslational modifications to histones [1,41].

One area of needed investigation is to better define the mechanisms that underlie the specificity of interactions between Hippo components and chromatin-remodeling complexes. *Drosophila* studies of Hippo and Brahma complexes highlight some of the complexity. Investigations into Yki recruitment of BAP–SWI/SNF chromatin-remodeling complexes demonstrated Yki–Sd interaction with the BAP–SWI/SNF subunit Brahma and regulation of Brahma protein stability by the Hippo kinase [29]. Brahma–Yki interactions were also shown to induce the expression of Yki target genes and influence tissue overgrowth [32]. Interestingly, different SWI/SNF subunits have been shown to be either augmentative or restrictive to Yki-mediated tissue overgrowth, indicating context specificity into the differential regulation of SWI/SNF on Yki activity [32,33]. More focused studies to uncover the basis of differential interactions are warranted, as are studies to address diverse roles of YAP/TAZ–SWI/SNF interactions in vertebrates.

Of relevance to this issue, recent work has demonstrated a role of the activator protein 1 (AP-1) in driving YAP/TAZ/TEAD activity at distal enhancers [75]. Activated target genes govern the onset of the S phase and cell mitosis, providing for a means of promoting tumorigenesis [75]. Furthermore, evidence of YAP–TEAD driving the expression of AP-1 targets has been shown to foster tumor growth [76], and AP-1 has been shown to co-occupy enhancer and promoter regions along with Tead4 in multiple cancer cell types [77]. AP-1 has also been documented to play a pertinent role in linking YAP/TAZ activity with regulation of the TGF-β/Smad3 signaling axis through promoting expression of Smad7 [78]. Interestingly, AP-1 has the capacity to recruit the BAF–SWI/SNF chromatin-remodeling complex to alter chromatin accessibility at enhancers [79]. These documented instances of YAP/TAZ/TEAD interaction with AP-1 to drive target gene expression provides for yet another avenue in which YAP/TAZ may work in conjunction with SWI/SNF chromatin-remodeling complexes to remodel and regulate target genes. Further investigation is needed to explore this possibility.

Interactions of Yki with the GAGA factor and the Mediator complex provide a coactivator function in facilitating Yki target transcription [28,30]. Despite the general conclusion that Yki and GAGA or Mediator protein binding are important in regulating target gene transcription, the underlying chromatin-remodeling mechanisms and protein complexes that mediate DNA reorganization remain unknown. As GAGA factor and the Mediator complex can associate with different chromatin-remodeling factors [57–63,65,66], investigations are needed to identify which specific complexes are involved and under what conditions they bind.

Gene repressive activity by YAP/TAZ has been shown to depend on NuRD complex recruitment. The NuRD complex, which has both nucleosome-remodeling and histone deacetylase activity, can be recruited by YAP/TAZ–TEAD to render chromatin inaccessible [23,27]. The interaction of YAP/TAZ with NuRD is commonly dependent on TEAD binding [27,73]. However, YAP/TAZ interactions with other DNA-bound factors provide the potential for NuRD recruitment under TEAD-independent contexts [18–21]. In support of this possibility, Yap overexpression in adult cardiomyocytes drives chromatin remodeling, promoting both open and closed states. While many of these remodeled regions showed TEAD consensus sites, loci that underwent compaction were not enriched for TEADs [22]. Future studies are needed to explore the role of non-TEAD partners of YAP/TAZ that may augment NuRD or other chromatin-modulating complexes to repress gene expression.

Additional questions regarding Hippo signaling and chromatin remodeling center on the relative timing of interactions between YAP/TAZ–TEAD (or other partners), chromatin-remodeling complexes, and DNA. Do chromatin-remodeling complexes recruit YAP/TAZ–TEAD/other to DNA or vice versa? Is enzymatic activity or target specificity altered by these interactions? What cell states facilitate the protein interactions? Importantly, can the relationships between Hippo components and chromatin remodeling be targeted for therapeutic purposes, particularly with regard to cancer and tissue repair following injury? Cumulatively, the documented collaborations of Yki/YAP/TAZ with chromatin-remodeling factors reviewed herein provides insight into the diverse mechanisms by which Hippo signaling controls gene regulation. With ongoing investigations probing cell-type and tissue-specific roles of the Hippo pathway in governing transcription, it is likely that additional interactions between Yki/YAP/TAZ and the machinery that manages chromatin structure will be discovered.

Author Contributions: Contributions for manuscript preparation is as follows: Conceptualization, B.A.L. and R.E.H.; Writing—Original Draft Preparation, R.E.H.; Writing—Review & Editing, B.A.L. and R.E.H.; Funding Acquisition, B.A.L.

Funding: This work was supported in part through NIH grant no. R01 EY029267 and funds provided by the Cardiovascular Center of the Medical College of Wisconsin.

Conflicts of Interest: The authors declare no conflict of interest.

Abbreviations

HUGO gene nomenclature for gene names and symbols was used within the text.

Abbreviation	Full Name
ARIDA	AT-Rich Interactive Domain-Containing Protein 1A
Arp	Actin-related protein
BAF	Brm/Brg1-Associated Factor
BAF53a	Brahma-Associated Factor 53a
BAP	Brahma-Associated Protein Complex
Brg1/BRG1	Brahma-Related Gene 1
Brm/BRM	Brahma
CHD3/4	Chromodomain Helicase DNA-Binding Protein 3/4
ChIP	Chromatin Immunoprecipitation
Co-IP	Coimmunoprecipitation
COMPASS	Complex of Proteins Associated with Set1
Crb	Crumbs
dE2f1	E2f Transcription Factor 1
diap1	death-associated inhibitor of apoptosis 1
dpp	decapentaplegic
FACT	Facilitates Chromatin Transcription Complex
FOXH1	Forkhead Box H1
GAF	GAGA factor
HDAC1/2	Histone Deacetylase 1/2
hESCs	Human Embryonic Stem Cells
HMT	Histone Methyltransferase
HNSCC	Head and Neck Squamous Cell Carcinoma
HRE	Hippo Response Element
ISC	Intestinal Stem Cell
Lats1/2	Large Tumor Suppressor Kinase 1/2
ME	Mesendoderm
MEC	Mammary Epithelial Cell

Med23	Mediator Complex Subunit 23
Mob1	MOB Kinase Activator 1
MS	Mass Spectrometry
Mst1/2	Mammalian STE20-like Kinase 1/2
Ncoa6	Nuclear Receptor Co-Activator 6
NuRD	Nucleosome-Remodeling and Deacetylase Complex
NURF	Nucleosome-Remodeling Factor Complex
OCT4	Octamer-Binding Transcription Factor 4
PBAP	Polybromo-Containing BAP Complex
RUNX1/2	Runt-Related Transcription Factor 1/2
Sav1	Salvador-Homolog 1
Sd	Scalloped
SMAD	*C. elegans* "Small" Worm Phenotype
	Drosophila Mothers Against Decapentaplegic
SWI/SNF	Switch/Sucrose Nonfermentable Complex
Taz	Transcriptional Coactivator with PDZ-Binding Motif
Tbx5	T-Box Transcription Factor 5
TEAD	TEA-Domain
TGF-β	Transforming Growth Factor β
Trl	Trithorax-like
Trr	Trithorax-related
Trx	Trithorax
TSO	TEAD–SMAD–OCT4 Complex
Wg	Wingless
WWC1/	WW Domain-Containing Protein 1
KIBRA	Kidney and Brain Expressed Protein
Yap	Yes-Associated Protein
Yki	Yorkie

References

1. Li, B.; Carey, M.; Workman, J.L. The Role of Chromatin during Transcription. *Cell* **2007**, *128*, 707–719. [CrossRef]

2. Tyagi, M.; Imam, N.; Verma, K.; Patel, A.K. Chromatin remodelers: We are the drivers! *Nucleus* **2016**, *7*, 388–404. [CrossRef]

3. Mohan, M.; Herz, H.-M.; Smith, E.R.; Zhang, Y.; Jackson, J.; Washburn, M.P.; Florens, L.; Eissenberg, J.C.; Shilatifard, A.; Washburn, M.P.; et al. The COMPASS Family of H3K4 Methylases in Drosophila. *Mol. Cell. Boil.* **2011**, *31*, 4310–4318. [CrossRef]

4. Malik, S.; Roeder, R.G. The metazoan Mediator co-activator complex as an integrative hub for transcriptional regulation. *Nat. Rev. Microbiol.* **2010**, *11*, 761–772. [CrossRef] [PubMed]

5. Misra, J.R.; Irvine, K.D. The Hippo Signaling Network and Its Biological Functions. *Annu. Genet.* **2018**, *52*, 65–87. [CrossRef]

6. Zhao, B.; Lei, Q.-Y.; Guan, K.-L. The Hippo-YAP pathway: new connections between regulation of organ size and cancer. *Cell Boil.* **2008**, *20*, 638–646. [CrossRef]

7. Oh, H.; Irvine, K.D. Yorkie: the final destination of Hippo signaling. *Trends Cell Biol.* **2010**, *20*, 410–417. [CrossRef]

8. Yu, F.-X.; Guan, K.-L. The Hippo pathway: regulators and regulations. *Genes Dev.* **2013**, *27*, 355–371. [CrossRef]

9. Callus, B.A.; Verhagen, A.M.; Vaux, D.L. Association of mammalian sterile twenty kinases, Mst1 and Mst2, with hSalvador via C-terminal coiled-coil domains, leads to its stabilization and phosphorylation. *FEBS J.* **2006**, *273*, 4264–4276. [CrossRef] [PubMed]

10. Chan, E.H.Y.; Nousiainen, M.; Chalamalasetty, R.B.; Schäfer, A.; A Nigg, E.; Silljé, H.H.W.; Sch, A. The Ste20-like kinase Mst2 activates the human large tumor suppressor kinase Lats1. *Oncogene* **2005**, *24*, 2076–2086. [CrossRef]

11. Praskova, M.; Khoklatchev, A.; Ortiz-Vega, S.; Avruch, J. Regulation of the MST1 kinase by autophosphorylation, by the growth inhibitory proteins, RASSF1 and NORE1, and by Ras. *Biochem. J.* **2004**, *381*, 453–462. [CrossRef] [PubMed]

12. Hao, Y.; Chun, A.; Cheung, K.; Rashidi, B.; Yang, X. Tumor Suppressor LATS1 Is a Negative Regulator of Oncogene YAP. *J. Biol. Chem.* **2008**, *283*, 5496–5509. [CrossRef]

13. Oka, T.; Mazack, V.; Sudol, M. Mst2 and Lats Kinases Regulate Apoptotic Function of Yes Kinase-associated Protein (YAP). *J. Boil. Chem.* **2008**, *283*, 27534–27546. [CrossRef]

14. Zhao, B.; Wei, X.; Li, W.; Udan, R.S.; Yang, Q.; Kim, J.; Xie, J.; Ikenoue, T.; Yu, J.; Li, L.; et al. Inactivation of YAP oncoprotein by the Hippo pathway is involved in cell contact inhibition and tissue growth control. *Genome Res.* **2007**, *21*, 2747–2761. [CrossRef]

15. Liu, C.-Y.; Zha, Z.-Y.; Zhou, X.; Zhang, H.; Huang, W.; Zhao, D.; Li, T.; Chan, S.W.; Lim, C.J.; Hong, W.; et al. The Hippo Tumor Pathway Promotes TAZ Degradation by Phosphorylating a Phosphodegron and Recruiting the SCFβ-TrCP E3 Ligase*. *J. Boil. Chem.* **2010**, *285*, 37159–37169. [CrossRef]

16. Zhao, B.; Li, L.; Tumaneng, K.; Wang, C.-Y.; Guan, K.-L. A coordinated phosphorylation by Lats and CK1 regulates YAP stability through SCFβ-TRCP. *Genes Dev.* **2010**, *24*, 72–85. [CrossRef]

17. Vassilev, A.; Shu, H.; Zhao, Y.; Kaneko, K.J.; Depamphilis, M.L. TEAD/TEF transcription factors utilize the activation domain of YAP65, a Src/Yes-associated protein localized in the cytoplasm. *Genes Dev.* **2001**, *15*, 1229–1241. [CrossRef]

18. Strano, S.; Monti, O.; Baccarini, A.; Sudol, M.; Sacchi, A.; Blandino, G. Physical interaction with yes-associated protein enhances p73 transcriptional activity. *Int. J. Biol. Chem.* **2001**, *37*, S279.

19. Rosenbluh, J.; Nijhawan, D.; Cox, A.G.; Li, X.; Neal, J.T.; Schafer, E.J.; Zack, T.I.; Wang, X.; Tsherniak, A.; Schinzel, A.C.; et al. β-catenin driven cancers require a YAP1 transcriptional complex for survival and tumorigenesis. *Cell* **2012**, *151*, 1457–1473. [CrossRef]

20. Grannas, K.; Arngården, L.; Lönn, P.; Mazurkiewicz, M.; Blokzijl, A.; Zieba, A.; Söderberg, O. Crosstalk between Hippo and TGFβ: Subcellular Localization of YAP/TAZ/Smad Complexes. *J. Mol. Boil.* **2015**, *427*, 3407–3415. [CrossRef] [PubMed]

21. Zaidi, S.K.; Sullivan, A.J.; Medina, R.; Ito, Y.; Van Wijnen, A.J.; Stein, J.L.; Lian, J.B.; Stein, G.S. Tyrosine phosphorylation controls Runx2-mediated subnuclear targeting of YAP to repress transcription. *EMBO J.* **2004**, *23*, 790–799. [CrossRef]

22. Monroe, T.O.; Hill, M.C.; Morikawa, Y.; Leach, J.P.; Heallen, T.; Cao, S.; Krijger, P.H.; De Laat, W.; Wehrens, X.H.; Rodney, G.G.; et al. YAP Partially Reprograms Chromatin Accessibility to Directly Induce Adult Cardiogenesis In Vivo. *Dev. Cell* **2019**, *48*, 765–779.e7. [CrossRef]

23. Beyer, T.A.; Weiss, A.; Khomchuk, Y.; Huang, K.; Ogunjimi, A.A.; Varelas, X.; Wrana, J.L. Switch Enhancers Interpret TGF-β and Hippo Signaling to Control Cell Fate in Human Embryonic Stem Cells. *Cell Rep.* **2013**, *5*, 1611–1624. [CrossRef] [PubMed]

24. Skibinski, A.; Breindel, J.L.; Prat, A.; Galván, P.; Smith, E.; Rolfs, A.; Gupta, P.B.; LaBaer, J.; Kuperwasser, C. The Hippo transducer TAZ interacts with the SWI/SNF complex to regulate breast epithelial lineage commitment. *Cell Rep.* **2014**, *6*, 1059–1072. [CrossRef]

25. Qing, Y.; Yin, F.; Wang, W.; Zheng, Y.; Guo, P.; Schozer, F.; Deng, H.; Pan, D. The Hippo effector Yorkie activates transcription by interacting with a histone methyltransferase complex through Ncoa6. *eLife* **2014**, *3*, 02564. [CrossRef]

26. Oh, H.; Slattery, M.; Ma, L.; White, K.P.; Mann, R.S.; Irvine, K.D. Yorkie promotes transcription by recruiting a Histone methyltransferase complex. *Cell Rep.* **2014**, *8*, 449–459. [CrossRef]

27. Kim, M.; Kim, T.; Johnson, R.L.; Lim, D.-S. Transcriptional Co-repressor Function of the Hippo Pathway Transducers YAP and TAZ. *Cell Rep.* **2015**, *11*, 270–282. [CrossRef]

28. Bayarmagnai, B.; Nicolay, B.N.; Islam, A.B.; Lopez-Bigas, N.; Frolov, M.V. Drosophila GAGA factor is required for full activation of the dE2f1-Yki/Sd transcriptional program. *Cell Cycle* **2012**, *11*, 4191–4202. [CrossRef]

29. Jin, Y.; Xu, J.; Yin, M.-X.; Lu, Y.; Hu, L.; Li, P.; Zhang, P.; Yuan, Z.; Ho, M.S.; Ji, H.; et al. Brahma is essential for Drosophila intestinal stem cell proliferation and regulated by Hippo signaling. *ELife* **2013**, *2*, e00999. [CrossRef]

30. Oh, H.; Slattery, M.; Ma, L.; Crofts, A.; White, K.P.; Mann, R.S.; Irvine, K.D. Genome-wide association of Yorkie with chromatin and chromatin remodeling complexes. *Cell Rep.* **2013**, *3*, 309–318. [CrossRef] [PubMed]

31. Saladi, S.V.; Ross, K.; Karaayvaz, M.; Tata, P.R.; Mou, H.; Rajagopal, J.; Ramaswamy, S.; Ellisen, L.W. ACTL6A Is Co-Amplified with *p63* in Squamous Cell Carcinoma to Drive YAP Activation, Regenerative Proliferation, and Poor Prognosis. *Cancer Cell* **2017**, *31*, 35–49. [CrossRef]
32. Zhu, Y.; Li, D.; Wang, Y.; Pei, C.; Liu, S.; Zhang, L.; Yuan, Z.; Zhang, P. Brahma regulates the Hippo pathway activity through forming complex with Yki–Sd and regulating the transcription of Crumbs. *Cell. Signal.* **2015**, *27*, 606–613. [CrossRef]
33. Song, S.; Herranz, H.; Cohen, S.M. The chromatin remodeling BAP complex limits tumor-promoting activity of the Hippo pathway effector Yki to prevent neoplastic transformation in Drosophila epithelia. *Model. Mech.* **2017**, *10*, 1201–1209. [CrossRef]
34. Chang, L.; Azzolin, L.; Di Biagio, D.; Zanconato, F.; Battilana, G.; Xiccato, R.L.; Aragona, M.; Giulitti, S.; Panciera, T.; Gandin, A.; et al. The SWI/SNF complex is a mechanoregulated inhibitor of YAP and TAZ. *Nat. Cell Boil.* **2018**, *563*, 265–269. [CrossRef]
35. Carlson, M.; Osmond, B.C.; Neigeborn, L.; Botstein, D. A Suppressor of snf1 Mutations Causes Constitutive High-Level Invertase Synthesis in Yeast. *Genetics* **1984**, *107*, 19–32.
36. Stern, M.; Jensen, R.; Herskowitz, I. Five SWI genes are required for expression of the HO gene in yeast. *J. Mol. Boil.* **1984**, *178*, 853–868. [CrossRef]
37. Elfring, L.K.; Daniel, C.; Papoulas, O.; Deuring, R.; Sarte, M.; Moseley, S.; Beek, S.J.; Waldrip, W.R.; Daubresse, G.; DePace, A.; et al. Genetic analysis of brahma: The Drosophila homolog of the yeast chromatin remodeling factor SWI2/SNF2. *Genetics* **1998**, *148*, 251–265.
38. Muchardt, C.; Yaniv, M. A human homologue of Saccharomyces cerevisiae SNF2/SWI2 and Drosophila brm genes potentiates transcriptional activation by the glucocorticoid receptor. *EMBO J.* **1993**, *12*, 4279–4290. [CrossRef] [PubMed]
39. Rando, O.J.; Zhao, K.; Janmey, P.; Crabtree, G.R. Phosphatidylinositol-dependent actin filament binding by the SWI/SNF-like BAF chromatin remodeling complex. *Proc. Natl. Acad. Sci. USA* **2002**, *99*, 2824–2829. [CrossRef]
40. Szerlong, H.; Saha, A.; Cairns, B.R. The nuclear actin-related proteins Arp7 and Arp9: a dimeric module that cooperates with architectural proteins for chromatin remodeling. *EMBO J.* **2003**, *22*, 3175–3187. [CrossRef] [PubMed]
41. Armstrong, J.; Emerson, B.M. Transcription of chromatin: these are complex times. *Genet. Dev.* **1998**, *8*, 165–172. [CrossRef]
42. Trouche, D.; Le Chalony, C.; Muchardt, C.; Yaniv, M.; Kouzarides, T. RB and hbrm cooperate to repress the activation functions of E2F1. *Proc. Natl. Acad. Sci. USA* **1997**, *94*, 11268–11273. [CrossRef] [PubMed]
43. Hargreaves, D.C.; Crabtree, G.R. ATP-dependent chromatin remodeling: genetics, genomics and mechanisms. *Cell Res.* **2011**, *21*, 396–420. [CrossRef] [PubMed]
44. Lickert, H.; Takeuchi, J.K.; Von Both, I.; Walls, J.R.; McAuliffe, F.; Adamson, S.L.; Henkelman, R.M.; Wrana, J.L.; Rossant, J.; Bruneau, B.G. Baf60c is essential for function of BAF chromatin remodelling complexes in heart development. *Nat. Cell Boil.* **2004**, *432*, 107–112. [CrossRef]
45. Zhan, X.; Shi, X.; Zhang, Z.; Chen, Y.; Wu, J.I. Dual role of Brg chromatin remodeling factor in Sonic hedgehog signaling during neural development. *Proc. Natl. Acad. Sci. USA* **2011**, *108*, 12758–12763. [CrossRef]
46. Ho, L.; Ronan, J.L.; Wu, J.; Staahl, B.T.; Chen, L.; Kuo, A.; Lessard, J.; Nesvizhskii, A.I.; Ranish, J.; Crabtree, G.R. An embryonic stem cell chromatin remodeling complex, esBAF, is essential for embryonic stem cell self-renewal and pluripotency. *Proc. Natl. Acad. Sci. USA* **2009**, *106*, 5181–5186. [CrossRef] [PubMed]
47. Ren, F.; Wang, B.; Yue, T.; Yun, E.-Y.; Ip, Y.T.; Jiang, J. Hippo signaling regulates Drosophila intestine stem cell proliferation through multiple pathways. *Proc. Natl. Acad. Sci. USA* **2010**, *107*, 21064–21069. [CrossRef] [PubMed]
48. Shaw, R.L.; Kohlmaier, A.; Polesello, C.; Veelken, C.; Edgar, B.A.; Tapon, N. The Hippo pathway regulates intestinal stem cell proliferation during Drosophila adult midgut regeneration. *J. Cell Sci.* **2010**, *123*, 4147–4158. [CrossRef]
49. Verghese, S.; Bedi, S.; Kango-Singh, M. Hippo signalling controls Dronc activity to regulate organ size in Drosophila. *Cell Death Differ.* **2012**, *19*, 1664–1676. [CrossRef] [PubMed]
50. Yu, F.-X.; Zhao, B.; Guan, K.-L. Hippo Pathway in Organ Size Control, Tissue Homeostasis, and Cancer. *Cell* **2015**, *163*, 811–828. [CrossRef] [PubMed]

51. Wilson, B.G.; Roberts, C.W.M. SWI/SNF nucleosome remodellers and cancer. *Nat. Rev. Cancer.* **2011**, *11*, 481–492. [CrossRef]

52. Beaucher, M.; Hersperger, E.; Page-McCaw, A.; Shearn, A. Metastatic ability of Drosophila tumors depends on MMP activity. *Dev. Boil.* **2007**, *303*, 625–634. [CrossRef] [PubMed]

53. Dekanty, A.; Milán, M. The interplay between morphogens and tissue growth. *EMBO Rep.* **2011**, *12*, 1003–1010. [CrossRef]

54. Yu, J.; Zheng, Y.; Dong, J.; Klusza, S.; Deng, W.-M.; Pan, D. Kibra functions as a tumor suppressor protein that regulates Hippo signaling in conjunction with Merlin and Expanded. *Dev. Cell* **2010**, *18*, 288–299. [CrossRef] [PubMed]

55. Farkas, G.; Gausz, J.; Galloni, M.; Reuter, G.; Gyurkovics, H.; Karch, F. The Trithorax-like gene encodes the Drosophila GAGA factor. *Nat. Cell Boil.* **1994**, *371*, 806–808. [CrossRef] [PubMed]

56. Okada, M.; Hirose, S. Chromatin Remodeling Mediated by Drosophila GAGA Factor and ISWI Activates fushi tarazu Gene Transcription In Vitro. *Mol. Cell. Boil.* **1998**, *18*, 2455–2461. [CrossRef] [PubMed]

57. Nakayama, T.; Shimojima, T.; Hirose, S. The PBAP remodeling complex is required for histone H3.3 replacement at chromatin boundaries and for boundary functions. *Development* **2012**, *139*, 4582–4590. [CrossRef]

58. Shimojima, T.; Okada, M.; Nakayama, T.; Ueda, H.; Okawa, K.; Iwamatsu, A.; Handa, H.; Hirose, S. Drosophila FACT contributes to Hox gene expression through physical and functional interactions with GAGA factor. *Genome Res.* **2003**, *17*, 1605–1616. [CrossRef] [PubMed]

59. Tsukiyama, T.; Wu, C. Purification and properties of an ATP-dependent nucleosome remodeling factor. *Cell* **1995**, *83*, 1011–1020. [CrossRef]

60. Xiao, H.; Sandaltzopoulos, R.; Wang, H.-M.; Hamiche, A.; Ranallo, R.; Lee, K.-M.; Fu, D.; Wu, C. Dual Functions of Largest NURF Subunit NURF301 in Nucleosome Sliding and Transcription Factor Interactions. *Mol. Cell* **2001**, *8*, 531–543. [CrossRef]

61. Hagstrom, K.; Muller, M.; Schedl, P. A Polycomb and GAGA dependent silencer adjoins the Fab-7 boundary in the Drosophila bithorax complex. *Genetics* **1997**, *146*, 1365–1380. [PubMed]

62. Horard, B.; Tatout, C.; Poux, S.; Pirrotta, V. Structure of a Polycomb Response Element and In Vitro Binding of Polycomb Group Complexes Containing GAGA Factor. *Mol. Cell. Boil.* **2000**, *20*, 3187–3197. [CrossRef]

63. Mishra, R.K.; Mihaly, J.; Barges, S.; Spierer, A.; Karch, F.; Hagstrom, K.; Schweinsberg, S.E.; Schedl, P. The iab-7 Polycomb Response Element Maps to a Nucleosome-Free Region of Chromatin and Requires Both GAGA and Pleiohomeotic for Silencing Activity. *Mol. Cell. Boil.* **2001**, *21*, 1311–1318. [CrossRef]

64. Fuda, N.J.; Guertin, M.J.; Sharma, S.; Danko, C.G.; Martins, A.L.; Siepel, A.; Lis, J.T. GAGA Factor Maintains Nucleosome-Free Regions and Has a Role in RNA Polymerase II Recruitment to Promoters. *PLoS Genet.* **2015**, *11*, e1005108. [CrossRef] [PubMed]

65. Malik, S.; Wallberg, A.E.; Kang, Y.K.; Roeder, R.G. TRAP/SMCC/Mediator-Dependent Transcriptional Activation from DNA and Chromatin Templates by Orphan Nuclear Receptor Hepatocyte Nuclear Factor 4. *Mol. Cell. Boil.* **2002**, *22*, 5626–5637. [CrossRef]

66. Acevedo, M.L.; Kraus, W.L. Mediator and p300/CBP-Steroid Receptor Coactivator Complexes Have Distinct Roles, but Function Synergistically, during Estrogen Receptor α-Dependent Transcription with Chromatin Templates. *Mol. Cell. Boil.* **2003**, *23*, 335–348. [CrossRef]

67. Eissenberg, J.C.; Shilatifard, A. Histone H3 lysine 4 (H3K4) methylation in development and differentiation. *Dev. Boil.* **2010**, *339*, 240–249. [CrossRef]

68. Shilatifard, A. The COMPASS Family of Histone H3K4 Methylases: Mechanisms of Regulation in Development and Disease Pathogenesis. *Annu. Biochem.* **2012**, *81*, 65–95. [CrossRef]

69. Ardehali, M.B.; Mei, A.; Zobeck, K.L.; Caron, M.; Lis, J.T.; Kusch, T. Drosophila Set1 is the major histone H3 lysine 4 trimethyltransferase with role in transcription. *EMBO J.* **2011**, *30*, 2817–2828. [CrossRef] [PubMed]

70. Mohan, M.; Liang, K.; Mickey, K.; Voets, O.; Shilatifard, A.; Herz, H.-M.; Takahashi, Y.-H.; Verrijzer, C.P.; Garruss, A.S. Enhancer-associated H3K4 monomethylation by Trithorax-related, the Drosophila homolog of mammalian Mll3/Mll4. *Genes Dev.* **2012**, *26*, 2604–2620.

71. Koontz, L.M.; Liu-Chittenden, Y.; Yin, F.; Zheng, Y.; Yu, J.; Huang, B.; Chen, Q.; Wu, S.; Pan, D. The Hippo effector Yorkie controls normal tissue growth by antagonizing Scalloped-mediated default repression. *Dev. Cell* **2013**, *25*, 388–401. [CrossRef]

72. Torchy, M.P.; Hamiche, A.; Klaholz, B.P. Structure and function insights into the NuRD chromatin remodeling complex. *Cell. Mol. Life Sci.* **2015**, *72*, 2491–2507. [CrossRef]

73. Xue, Y.; Wong, J.; Moreno, G.; Young, M.K.; Côté, J.; Wang, W. NURD, a Novel Complex with Both ATP-Dependent Chromatin-Remodeling and Histone Deacetylase Activities. *Mol. Cell* **1998**, *2*, 851–861. [CrossRef]

74. Zhang, T.; Wei, G.; Millard, C.J.; Fischer, R.; Konietzny, R.; Kessler, B.M.; Schwabe, J.W.R.; Brockdorff, N. A variant NuRD complex containing PWWP2A/B excludes MBD2/3 to regulate transcription at active genes. *Nat. Commun.* **2018**, *9*, 3798. [CrossRef]

75. Zanconato, F.; Forcato, M.; Battilana, G.; Azzolin, L.; Quaranta, E.; Bodega, B.; Rosato, A.; Bicciato, S.; Cordenonsi, M.; Piccolo, S. Genome-wide association between YAP/TAZ/TEAD and AP-1 at enhancers drives oncogenic growth. *Nat. Cell Boil.* **2015**, *17*, 1218–1227. [CrossRef]

76. Maglic, D.; Schlegelmilch, K.; Dost, A.F.; Panero, R.; Dill, M.T.; A Calogero, R.; Camargo, F.D. YAP-TEAD signaling promotes basal cell carcinoma development via a c-JUN/AP1 axis. *EMBO J.* **2018**, *37*, e98642. [CrossRef]

77. Liu, X.; Li, H.; Rajurkar, M.; Li, Q.; Cotton, J.L.; Ou, J.; Zhu, L.J.; Goel, H.L.; Mercurio, A.M.; Park, J.-S.; et al. Tead and AP1 coordinate transcription and motility. *Cell Rep.* **2016**, *14*, 1169–1180. [CrossRef]

78. Qin, Z.; Xia, W.; Fisher, G.J.; Voorhees, J.J.; Quan, T. YAP/TAZ regulates TGF-β/Smad3 signaling by induction of Smad7 via AP-1 in human skin dermal fibroblasts. *Cell Commun. Signal.* **2018**, *16*, 18. [CrossRef]

79. Vierbuchen, T.; Ling, E.; Cowley, C.J.; Couch, C.H.; Wang, X.; Harmin, D.A.; Roberts, C.W.; Greenberg, M.E. AP-1 transcription factors and the SWI/SNF complex mediate signal-dependent enhancer selection. *Mol. Cell* **2017**, *68*, 1067–1082.e12. [CrossRef]

cells

MDPI

Review
GPCR-Hippo Signaling in Cancer

Jiaqian Luo and Fa-Xing Yu *

Children's Hospital and Institutes of Biomedical Sciences, Fudan University, Shanghai 200032, China; jqluo15@fudan.edu.cn
* Correspondence: fxyu@fudan.edu.cn; Tel.: +86-21-5423-7304

Received: 29 March 2019; Accepted: 7 May 2019; Published: 8 May 2019

Abstract: The Hippo signaling pathway is involved in tissue size regulation and tumorigenesis. Genetic deletion or aberrant expression of some Hippo pathway genes lead to enhanced cell proliferation, tumorigenesis, and cancer metastasis. Recently, multiple studies have identified a wide range of upstream regulators of the Hippo pathway, including mechanical cues and ligands of G protein-coupled receptors (GPCRs). Through the activation related G proteins and possibly rearrangements of actin cytoskeleton, GPCR signaling can potently modulate the phosphorylation states and activity of YAP and TAZ, two homologous oncogenic transcriptional co-activators, and major effectors of the Hippo pathway. Herein, we summarize the network, regulation, and functions of GPCR-Hippo signaling, and we will also discuss potential anti-cancer therapies targeting GPCR-YAP signaling.

Keywords: G protein-coupled receptor; GPCR; Hippo pathway; YAP/TAZ; signal transduction; cancer; tumorigenesis; anti-cancer therapy

1. The Hippo Signaling Network

The Hippo pathway is initially established in *Drosophila melanogaster* (fruit flies), following extensive genetic screens for tumor suppressors, and is highly conserved in mammals [1,2]. The Hippo pathway plays a crucial role in regulating cell survival, proliferation, differentiation, and organ size [1–4]. The core Hippo pathway in mammals can be represented by a kinase cascade consisting of Ste20-like kinases 1/2 (MST1/2), MAP kinase kinase kinase kinases (MAP4K1-7), Large tumor suppressor 1/2 (LATS1/2), Salvador 1 (SAV1, also known as WW45), MOB kinase activator 1A/B (MOB1A/B), Yes-association protein (YAP), and Transcriptional coactivator with PDZ-binding motif (TAZ, also known as WWTR1), TEA domain family members (TEAD1-4), and Vestigial-like family member 4 (VGLL4). Mechanistically, MST1/2 in complex with SAV1 phosphorylate and activate LATS1/2, and MAP4K proteins plays overlapping, yet non-redundant, roles in activating LATS1/2 [5]. LATS1/2 subsequently phosphorylate multiple serine residues of YAP (including S127 and S318) and TAZ (including S89 and S311). Phosphorylation of YAP/TAZ lead to 14-3-3 mediated cytoplasmic retention and ubiquitination-dependent proteasome degradation [6–8]. When upstream kinases are inactivated, dephosphorylated YAP/TAZ translocate into the nucleus, bind with TEAD1-4 and induce the expression of target genes such as connective tissue growth factor (CTFG) and cysteine-rich angiogenic inducer 61(CYR61) [7,9]. Without nuclear YAP/TAZ, TEAD1-4 interact with VGLL4, which may repress transcription of target genes [10,11]. The Hippo signaling output is dependent on transcriptional activity of YAP/TAZ and the latter is mainly inhibited by Hippo pathway kinases.

Several proteins may interpret and transmit physiological signals to core components of the Hippo pathway. The apical membrane-associated FERM-domain protein Neurofibromin 2 (NF2, also known as Merlin) is an activator of the Hippo pathway [12]. NF2 functions by forming a complex with Kidney and brain (KIBRA, also known as WWC1) to activate MST1/2 or recruiting LATS1/2 to plasma membrane for activation by MST1/2 [13]. KIBRA may activate LATS1/2 in a MST1/2-dependent or

-independent manner [14]. AMOT family proteins (AMOTp130, AMOTL1, and AMOTL2) interact with YAP/TAZ and enhance cytoplasmic or junctional localization of YAP/TAZ, and LATS1/2 activity is also mildly induced by AMOT proteins, both lead to an inhibition of YAP/TAZ activity [15–18]. Ras association domain family (RASSF) proteins interact with MST1/2 or SAV1, and may mediate RAS signaling to the Hippo pathway [19–21]. Further studies are required to understand how these proteins serve as bridges linking mechanical or biochemical cues and Hippo pathway kinases.

To date, an array of environmental stimuli has been shown to regulate YAP/TAZ activity. Cell-cell contact is well-known to suppress YAP/TAZ activity by promoting LATS1/2 activation [7]. The mechanical force, such as stiffness of extracellular matrix (ECM), cell geometry and shear stress, regulates phosphorylation and subcellular localization of YAP/TAZ, and recently small GTPase RAP2 has been shown to mediate matrix stiffness signals to LATS1/2 [22,23]. Various stress signals, including oxidative stress, hypoxia, energy stress, endoplasmic reticulum (ER) stress, and osmotic stress modulate YAP/TAZ activity as well [24–29]. Moreover, G protein-coupled receptors (GPCRs) can mediate diverse diffusive signals to modulate Hippo pathway activity, the regulation of the Hippo pathway by GPCR signaling will be further discussed below [30–32].

2. The Hippo Pathway in Tumorigenesis

The link between the Hippo pathway and cancer development has been recently reviewed elsewhere [33,34]. Among components of the Hippo pathway, YAP and TAZ are considered as oncoproteins, whereas most upstream regulators are with tumor suppressor functions. In mouse models, transgenic expression of *Yap*, or genetic ablation of *Nf2*, *Sav1*, *Mst1/2*, *Lats1/2*, *Mob1*, *Wwc1/2*, and *Rassf1a* all lead to tumorigenesis [35–44].

As oncoproteins, YAP/TAZ are able to promote cell proliferation, cell transformation, and cancer cell stemness. YAP/TAZ can induce cell proliferation and reduce cell death, which together lead to increased cell numbers [8,45]. YAP/TAZ may also promote cell transformation, as overexpression of YAP in human non-transformed mammary epithelial cells induces epithelial-to-mesenchymal transition (EMT), and increased TAZ expression in mammary cells leads to the acquisition of a spindle-shaped morphology and increased cell migration and invasion [46–48]. Recently, multiple studies have shown that YAP/TAZ play a role in regulating cancer stem cells (CSCs) [49]. YAP activation leads to dedifferentiate of matured cells and expands undifferentiated liver, epidermal, neural, cardiac, muscle, and intestinal stem/progenitor cells [37,50–56]. In breast cancer, TAZ expression is enriched in CSCs with high self-renewal and tumor initiating capacities [48]. YAP also induces esophageal CSC properties via upregulation of SOX9 [57]. Together, enhanced YAP/TAZ activity may promote cancer development by multiple approaches, such as modulating cell proliferation, movement, and stemness.

YAP/TAZ are activated in diverse human cancers and may serve as an indicator of poor prognosis. Elevated expression of YAP/TAZ is frequently observed in human cancers, including liver, breast, prostate, colorectal, gastric, lung, and brain tumors, especially in high-grade or metastatic tumors [56,58–64]. The expression of YAP/TAZ also functions as a prognostic marker, for instance, YAP/TAZ expression is associated with poor prognosis in hepatocellular carcinoma (HCC), cholangiocarcinoma patients, lung, and colorectal cancers [65–68]. Moreover, high YAP/TAZ expression in cancer may also predict resistance to therapies and cancer relapse [69,70]. Taken together, these results demonstrate that Hippo pathway, especially YAP/TAZ activity, is involved in human cancer development and may function as a molecular target for cancer diagnosis and therapy.

Even though activation of YAP/TAZ occurs frequently in human cancers, the mutation rates of Hippo pathway genes are unexpectedly modest. However, one exception is *NF2*. Inactivating mutation of *NF2* is observed in multiple cancers including meningiomas, schwannomas, and mesotheliomas [71,72]. Additional genetic alterations of Hippo pathway genes in cancer have also been reported, for instance, WWTR1-CAMTA and YAP1-TFE3 fusion are found in epithelioid hemangioma [73,74]. In addition, silencing of *MST1/2*, *LATS1/2*, or *RASSF1* due to promotor hypermethylation are reported in soft tissue sarcomas, breast cancer, and lower-grade glioma [20,75–79].

However, these genetic or epigenetic alterations are not sufficient to explain the widespread YAP/TAZ activation in cancers, especially in cancers with high incidences, and additional molecular mechanisms may contribute to YAP/TAZ activation in cancer.

3. Regulation of Hippo Pathway by GPCRs

GPCRs represent the largest family of cell surface receptors in human genome, and they are involved in a wide range of physiological processes by transmitting diverse extracellular signals into cells. Recent studies suggest that the Hippo pathway is a downstream branch of GPCR signaling. Many GPCRs mediated signals can modulate YAP/TAZ activity, either positively or negatively, dependent on the nature of signals, receptors, and adaptor proteins [4,31].

Following the initial discovery that sphingosine-1-phosphate (S1P) and lysophosphatidic acid (LPA) can induce YAP/TAZ activity, diverse GPCR related signals have been shown to modulate YAP/TAZ activity [30,32] (Table 1). For instance, simple molecules such as protons, which are associated with extracellular pH, can induce YAP activity [80]. Metabolites such purines, adenosine, epinephrine, glutamate, fatty acids, and bile acids activate YAP through GPCRs stimulation [30,32,81–83]. Polypeptides and secreting proteins, such as thrombin, glucagon, Angiotensin II, and Endothelin, also modulate YAP/TAZ activity via GPCR signaling [32,84–87]. These signals, either locally or from a long range, represent major constitutes of cell niche or microenvironment, suggesting that the Hippo pathway is regulated collectively by signals surrounding a given cell. Adhesion GPCRs link cells to their neighbors and probably cell matrix [88], these receptors may also link physical signals to the Hippo pathway.

Table 1. The regulation of YAP/TAZ activity by various G-protein coupled receptors in human cancers. Representative GPCRs that are most frequently implicated in human cancer are shown and their regulation on YAP/TAZ activation are listed in the following table.

GPCRs	Ligand	Coupling Protein	YAP/TAZ Activation	Associated Cancer Type	References
GPER	Estrogen	Gαq/11	↑	Breast cancer	[89]
LPA receptors	LPA	Gα12/13, Gαq/11	↑	Colon cancer Ovarian cancer Prostate cancer Breast caner	[32,90]
S1P receptors	S1P	Gα12/13	↑	Hepatocellular carcinoma	[91]
Protease-activated receptors (PARs)	Thrombin	Gαq, Gα12/13, Gαi	↑	Melanoma Colon cancer Breast cancer Lung cancer Pancreatic cancer Prostate cancer Squamous cell carcinoma of the head and neck	[87,92,93]
ETAR	Endothelin-1	Gαq/11	↑	Colorectal cancer	[86]
EP2, EP4	PGE2	Gαq/11	↑	Colon caner Hepatocellular Carcinoma Head and neck cancer Non-small-cell lung cancer	[94,95]
Frizzleds (FZD)	Wnts	Gα12/13	↑	Colorectal cancer Prostate cancer Hepatocellular carcinoma	[96]
Chemokine (C-X-C motif) receptor 4	SDF1/CXCL12	Gα12/13, Gαq/11, Gαi/o	↑	Breast cancer Non-small cell and small cell lung cancer Oral squamous carcinoma Chronic Myelogenous Leukemia	[97–99]

Table 1. *Cont.*

GPCRs	Ligand	Coupling Protein	YAP/TAZ Activation	Associated Cancer Type	References
Chemokine (C-X-C motif) receptor 2	IL8, CXCL5	Gαi	↑	Head and neck squamous cell carcinoma Non-small cell lung cancer (NSCLC) Ovarian cancer	[100–102]
Angiotensin II receptor AT1	Angiotensin II	Gαq/11	↑	Prostate cancer Cholangiocarcinoma	[103,104]
Free Fatty Acid receptor 1(FFAR1)	Fatty acids	Gαq/11, Gαi/o	↑	Prostate cancer	[105]
β1- and β2-adrenergic receptors	Catecholamines (e.g., dobutamine)	Gαs	↓	Breast cancer	[106]

GPCR ligands regulate the Hippo pathway differentially (Figure 1). It has been established that the effect of GPCR ligands on YAP/TAZ activity is dependent on the type of downstream G proteins activated [32]. GPCRs coupled with $G\alpha_{12/13}$, $G\alpha_{q/11}$, or $G\alpha_{i/o}$, such as LPA and thrombin receptors, will activate YAP/TAZ; on the contrary, GPCRs coupled with $G\alpha_s$, such as epinephrine and glucagon receptors, will inhibit YAP/TAZ [32]. The function of GPCRs and G proteins on the Hippo signaling is most likely depended on protein kinases (such as PKA and PKC), Rho GTPases, and remodeling of the actin cytoskeleton [4]. PKA has been proposed to mediate upstream signals by repressing actin fiber formation or phosphorylating LATS1/2 directly [107–109]. The effect of PKC appears to be isoform-specific, for instance canonical PKC isoforms induce YAP/TAZ activity, whereas novel PKC isoforms repress YAP/TAZ activity [110]. The isoform-specific effect of PKC towards YAP/TAZ may explain cell type-dependent response to PKC activation, as the expression of PKC isoforms vary across different cell types. It seems like MST1/2 is not a direct target of GPCR signaling but the phosphorylation level of LATS1/2 is sensitive to different GPCR ligands [32]. In the absence of MST1/2, MAP4Ks may be responsible for LATS1/2 phosphorylation, as the deletion of MST1/2 and MAP4Ks together abolished the regulation of LATS1/2 phosphorylation by GPCR signaling [5]. Collectively, G proteins and related kinases relay the GPCR signaling to regulate dynamics of the actin cytoskeleton, which, in turn, can be sensed by the Hippo pathway. How different states of actin cytoskeleton sensed by Hippo pathway components remains unclear.

Some seven-(pass)-transmembrane domain receptors, such as Frizzled and smoothened (SMO), are not considered as typical GPCRs. Recent evidence suggests that these atypical GPCRs also regulate the Hippo pathway in a G protein-dependent manner. For instance, Wnt ligands and their receptors (Frizzle proteins) can repress LATS1/2 activity and lead to enhanced YAP/TAZ activity [111]. In addition, Hedgehog (Hh) ligands, via SMO -$G\alpha_s$-cAMP-PKA signaling axis, lead to repression of YAP/TAZ [112]. Thus, atypical GPCRs can regulate the Hippo pathway and also contribute to the crosstalk between Hippo and other important pathways (such as Wnt and Hedgehog) in development and cancer.

The regulation of Hippo pathway by GPCR signaling can also be fine-tuned by additional signals. For instance, it has been shown that the effect of GPCR on YAP/TAZ activity is enhanced when insulin is present, and PI3K and PKD downstream of insulin receptor are involved in this regulation [113]. Moreover, MAPK signaling has also been shown to modulate the Hippo pathway [114]. Hence, the crosstalk between the GPCR-Hippo signaling axis with other pathways should be explored in the future.

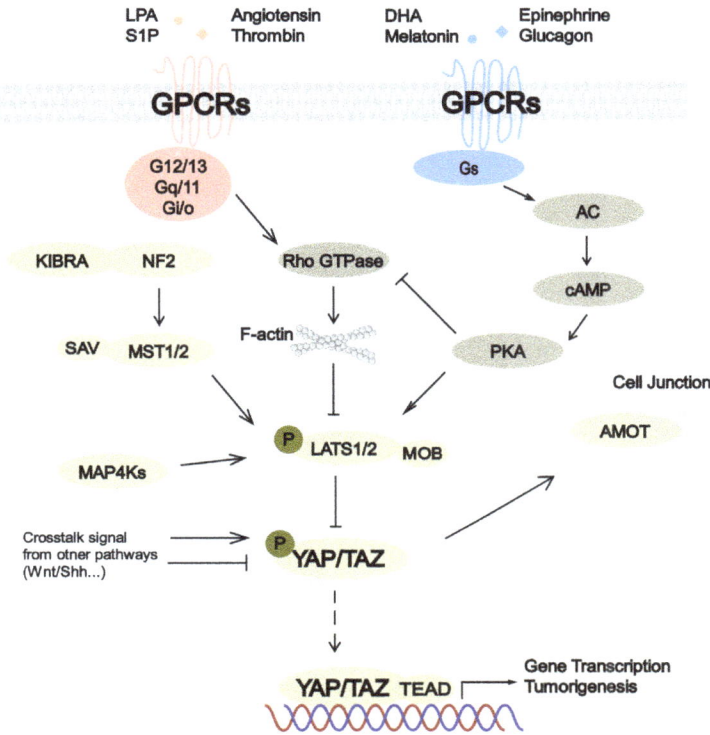

Figure 1. GPCRs modulate Hippo-YAP signaling via GPCRs–G-protein–cytoskeleton axis. The molecular scheme of GPCR-Hippo signaling is shown, including activating and inhibitory regulation of YAP/TAZ. GPCRs and G proteins activating YAP/TAZ are marked with red and those inhibiting YAP/TAZ are in blue.

4. Widespread Alternations of GPCR-YAP Signaling Axis in Cancer

Aberrant GPCR signaling is an important mechanism in cancer development [115]. Different effectors, such as MAPK signaling, mediate the aberrant GPCR signaling to promote tumorigenesis [116]. As a new downstream branch of GPCR signaling, the Hippo tumor suppressor pathway may also play a role in this process. Moreover, aberrant GPCR signaling represents a potential mechanism responsible for prevalent YAP/TAZ activation in human cancers (Table 1).

Cancer genome sequencing analyses revealed that mutations in GPCRs and G proteins are widespread and frequent in multiple tumor types [117]. It has been reported that *GNAQ* and *GNA11* encoding the alpha subunits of $G\alpha_q$ and $G\alpha_{11}$, respectively, are frequently mutated at Arg183 or Gly209 in uveal melanoma and blue nevi [118,119], the mutations at Arg183 and Gly209 result in constitutive activation of $G\alpha_{q/11}$, and several downstream effectors including YAP/TAZ are activated, leading to tumorigenesis [120,121]. Mutations in *GNAS* has been discovered in human medulloblastoma [122] and it has been demonstrated in mice that *GNAS* loss leads to YAP activation and tumorigenesis [109,123]. Thus, YAP activation might be a common mechanism underlying cancer associated mutations on G proteins. However, mutations at Gly227 on *GNAS* contribute to the development of hormone-secreting pituitary tumors and thyroid adenomas [124,125], in principle this mutation will inactivate YAP/TAZ [126].

Mutations in genes encoding GPCRs are observed in approximately 20% of cancers, including mutations in *TSHR* in thyroid cancer, luteinizing hormone receptor (*LHCGR*), and follicle stimulating hormone receptor (*FSHR*) in breast, lung, and colon cancers [127]. Smoothened (*SMO*) is also frequently mutated in cancers arising at the colon, central nervous system, and many other cancers types [128]. Mutations observed in the family of GPCR adhesion receptors, the majority of which are still orphan, resulted in constitutive activation of the receptors, leading to pathological conditions [129]. Moreover, mutated glutamate receptors, such as GRM8, GRM1, and GRM3, have been implicated in squamous non-small cell lung cancer (NSCLC) and melanomas [126]. Currently, the effect of these mutations on the Hippo pathway has not been systematically examined.

Aberrant expression of GPCRs and their ligands may also contribute to tumorigenesis. Elevated *PAR1* (a thrombin receptor) expression is associated with poor differentiation and metastasis of breast cancer [130]. Increased expression of G protein–coupled estrogen receptors (*GPER*) and TAZ activation are detected at the early stage of breast tumor development [89]. Additionally, aberrant expression of LPA receptors may elicit cancer initiation and progression in breast cancer and ovarian cancer via activation of the Rho-dependent transduction pathway [131]. Abnormal G-protein coupled hormone receptor also involved in several adrenal diseases, including tumor and hyperplasia [132]. Kaposi's sarcoma (KS) is caused by infection of human herpesvirus 8 (HHV-8, also known as KSHV) and a viral GPCR (vGPCR) encoded in HHV-8 genome can activate TAZ, which is essential for the development of KS [133]. Recently, an orphan G protein-coupled receptor GPRC5A has been identified as a hypoxia-induced protein, which protects hypoxic tumor cells from apoptosis via the HIF-GPRC5A-RhoA-YAP axis [134]. For GPCR ligands, the levels of LPA has been considered as a marker for ovarian cancer, and high circulating angiotensin II is associated with carcinogenesis, prognosis, and drug resistance in several malignancies, such as colorectal cancer, hepatocellular carcinoma, melanoma, ovarian cancer, and breast cancer [135–141]. Together, abnormal expression of GPCRs and GPCR ligands are associated with the development of different cancers and YAP/TAZ activation may, at least in part, participate in the tumorigenic process.

5. Potential Cancer Therapies Targeting GPCR-Hippo Signaling Axis

Given the frequent dysregulation of GPCR signaling and Hippo pathway in cancer, the GPCR-Hippo signaling axis may serve as a therapeutic target for cancer treatment. As the Hippo pathway is a downstream branch of the GPCR signaling, cancers initiated by aberrant GPCR signaling might be treated by modulating the Hippo pathway, especially YAP/TAZ activity. Meanwhile, for cancers with a dependency on high YAP/TAZ activity, drugs targeting GPCRs and G proteins may reduce YAP/TAZ activation and slowdown cancer progression.

The Hippo pathway can be modulated at multiple levels [69]. As the function of Hippo pathway is mediated mainly by YAP/TAZ and associated transcription factors TEAD1-4, different approaches have been developed to disrupt YAP/TAZ-TEAD interaction. Liu and colleagues discovered the porphyrin family compounds, such as verteporfin (VP), can effectively block YAP-TEAD interaction [142]. A VGLL4-mimicking peptide has also been developed to displace YAP from TEAD [61]. Recently, it has been reported that the pocket in TEAD critical for YAP/TAZ binding is palmitoylated and targeting TEAD palmitoylation is a new approach to repress YAP/TAZ activity [143–146]. Both VP and VGLL4-mimicking peptides have been used in vitro and in vivo to repress YAP/TAZ activity, tissue growth, and tumorigenesis, whereas further improvement of these drugs might be required for clinical use.

GPCRs represent a major target of currently available drugs, and some GPCR-based drugs could be repositioned to block YAP/TAZ activity. For instance, $G\alpha_s$-targeted molecules may repress YAP/TAZ activity in a way that is similar to epinephrine, dobutamine, and glucagon [106,147]. In contrast, antagonizing or depleting $G\alpha_{12/13}$-, $G\alpha_{q/11}$-, or $G\alpha_{i/o}$-mediated signals, such as using phosphatase-resistant LPA analogues and monoclonal antibodies specific for LPA or S1P [117,148], may also limit YAP/TAZ activity. It is noteworthy that some GPCR-based drugs, like β-blockers and dopamine, will significantly affect heart and psychiatric functions, thus, side effects must be considered before using these drugs in cancer therapies [149,150].

Proteins transmitting GPCR signaling can also be targeted to manipulate the Hippo pathway. Recently, cyclic depsipeptide FR900359 has been shown to target mutant $G\alpha_{q/11}$ and repress downstream effectors MAPK and YAP [151,152]. The activity of PKA is associated with cellular cyclic-AMP (cAMP) levels, Forskolin or phosphodiesterase inhibitors, such as Rolipram, have been shown to induce PKA activity and repress YAP/TAZ [112,148]. PKC inhibitors can also repress YAP/TAZ activity in a cell type-dependent manner [121]. Rho GTPases are central for the regulation of Hippo pathway by GPCR signaling, it has been shown that statins, inhibitors of HMG-CoA reductase (HMGCR), can indirectly inactivate Rho GTPases and reduce YAP/TAZ nuclear localization [153,154]. As GPCR-Hippo signaling is a complex signaling network, drugs targeting GPCRs, G proteins, or downstream signaling nodes may affect effectors other than YAP/TAZ, hence, the specificity towards the Hippo pathway will be compromised when using these drugs.

6. Conclusions

In the last few years, Hippo pathway has been under the spotlight due to its function in organ size control and tumorigenesis. GPCRs, the largest cell membrane receptor family, regulate an array of downstream signal pathways, including Hippo pathway effectors YAP/TAZ. Aberrant GPCR and YAP/TAZ activation have been observed in pathogenesis of several types of cancer. Although further in-depth studies are still required to address issues relevant to side effects and pharmacodynamics, it is clear that GPCR-Hippo signaling axis will be a promising target for anticancer therapy.

Funding: F.-X.Y. would like to gratefully acknowledge financial support from the National Natural Science Foundation of China (81622038, 81772965, and 31571479) and Shanghai Municipal Commission of Health and Family Planning (2017BR018).

Conflicts of Interest: The authors declare no conflict of interest.

References

1. Johnson, R.; Halder, G. The two faces of Hippo: targeting the Hippo pathway for regenerative medicine and cancer treatment. *Nat. Rev. Drug Discov.* **2014**, *13*, 63–79. [CrossRef] [PubMed]
2. Pan, D. The hippo signaling pathway in development and cancer. *Dev. Cell* **2010**, *19*, 491–505. [CrossRef] [PubMed]
3. Harvey, K.F.; Zhang, X.; Thomas, D.M. The Hippo pathway and human cancer. *Nat. Rev. Cancer* **2013**, *13*, 246–257. [CrossRef] [PubMed]
4. Yu, F.X.; Zhao, B.; Guan, K.L. Hippo Pathway in Organ Size Control, Tissue Homeostasis, and Cancer. *Cell* **2015**, *163*, 811–828. [CrossRef]
5. Meng, Z.; Moroishi, T.; Mottier-Pavie, V.; Plouffe, S.W.; Hansen, C.G.; Hong, A.W.; Park, H.W.; Mo, J.S.; Lu, W.; Lu, S.; et al. MAP4K family kinases act in parallel to MST1/2 to activate LATS1/2 in the Hippo pathway. *Nat. Commun.* **2015**, *6*, 8357. [CrossRef]
6. Zhao, B.; Li, L.; Tumaneng, K.; Wang, C.-Y.; Guan, K.-L. A coordinated phosphorylation by Lats and CK1 regulates YAP stability through SCF(beta-TRCP). *Genes Dev.* **2010**, *24*, 72–85. [CrossRef]
7. Zhao, B.; Wei, X.; Li, W.; Udan, R.S.; Yang, Q.; Kim, J.; Xie, J.; Ikenoue, T.; Yu, J.; Li, L.; et al. Inactivation of YAP oncoprotein by the Hippo pathway is involved in cell contact inhibition and tissue growth control. *Genes Dev.* **2007**, *21*, 2747–2761. [CrossRef]
8. Lei, Q.Y.; Zhang, H.; Zhao, B.; Zha, Z.Y.; Bai, F.; Pei, X.H.; Zhao, S.; Xiong, Y.; Guan, K.L. TAZ promotes cell proliferation and epithelial-mesenchymal transition and is inhibited by the hippo pathway. *Mol. Cell. Biol.* **2008**, *28*, 2426–2436. [CrossRef]
9. Lamar, J.M.; Stern, P.; Liu, H.; Schindler, J.W.; Jiang, Z.-G.; Hynes, R.O. The Hippo pathway target, YAP, promotes metastasis through its TEAD-interaction domain. *Proc. Natl. Acad. Sci. USA* **2012**, *109*, E2441–E2450. [CrossRef]

10. Zhang, W.; Gao, Y.; Li, P.; Shi, Z.; Guo, T.; Li, F.; Han, X.; Feng, Y.; Zheng, C.; Wang, Z.; et al. VGLL4 functions as a new tumor suppressor in lung cancer by negatively regulating the YAP-TEAD transcriptional complex. *Cell Res.* **2014**, *24*, 331–343. [CrossRef]

11. Koontz, L.M.; Liu-Chittenden, Y.; Yin, F.; Zheng, Y.G.; Yu, J.Z.; Huang, B.; Chen, Q.; Wu, S.; Pan, D.J. The Hippo Effector Yorkie Controls Normal Tissue Growth by Antagonizing Scalloped-Mediated Default Repression. *Dev. Cell* **2013**, *25*, 388–401. [CrossRef] [PubMed]

12. Striedinger, K.; VandenBerg, S.R.; Baia, G.S.; McDermott, M.W.; Gutmann, D.H.; Lal, A. The neurofibromatosis 2 tumor suppressor gene product, merlin, regulates human meningioma cell growth by signaling through YAP. *Neoplasia* **2008**, *10*, 1204–1212. [CrossRef] [PubMed]

13. Yin, F.; Yu, J.; Zheng, Y.; Chen, Q.; Zhang, N.; Pan, D. Spatial organization of Hippo signaling at the plasma membrane mediated by the tumor suppressor Merlin/NF2. *Cell* **2013**, *154*, 1342–1355. [CrossRef] [PubMed]

14. Yu, J.; Zheng, Y.; Dong, J.; Klusza, S.; Deng, W.-M.; Pan, D. Kibra functions as a tumor suppressor protein that regulates Hippo signaling in conjunction with Merlin and Expanded. *Dev. Cell* **2010**, *18*, 288–299. [CrossRef]

15. Chan, S.W.; Lim, C.J.; Chong, Y.F.; Pobbati, A.V.; Huang, C.; Hong, W. Hippo pathway-independent restriction of TAZ and YAP by angiomotin. *J. Biol. Chem.* **2011**, *286*, 7018–7026. [CrossRef] [PubMed]

16. Paramasivam, M.; Sarkeshik, A.; Yates, J.R., 3rd; Fernandes, M.J.; McCollum, D. Angiomotin family proteins are novel activators of the LATS2 kinase tumor suppressor. *Mol. Biol. Cell* **2011**, *22*, 3725–3733. [CrossRef] [PubMed]

17. Wang, W.; Huang, J.; Chen, J. Angiomotin-like proteins associate with and negatively regulate YAP1. *J. Biol. Chem.* **2011**, *286*, 4364–4370. [CrossRef] [PubMed]

18. Zhao, B.; Li, L.; Lu, Q.; Wang, L.H.; Liu, C.Y.; Lei, Q.; Guan, K.L. Angiomotin is a novel Hippo pathway component that inhibits YAP oncoprotein. *Genes Dev.* **2011**, *25*, 51–63. [CrossRef]

19. Polesello, C.; Huelsmann, S.; Brown, N.H.; Tapon, N. The Drosophila RASSF homolog antagonizes the hippo pathway. *Curr. Biol.* **2006**, *16*, 2459–2465. [CrossRef]

20. Vlahov, N.; Scrace, S.; Soto, M.S.; Grawenda, A.M.; Bradley, L.; Pankova, D.; Papaspyropoulos, A.; Yee, K.S.; Buffa, F.; Goding, C.R.; et al. Alternate RASSF1 Transcripts Control SRC Activity, E-Cadherin Contacts, and YAP-Mediated Invasion. *Curr. Biol.* **2015**, *25*, 3019–3034. [CrossRef]

21. Khokhlatchev, A.; Rabizadeh, S.; Xavier, R.; Nedwidek, M.; Chen, T.; Zhang, X.F.; Seed, B.; Avruch, J. Identification of a novel Ras-regulated proapoptotic pathway. *Curr. Biol.* **2002**, *12*, 253–265. [CrossRef]

22. Meng, Z.; Qiu, Y.; Lin, K.C.; Kumar, A.; Placone, J.K.; Fang, C.; Wang, K.-C.; Lu, S.; Pan, M.; Hong, A.W.; et al. RAP2 mediates mechanoresponses of the Hippo pathway. *Nature* **2018**, *560*, 655–660. [CrossRef]

23. Dupont, S.; Morsut, L.; Aragona, M.; Enzo, E.; Giulitti, S.; Cordenonsi, M.; Zanconato, F.; Le Digabel, J.; Forcato, M.; Bicciato, S.; et al. Role of YAP/TAZ in mechanotransduction. *Nature* **2011**, *474*, 179–183. [CrossRef]

24. Shao, D.; Zhai, P.; Del Re, D.P.; Sciarretta, S.; Yabuta, N.; Nojima, H.; Lim, D.-S.; Pan, D.; Sadoshima, J. A functional interaction between Hippo-YAP signalling and FoxO1 mediates the oxidative stress response. *Nat. Commun.* **2014**, *5*, 3315. [CrossRef]

25. Ma, B.; Chen, Y.; Chen, L.; Cheng, H.; Mu, C.; Li, J.; Gao, R.; Zhou, C.; Cao, L.; Liu, J.; et al. Hypoxia regulates Hippo signalling through the SIAH2 ubiquitin E3 ligase. *Nat. Cell Biol.* **2014**, *17*, 95–103. [CrossRef]

26. Ma, B.; Cheng, H.; Gao, R.; Mu, C.; Chen, L.; Wu, S.; Chen, Q.; Zhu, Y. Zyxin-Siah2-Lats2 axis mediates cooperation between Hippo and TGF-β signalling pathways. *Nat. Commun.* **2016**, *7*, 11123. [CrossRef]

27. Lin, K.C.; Moroishi, T.; Meng, Z.; Jeong, H.-S.; Plouffe, S.W.; Sekido, Y.; Han, J.; Park, H.W.; Guan, K.-L. Regulation of Hippo pathway transcription factor TEAD by p38 MAPK-induced cytoplasmic translocation. *Nat. Cell Biol.* **2017**, *19*, 996–1002. [CrossRef]

28. Wu, H.; Wei, L.; Fan, F.; Ji, S.; Zhang, S.; Geng, J.; Hong, L.; Fan, X.; Chen, Q.; Tian, J.; et al. Integration of Hippo signalling and the unfolded protein response to restrain liver overgrowth and tumorigenesis. *Nat. Commun.* **2015**, *6*, 6239. [CrossRef]

29. Hong, A.W.; Meng, Z.; Yuan, H.-X.; Plouffe, S.W.; Moon, S.; Kim, W.; Jho, E.-H.; Guan, K.-L. Osmotic stress-induced phosphorylation by NLK at Ser128 activates YAP. *EMBO Rep.* **2017**, *18*, 72–86. [CrossRef]

30. Miller, E.; Yang, J.; DeRan, M.; Wu, C.; Su, A.I.; Bonamy, G.M.C.; Liu, J.; Peters, E.C.; Wu, X. Identification of Serum-Derived Sphingosine-1-Phosphate as a Small Molecule Regulator of YAP. *Chem. Biol.* **2012**, *19*, 955–962. [CrossRef]

31. Yu, F.-X.; Mo, J.-S.; Guan, K.-L. Upstream regulators of the Hippo pathway. *Cell Cycle* **2012**, *11*, 4097–4098. [CrossRef]

32. Yu, F.-X.; Zhao, B.; Panupinthu, N.; Jewell, J.L.; Lian, I.; Wang, L.H.; Zhao, J.; Yuan, H.; Tumaneng, K.; Li, H.; et al. Regulation of the Hippo-YAP pathway by G-protein-coupled receptor signaling. *Cell* **2012**, *150*, 780–791. [CrossRef]

33. Zanconato, F.; Cordenonsi, M.; Piccolo, S. YAP/TAZ at the Roots of Cancer. *Cancer Cell* **2016**, *29*, 783–803. [CrossRef]

34. Moroishi, T.; Hansen, C.G.; Guan, K.L. The emerging roles of YAP and TAZ in cancer. *Nat. Rev. Cancer* **2015**, *15*, 73–79. [CrossRef]

35. Yu, F.-X.; Meng, Z.; Plouffe, S.W.; Guan, K.-L. Hippo Pathway Regulation of Gastrointestinal Tissues. *Annu. Rev. Physiol.* **2015**, *77*, 201–227. [CrossRef]

36. Zhang, N.; Bai, H.; David, K.K.; Dong, J.; Zheng, Y.; Cai, J.; Giovannini, M.; Liu, P.; Anders, R.A.; Pan, D. The Merlin/NF2 Tumor Suppressor Functions through the YAP Oncoprotein to Regulate Tissue Homeostasis in Mammals. *Dev. Cell* **2010**, *19*, 27–38. [CrossRef]

37. Lee, K.-P.; Lee, J.-H.; Kim, T.-S.; Kim, T.-H.; Park, H.-D.; Byun, J.-S.; Kim, M.-C.; Jeong, W.-I.; Calvisi, D.F.; Kim, J.-M.; et al. The Hippo-Salvador pathway restrains hepatic oval cell proliferation, liver size, and liver tumorigenesis. *Proc. Natl. Acad. Sci. USA* **2010**, *107*, 8248–8253. [CrossRef]

38. Zhou, D.; Conrad, C.; Xia, F.; Park, J.-S.; Payer, B.; Yin, Y.; Lauwers, G.Y.; Thasler, W.; Lee, J.T.; Avruch, J.; et al. Mst1 and Mst2 maintain hepatocyte quiescence and suppress hepatocellular carcinoma development through inactivation of the Yap1 oncogene. *Cancer Cell* **2009**, *16*, 425–438. [CrossRef] [PubMed]

39. St John, M.A.; Tao, W.; Fei, X.; Fukumoto, R.; Carcangiu, M.L.; Brownstein, D.G.; Parlow, A.F.; McGrath, J.; Xu, T. Mice deficient of Lats1 develop soft-tissue sarcomas, ovarian tumours and pituitary dysfunction. *Nat.Genet.* **1999**, *21*, 182–186. [CrossRef] [PubMed]

40. Nishio, M.; Hamada, K.; Kawahara, K.; Sasaki, M.; Noguchi, F.; Chiba, S.; Mizuno, K.; Suzuki, S.O.; Dong, Y.; Tokuda, M.; et al. Cancer susceptibility and embryonic lethality in Mob1a/1b double-mutant mice. *J. Clin. Invest.* **2012**, *122*, 4505–4518. [CrossRef]

41. Knight, J.F.; Sung, V.Y.C.; Kuzmin, E.; Couzens, A.L.; de Verteuil, D.A.; Ratcliffe, C.D.H.; Coelho, P.P.; Johnson, R.M.; Samavarchi-Tehrani, P.; Gruosso, T.; et al. KIBRA (WWC1) Is a Metastasis Suppressor Gene Affected by Chromosome 5q Loss in Triple-Negative Breast Cancer. *Cell Rep.* **2018**, *22*, 3191–3205. [CrossRef] [PubMed]

42. Van der Weyden, L.; Tachibana, K.K.; Gonzalez, M.A.; Adams, D.J.; Ng, B.L.; Petty, R.; Venkitaraman, A.R.; Arends, M.J.; Bradley, A. The RASSF1A isoform of RASSF1 promotes microtubule stability and suppresses tumorigenesis. *Mol. Cell. Biol.* **2005**, *25*, 8356–8367. [CrossRef]

43. Van der Weyden, L.; Arends, M.J.; Dovey, O.M.; Harrison, H.L.; Lefebvre, G.; Conte, N.; Gergely, F.V.; Bradley, A.; Adams, D.J. Loss of Rassf1a cooperates with Apc(Min) to accelerate intestinal tumourigenesis. *Oncogene* **2008**, *27*, 4503–4508. [CrossRef] [PubMed]

44. Van der Weyden, L.; Papaspyropoulos, A.; Poulogiannis, G.; Rust, A.G.; Rashid, M.; Adams, D.J.; Arends, M.J.; O'Neill, E. Loss of RASSF1A synergizes with deregulated RUNX2 signaling in tumorigenesis. *Cancer Res.* **2012**, *72*, 3817–3827. [CrossRef] [PubMed]

45. Dong, J.; Feldmann, G.; Huang, J.; Wu, S.; Zhang, N.; Comerford, S.A.; Gayyed, M.F.; Anders, R.A.; Maitra, A.; Pan, D. Elucidation of a universal size-control mechanism in Drosophila and mammals. *Cell* **2007**, *130*, 1120–1133. [CrossRef]

46. Overholtzer, M.; Zhang, J.; Smolen, G.A.; Muir, B.; Li, W.; Sgroi, D.C.; Deng, C.-X.; Brugge, J.S.; Haber, D.A. Transforming properties of YAP, a candidate oncogene on the chromosome 11q22 amplicon. *Proc. Natl. Acad. Sci. USA* **2006**, *103*, 12405–12410. [CrossRef]

47. Chan, S.W.; Lim, C.J.; Guo, K.; Ng, C.P.; Lee, I.; Hunziker, W.; Zeng, Q.; Hong, W. A Role for TAZ in Migration, Invasion, and Tumorigenesis of Breast Cancer Cells. *Cancer Res.* **2008**, *68*, 2592–2598. [CrossRef]

48. Cordenonsi, M.; Zanconato, F.; Azzolin, L.; Forcato, M.; Rosato, A.; Frasson, C.; Inui, M.; Montagner, M.; Parenti, A.R.; Poletti, A.; et al. The Hippo Transducer TAZ Confers Cancer Stem Cell-Related Traits on Breast Cancer Cells. *Cell* **2011**, *147*, 759–772. [CrossRef]

49. Dawood, S.; Austin, L.; Cristofanilli, M. Cancer stem cells: implications for cancer therapy. *Oncology* **2014**, *28*, 1101–1107, 1110.

50. Schlegelmilch, K.; Mohseni, M.; Kirak, O.; Pruszak, J.; Rodriguez, J.R.; Zhou, D.; Kreger, B.T.; Vasioukhin, V.; Avruch, J.; Brummelkamp, T.R.; et al. Yap1 acts downstream of α-catenin to control epidermal proliferation. *Cell* **2011**, *144*, 782–795. [CrossRef]

51. Van Hateren, N.J.; Das, R.M.; Hautbergue, G.M.; Borycki, A.-G.; Placzek, M.; Wilson, S.A. FatJ acts via the Hippo mediator Yap1 to restrict the size of neural progenitor cell pools. *Development* **2011**, *138*, 1893–1902. [CrossRef]

52. Heallen, T.; Zhang, M.; Wang, J.; Bonilla-Claudio, M.; Klysik, E.; Johnson, R.L.; Martin, J.F. Hippo pathway inhibits Wnt signaling to restrain cardiomyocyte proliferation and heart size. *Science* **2011**, *332*, 458–461. [CrossRef]

53. Judson, R.N.; Tremblay, A.M.; Knopp, P.; White, R.B.; Urcia, R.; De Bari, C.; Zammit, P.S.; Camargo, F.D.; Wackerhage, H. The Hippo pathway member Yap plays a key role in influencing fate decisions in muscle satellite cells. *J. Cell Sci.* **2012**, *125*, 6009–6019. [CrossRef]

54. Cai, J.; Zhang, N.; Zheng, Y.; de Wilde, R.F.; Maitra, A.; Pan, D. The Hippo signaling pathway restricts the oncogenic potential of an intestinal regeneration program. *Genes Dev.* **2010**, *24*, 2383–2388. [CrossRef]

55. Fernandez-L, A.; Northcott, P.A.; Dalton, J.; Fraga, C.; Ellison, D.; Angers, S.; Taylor, M.D.; Kenney, A.M. YAP1 is amplified and up-regulated in hedgehog-associated medulloblastomas and mediates Sonic hedgehog-driven neural precursor proliferation. *Genes Dev.* **2009**, *23*, 2729–2741. [CrossRef]

56. Bhat, K.P.L.; Salazar, K.L.; Balasubramaniyan, V.; Wani, K.; Heathcock, L.; Hollingsworth, F.; James, J.D.; Gumin, J.; Diefes, K.L.; Kim, S.H.; et al. The transcriptional coactivator TAZ regulates mesenchymal differentiation in malignant glioma. *Genes Dev.* **2011**, *25*, 2594–2609. [CrossRef]

57. Song, S.; Ajani, J.A.; Honjo, S.; Maru, D.M.; Chen, Q.; Scott, A.W.; Heallen, T.R.; Xiao, L.; Hofstetter, W.L.; Weston, B.; et al. Hippo Coactivator YAP1 Upregulates SOX9 and Endows Esophageal Cancer Cells with Stem-like Properties. *Cancer Res.* **2014**, *74*, 4170–4182. [CrossRef]

58. Li, H.; Wolfe, A.; Septer, S.; Edwards, G.; Zhong, X.; Bashar Abdulkarim, A.; Ranganathan, S.; Apte, U. Deregulation of Hippo kinase signalling in Human hepatic malignancies. *Liver Int.* **2012**, *32*, 38–47. [CrossRef]

59. Díaz-Martín, J.; López-García, M.Á.; Romero-Pérez, L.; Atienza-Amores, M.R.; Pecero, M.L.; Castilla, M.Á.; Biscuola, M.; Santón, A.; Palacios, J. Nuclear TAZ expression associates with the triple-negative phenotype in breast cancer. *Endocr. Relat. Cancer* **2015**, *22*, 443–454. [CrossRef]

60. Liu, C.-Y.; Yu, T.; Huang, Y.; Cui, L.; Hong, W. ETS (E26 transformation-specific) up-regulation of the transcriptional co-activator TAZ promotes cell migration and metastasis in prostate cancer. *J. Biol. Chem.* **2017**, *292*, 9420–9430. [CrossRef]

61. Jiao, S.; Wang, H.; Shi, Z.; Dong, A.; Zhang, W.; Song, X.; He, F.; Wang, Y.; Zhang, Z.; Wang, W.; et al. A peptide mimicking VGLL4 function acts as a YAP antagonist therapy against gastric cancer. *Cancer Cell* **2014**, *25*, 166–180. [CrossRef]

62. Noguchi, S.; Saito, A.; Horie, M.; Mikami, Y.; Suzuki, H.I.; Morishita, Y.; Ohshima, M.; Abiko, Y.; Mattsson, J.S.M.; Konig, H.; et al. An Integrative Analysis of the Tumorigenic Role of TAZ in Human Non-Small Cell Lung Cancer. *Clin.Cancer Res.* **2014**, *20*, 4660–4672. [CrossRef]

63. Cheng, H.; Zhang, Z.; Rodriguez-Barrueco, R.; Borczuk, A.; Liu, H.; Yu, J.; Silva, J.M.; Cheng, S.K.; Perez-Soler, R.; Halmos, B. Functional genomics screen identifies YAP1 as a key determinant to enhance treatment sensitivity in lung cancer cells. *Oncotarget* **2016**, *7*, 28976–28988. [CrossRef]

64. Orr, B.A.; Bai, H.; Odia, Y.; Jain, D.; Anders, R.A.; Eberhart, C.G. Yes-associated protein 1 is widely expressed in human brain tumors and promotes glioblastoma growth. *J. Neuropathol. Exp. Neurol.* **2011**, *70*, 568–577. [CrossRef]

65. Xu, M.Z.; Yao, T.-J.; Lee, N.P.Y.; Ng, I.O.L.; Chan, Y.-T.; Zender, L.; Lowe, S.W.; Poon, R.T.P.; Luk, J.M. Yes-associated protein is an independent prognostic marker in hepatocellular carcinoma. *Cancer* **2009**, *115*, 4576–4585. [CrossRef]

66. Lee, K.; Lee, K.-B.; Jung, H.Y.; Yi, N.-J.; Lee, K.-W.; Suh, K.-S.; Jang, J.-J. The correlation between poor prognosis and increased yes-associated protein 1 expression in keratin 19 expressing hepatocellular carcinomas and cholangiocarcinomas. *BMC Cancer* **2017**, *17*, 441. [CrossRef]

67. Liu, X.-L.; Zuo, R.; Ou, W.-B. The hippo pathway provides novel insights into lung cancer and mesothelioma treatment. *J. Cancer Res. Clin. Oncol.* **2018**, *144*, 2097–2106. [CrossRef]

68. Zhang, L.; Song, X.; Li, X.; Wu, C.; Jiang, J. Yes-Associated Protein 1 as a Novel Prognostic Biomarker for Gastrointestinal Cancer: A Meta-Analysis. *BioMed Res. Int.* **2018**, *2018*, 4039173. [CrossRef]

69. Gong, R.; Yu, F.-X. Targeting the Hippo Pathway for Anti-cancer Therapies. *Curr. Med. Chem.* **2015**, *22*, 4104–4117. [CrossRef]

70. Kim, M.H.; Kim, J. Role of YAP/TAZ transcriptional regulators in resistance to anti-cancer therapies. *Cell. Mol. Life Sci.* **2017**, *74*, 1457–1474. [CrossRef]

71. Evans, D.G.R. Neurofibromatosis 2 [Bilateral acoustic neurofibromatosis, central neurofibromatosis, NF2, neurofibromatosis type II]. *Genet. Med.* **2009**, *11*, 599–610. [CrossRef] [PubMed]

72. Thurneysen, C.; Opitz, I.; Kurtz, S.; Weder, W.; Stahel, R.A.; Felley-Bosco, E. Functional inactivation of NF2/merlin in human mesothelioma. *Lung Cancer* **2009**, *64*, 140–147. [CrossRef] [PubMed]

73. Tanas, M.R.; Ma, S.; Jadaan, F.O.; Ng, C.K.; Weigelt, B.; Reis-Filho, J.S.; Rubin, B.P. Mechanism of action of a WWTR1(TAZ)-CAMTA1 fusion oncoprotein. *Oncogene* **2016**, *35*, 929–938. [CrossRef] [PubMed]

74. Errani, C.; Zhang, L.; Sung, Y.S.; Hajdu, M.; Singer, S.; Maki, R.G.; Healey, J.H.; Antonescu, C.R. A novel WWTR1-CAMTA1 gene fusion is a consistent abnormality in epithelioid hemangioendothelioma of different anatomic sites. *Genes Chromosomes Cancer* **2011**, *50*, 644–653. [CrossRef] [PubMed]

75. Seidel, C.; Schagdarsurengin, U.; Blümke, K.; Würl, P.; Pfeifer, G.P.; Hauptmann, S.; Taubert, H.; Dammann, R. Frequent hypermethylation of MST1 andMST2 in soft tissue sarcoma. *Mol. Carcinog.* **2007**, *46*, 865–871. [CrossRef]

76. Takahashi, Y.; Miyoshi, Y.; Takahata, C.; Irahara, N.; Taguchi, T.; Tamaki, Y.; Noguchi, S. Down-regulation of LATS1 and LATS2 mRNA expression by promoter hypermethylation and its association with biologically aggressive phenotype in human breast cancers. *Clin. Cancer Res.* **2005**, *11*, 1380–1385. [CrossRef]

77. Jiang, Z.; Li, X.; Hu, J.; Zhou, W.; Jiang, Y.; Li, G.; Lu, D. Promoter hypermethylation-mediated down-regulation of LATS1 and LATS2 in human astrocytoma. *Neurosci. Res.* **2006**, *56*, 450–458. [CrossRef]

78. Sanchez-Vega, F.; Mina, M.; Armenia, J.; Chatila, W.K.; Luna, A.; La, K.C.; Dimitriadoy, S.; Liu, D.L.; Kantheti, H.S.; Saghafinia, S.; et al. Oncogenic Signaling Pathways in The Cancer Genome Atlas. *Cell* **2018**, *173*, 321–337. [CrossRef]

79. Grawenda, A.M.; O'Neill, E. Clinical utility of RASSF1A methylation in human malignancies. *Br. J. Cancer* **2015**, *113*, 372–381. [CrossRef]

80. Zhu, H.; Cheng, X.; Niu, X.; Zhang, Y.; Guan, J.; Liu, X.; Tao, S.; Wang, Y.; Zhang, C. Proton-sensing GPCR-YAP Signalling Promotes Cell Proliferation and Survival. *Int. J. Biol. Sci.* **2015**, *11*, 1181–1189. [CrossRef]

81. Anakk, S.; Bhosale, M.; Schmidt, V.A.; Johnson, R.L.; Finegold, M.J.; Moore, D.D. Bile acids activate YAP to promote liver carcinogenesis. *Cell Rep.* **2013**, *5*, 1060–1069. [CrossRef]

82. Thirunavukkarasan, M.; Wang, C.; Rao, A.; Hind, T.; Teo, Y.R.; Siddiquee, A.A.-M.; Goghari, M.A.I.; Kumar, A.P.; Herr, D.R. Short-chain fatty acid receptors inhibit invasive phenotypes in breast cancer cells. *PLoS ONE* **2017**, *12*, e0186334. [CrossRef]

83. Koo, J.H.; Guan, K.-L. Interplay between YAP/TAZ and Metabolism. *Cell Metab.* **2018**, *28*, 196–206. [CrossRef]

84. Yu, O.M.; Miyamoto, S.; Brown, J.H. Myocardin-Related Transcription Factor A and Yes-Associated Protein Exert Dual Control in G Protein-Coupled Receptor- and RhoA-Mediated Transcriptional Regulation and Cell Proliferation. *Mol. Cell. Biol.* **2016**, *36*, 39–49. [CrossRef]

85. Wennmann, D.O.; Vollenbröker, B.; Eckart, A.K.; Bonse, J.; Erdmann, F.; Wolters, D.A.; Schenk, L.K.; Schulze, U.; Kremerskothen, J.; Weide, T.; et al. The Hippo pathway is controlled by Angiotensin II signaling and its reactivation induces apoptosis in podocytes. *Cell Death Dis.* **2014**, *5*, e1519. [CrossRef]

86. Wang, Z.; Liu, P.; Zhou, X.; Wang, T.; Feng, X.; Sun, Y.-P.; Xiong, Y.; Yuan, H.-X.; Guan, K.-L. Endothelin Promotes Colorectal Tumorigenesis by Activating YAP/TAZ. *Cancer Res.* **2017**, *77*, 2413–2423. [CrossRef]

87. Mo, J.-S.; Yu, F.-X.; Gong, R.; Brown, J.H.; Guan, K.-L. Regulation of the Hippo-YAP pathway by protease-activated receptors (PARs). *Genes Dev.* **2012**, *26*, 2138–2143. [CrossRef] [PubMed]

88. Langenhan, T.; Aust, G.; Hamann, J. Sticky Signaling—Adhesion Class G Protein-Coupled Receptors Take the Stage. *Sci. Signal.* **2013**, *6*. [CrossRef] [PubMed]

89. Zhou, X.; Wang, S.; Wang, Z.; Feng, X.; Liu, P.; Lv, X.-B.; Li, F.; Yu, F.-X.; Sun, Y.; Yuan, H.; et al. Estrogen regulates Hippo signaling via GPER in breast cancer. *J. Clin. Invest.* **2015**, *125*, 2123–2135. [CrossRef]

90. Cai, H.; Xu, Y. The role of LPA and YAP signaling in long-term migration of human ovarian cancer cells. *Cell Commun. Signal.* **2013**, *11*, 31. [CrossRef]

91. Cheng, J.-C.; Wang, E.Y.; Yi, Y.; Thakur, A.; Tsai, S.-H.; Hoodless, P.A. S1P Stimulates Proliferation by Upregulating CTGF Expression through S1PR2-Mediated YAP Activation. *Mol. Cancer Res.* **2018**, *16*, 1543–1555. [CrossRef] [PubMed]

92. Zhou, P.-J.; Xue, W.; Peng, J.; Wang, Y.; Wei, L.; Yang, Z.; Zhu, H.H.; Fang, Y.-X.; Gao, W.-Q. Elevated expression of Par3 promotes prostate cancer metastasis by forming a Par3/aPKC/KIBRA complex and inactivating the hippo pathway. *J. Exp. Clin. Cancer Res.* **2017**, *36*, 139. [CrossRef] [PubMed]

93. Nag, J.; Bar-Shavit, R. Transcriptional Landscape of PARs in Epithelial Malignancies. *Int. J. Mol. Sci.* **2018**, *19*, 3451. [CrossRef]

94. Kim, H.-B.; Kim, M.; Park, Y.-S.; Park, I.; Kim, T.; Yang, S.-Y.; Cho, C.J.; Hwang, D.; Jung, J.-H.; Markowitz, S.D.; et al. Prostaglandin E2 Activates YAP and a Positive-Signaling Loop to Promote Colon Regeneration After Colitis but Also Carcinogenesis in Mice. *Gastroenterology* **2017**, *152*, 616–630. [CrossRef] [PubMed]

95. Xu, G.; Wang, Y.; Li, W.; Cao, Y.; Xu, J.; Hu, Z.; Hao, Y.; Hu, L.; Sun, Y. COX-2 Forms Regulatory Loop with YAP to Promote Proliferation and Tumorigenesis of Hepatocellular Carcinoma Cells. *Neoplasia* **2018**, *20*, 324–334. [CrossRef] [PubMed]

96. Katoh, M.; Katoh, M. Molecular genetics and targeted therapy of WNT-related human diseases (Review). *Int. J. Mol. Med.* **2017**, *40*, 587–606. [CrossRef] [PubMed]

97. Zheng, C.-H.; Chen, X.-M.; Zhang, F.-B.; Zhao, C.; Tu, S.-S. Inhibition of CXCR4 regulates epithelial mesenchymal transition of NSCLC via the Hippo-YAP signaling pathway. *Cell Biol. Int.* **2018**, *42*, 1386–1394. [CrossRef]

98. Burger, J.A.; Kipps, T.J. CXCR4: A key receptor in the crosstalk between tumor cells and their microenvironment. *Blood* **2006**, *107*, 1761–1767. [CrossRef]

99. Almofti, A.; Uchida, D.; Begum, N.M.; Tomizuka, Y.; Iga, H.; Yoshida, H.; Sato, M. The clinicopathological significance of the expression of CXCR4 protein in oral squamous cell carcinoma. *Int. J. Oncol.* **2004**, *25*, 65–71. [CrossRef] [PubMed]

100. Raghuwanshi, S.K.; Su, Y.; Singh, V.; Haynes, K.; Richmond, A.; Richardson, R.M. The Chemokine Receptors CXCR1 and CXCR2 Couple to Distinct G Protein-Coupled Receptor Kinases To Mediate and Regulate Leukocyte Functions. *J. Immunol.* **2012**, *189*, 2824–2832. [CrossRef]

101. Lee, Z.; Swaby, R.F.; Liang, Y.; Yu, S.; Liu, S.; Lu, K.H.; Bast, R.C.; Mills, G.B.; Fang, X. Lysophosphatidic Acid Is a Major Regulator of Growth-Regulated Oncogene α in Ovarian Cancer. *Cancer Res.* **2006**, *66*, 2740–2748. [CrossRef]

102. Sharif, G.M.; Schmidt, M.O.; Yi, C.; Hu, Z.; Haddad, B.R.; Glasgow, E.; Riegel, A.T.; Wellstein, A. Cell growth density modulates cancer cell vascular invasion via Hippo pathway activity and CXCR2 signaling. *Oncogene* **2015**, *34*, 5879–5889. [CrossRef] [PubMed]

103. Uemura, H.; Hasumi, H.; Ishiguro, H.; Teranishi, J.-I.; Miyoshi, Y.; Kubota, Y. Renin-angiotensin system is an important factor in hormone refractory prostate cancer. *Prostate* **2006**, *66*, 822–830. [CrossRef] [PubMed]

104. Saikawa, S.; Kaji, K.; Nishimura, N.; Seki, K.; Sato, S.; Nakanishi, K.; Kitagawa, K.; Kawaratani, H.; Kitade, M.; Moriya, K.; et al. Angiotensin receptor blockade attenuates cholangiocarcinoma cell growth by inhibiting the oncogenic activity of Yes-associated protein. *Cancer Lett.* **2018**, *434*, 120–129. [CrossRef] [PubMed]

105. Wang, J.; Hong, Y.; Shao, S.; Zhang, K.; Hong, W. FFAR1-and FFAR4-dependent activation of Hippo pathway mediates DHA-induced apoptosis of androgen-independent prostate cancer cells. *Biochem. Biophys. Res. Commun.* **2018**, *506*, 590–596. [CrossRef] [PubMed]

106. Dethlefsen, C.; Hansen, L.S.; Lillelund, C.; Andersen, C.; Gehl, J.; Christensen, J.F.; Pedersen, B.K.; Hojman, P. Exercise-Induced Catecholamines Activate the Hippo Tumor Suppressor Pathway to Reduce Risks of Breast Cancer Development. *Cancer Res.* **2017**, *77*, 4894–4904. [CrossRef] [PubMed]

107. Yu, F.-X.; Zhang, Y.; Park, H.W.; Jewell, J.L.; Chen, Q.; Deng, Y.; Pan, D.; Taylor, S.S.; Lai, Z.-C.; Guan, K.-L. Protein kinase A activates the Hippo pathway to modulate cell proliferation and differentiation. *Genes Dev.* **2013**, *27*, 1223–1232. [CrossRef]

108. Kim, M.; Kim, M.; Lee, S.; Kuninaka, S.; Saya, H.; Lee, H.; Lee, S.; Lim, D.-S. cAMP/PKA signalling reinforces the LATS-YAP pathway to fully suppress YAP in response to actin cytoskeletal changes. *EMBO J.* **2013**, *32*, 1543–1555. [CrossRef] [PubMed]

109. Iglesias-Bartolome, R.; Torres, D.; Marone, R.; Feng, X.; Martin, D.; Simaan, M.; Chen, M.; Weinstein, L.S.; Taylor, S.S.; Molinolo, A.A.; et al. Inactivation of a Gα(s)-PKA tumour suppressor pathway in skin stem cells initiates basal-cell carcinogenesis. *Nat. Cell Biol.* **2015**, *17*, 793–803. [CrossRef]

110. Gong, R.; Hong, A.W.; Plouffe, S.W.; Zhao, B.; Liu, G.; Yu, F.-X.; Xu, Y.; Guan, K.-L. Opposing roles of conventional and novel PKC isoforms in Hippo-YAP pathway regulation. *Cell Res.* **2015**, *25*, 985–988. [CrossRef]

111. Park, H.W.; Kim, Y.C.; Yu, B.; Moroishi, T.; Mo, J.S.; Plouffe, S.W.; Meng, Z.; Lin, K.C.; Yu, F.X.; Alexander, C.M.; et al. Alternative Wnt Signaling Activates YAP/TAZ. *Cell* **2015**, *162*, 780–794. [CrossRef] [PubMed]

112. Rao, R.; Salloum, R.; Xin, M.; Lu, Q.R. The G protein Gα$_s$ acts as a tumor suppressor in sonic hedgehog signaling-driven tumorigenesis. *Cell Cycle* **2016**, *15*, 1325–1330. [CrossRef]

113. Hao, F.; Xu, Q.; Zhao, Y.; Stevens, J.V.; Young, S.H.; Sinnett-Smith, J.; Rozengurt, E. Insulin Receptor and GPCR Crosstalk Stimulates YAP via PI3K and PKD in Pancreatic Cancer Cells. *Mol. Cancer Res.* **2017**, *15*, 929–941. [CrossRef] [PubMed]

114. Feng, R.; Gong, J.; Wu, L.; Wang, L.; Zhang, B.; Liang, G.; Zheng, H.; Xiao, H. MAPK and Hippo signaling pathways crosstalk via the RAF-1/MST-2 interaction in malignant melanoma. *Oncol. Rep.* **2017**, *38*, 1199–1205. [CrossRef] [PubMed]

115. Dorsam, R.T.; Gutkind, J.S. G-protein-coupled receptors and cancer. *Nat. Rev. Cancer* **2007**, *7*, 79–94. [CrossRef] [PubMed]

116. Gutkind, J.S. The pathways connecting G protein-coupled receptors to the nucleus through divergent mitogen-activated protein kinase cascades. *J. Biol. Chem.* **1998**, *273*, 1839–1842. [CrossRef]

117. Nieto Gutierrez, A.; McDonald, P.H. GPCRs: Emerging anti-cancer drug targets. *Cell. Signal.* **2018**, *41*, 65–74. [CrossRef]

118. Van Raamsdonk, C.D.; Bezrookove, V.; Green, G.; Bauer, J.; Gaugler, L.; O'Brien, J.M.; Simpson, E.M.; Barsh, G.S.; Bastian, B.C. Frequent somatic mutations of GNAQ in uveal melanoma and blue naevi. *Nature* **2009**, *457*, 599–602. [CrossRef]

119. Van Raamsdonk, C.D.; Griewank, K.G.; Crosby, M.B.; Garrido, M.C.; Vemula, S.; Wiesner, T.; Obenauf, A.C.; Wackernagel, W.; Green, G.; Bouvier, N.; et al. Mutations in *GNA11* in Uveal Melanoma. *N. Eng. J. Med.* **2010**, *363*, 2191–2199. [CrossRef]

120. Yu, F.X.; Luo, J.; Mo, J.S.; Liu, G.; Kim, Y.C.; Meng, Z.; Zhao, L.; Peyman, G.; Ouyang, H.; Jiang, W.; et al. Mutant Gq/11 promote uveal melanoma tumorigenesis by activating YAP. *Cancer Cell* **2014**, *25*. [CrossRef]

121. Chen, X.; Wu, Q.; Depeille, P.; Chen, P.; Thornton, S.; Kalirai, H.; Coupland, S.E.; Roose, J.P.; Bastian, B.C. RasGRP3 Mediates MAPK Pathway Activation in GNAQ Mutant Uveal Melanoma. *Cancer Cell* **2017**, *31*, 685–696. [CrossRef] [PubMed]

122. Huh, J.Y.; Kwon, M.J.; Seo, K.Y.; Kim, M.K.; Chae, K.Y.; Kim, S.H.; Ki, C.S.; Yoon, M.S.; Kim, D.H. Novel nonsense GNAS mutation in a 14-month-old boy with plate-like osteoma cutis and medulloblastoma. *J. Dermatol.* **2014**, *41*, 319–321. [CrossRef]

123. Deng, Y.; Wu, L.M.N.; Bai, S.; Zhao, C.; Wang, H.; Wang, J.; Xu, L.; Sakabe, M.; Zhou, W.; Xin, M.; et al. A reciprocal regulatory loop between TAZ/YAP and G-protein Gαs regulates Schwann cell proliferation and myelination. *Nat. Commun.* **2017**, *8*, 15161. [CrossRef] [PubMed]

124. Landis, C.A.; Masters, S.B.; Spada, A.; Pace, A.M.; Bourne, H.R.; Vallar, L. GTPase inhibiting mutations activate the alpha chain of Gs and stimulate adenylyl cyclase in human pituitary tumours. *Nature* **1989**, *340*, 692–696. [CrossRef]

125. Weinstein, L.S.; Shenker, A.; Gejman, P.V.; Merino, M.J.; Friedman, E.; Spiegel, A.M. Activating mutations of the stimulatory G protein in the McCune-Albright syndrome. *N. Eng. J. Med.* **1991**, *325*, 1688–1695. [CrossRef]

126. O'Hayre, M.; Degese, M.S.; Gutkind, J.S. Novel insights into G protein and G protein-coupled receptor signaling in cancer. *Curr. Opin. Cell Biol.* **2014**, *27*, 126–135. [CrossRef] [PubMed]

127. O'Hayre, M.; Vázquez-Prado, J.; Kufareva, I.; Stawiski, E.W.; Handel, T.M.; Seshagiri, S.; Gutkind, J.S. The Emerging Mutational Landscape of G-proteins and G-protein Coupled Receptors in Cancer. *Nat. Rev. Cancer* **2013**, *13*, 412–424. [CrossRef]

128. Scales, S.J.; de Sauvage, F.J. Mechanisms of Hedgehog pathway activation in cancer and implications for therapy. *Trends Pharmacol. Sci.* **2009**, *30*, 303–312. [CrossRef]

129. Araç, D.; Boucard, A.A.; Bolliger, M.F.; Nguyen, J.; Soltis, S.M.; Südhof, T.C.; Brunger, A.T. A novel evolutionarily conserved domain of cell-adhesion GPCRs mediates autoproteolysis. *EMBO J.* **2012**, *31*, 1364–1378. [CrossRef]

130. Hernández, N.A.; Correa, E.; Avila, E.P.; Vela, T.A.; Pérez, V.M. PAR1 is selectively over expressed in high grade breast cancer patients: A cohort study. *J. Transl. Med.* **2009**, *7*, 47. [CrossRef]

131. Bar-Shavit, R.; Maoz, M.; Kancharla, A.; Nag, J.K.; Agranovich, D.; Grisaru-Granovsky, S.; Uziely, B. G Protein-Coupled Receptors in Cancer. *Int. J. Mol. Sci.* **2016**, *17*. [CrossRef]

132. St-Jean, M.; Ghorayeb, N.E.; Bourdeau, I.; Lacroix, A. Aberrant G-protein coupled hormone receptor in adrenal diseases. *Best Pract. Res. Clin. Endocrinol. Metab.* **2018**, *32*, 165–187. [CrossRef]

133. Liu, G.; Yu, F.-X.; Kim, Y.C.; Meng, Z.; Naipauer, J.; Looney, D.J.; Liu, X.; Gutkind, J.S.; Mesri, E.A.; Guan, K.-L. Kaposi sarcoma-associated herpesvirus promotes tumorigenesis by modulating the Hippo pathway. *Oncogene* **2015**, *34*, 3536–3546. [CrossRef]

134. Greenhough, A.; Bagley, C.; Heesom, K.J.; Gurevich, D.B.; Gay, D.; Bond, M.; Collard, T.J.; Paraskeva, C.; Martin, P.; Sansom, O.J.; et al. Cancer cell adaptation to hypoxia involves a HIF-GPRC5A-YAP axis. *EMBO Mol. Med.* **2018**, *10*. [CrossRef]

135. Wang, Y.; Pei, H.; Jia, Y.; Liu, J.; Li, Z.; Ai, K.; Lu, Z.; Lu, L. Synergistic Tailoring of Electrostatic and Hydrophobic Interactions for Rapid and Specific Recognition of Lysophosphatidic Acid, an Early-Stage Ovarian Cancer Biomarker. *J. Am. Chem. Soc.* **2017**, *139*, 11616–11621. [CrossRef] [PubMed]

136. Yu, D.; Cai, Y.; Zhou, W.; Sheng, J.; Xu, Z. The Potential of Angiogenin as a Serum Biomarker for Diseases: Systematic Review and Meta-Analysis. *Dis. Markers* **2018**, *2018*, 1984718. [CrossRef] [PubMed]

137. Ramcharan, S.K.; Lip, G.Y.; Stonelake, P.S.; Blann, A.D. Angiogenin outperforms VEGF, EPCs and CECs in predicting Dukes' and AJCC stage in colorectal cancer. *Eur. J. Clin. Invest.* **2013**, *43*, 801–808. [CrossRef]

138. Hisai, H.; Kato, J.; Kobune, M.; Murakami, T.; Miyanishi, K.; Takahashi, M.; Yoshizaki, N.; Takimoto, R.; Terui, T.; Niitsu, Y. Increased expression of angiogenin in hepatocellular carcinoma in correlation with tumor vascularity. *Clin. Cancer Res.* **2003**, *9*, 4852–4859.

139. Tas, F.; Duranyildiz, D.; Oguz, H.; Camlica, H.; Yasasever, V.; Topuz, E. Circulating serum levels of angiogenic factors and vascular endothelial growth factor receptors 1 and 2 in melanoma patients. *Melanoma Res.* **2006**, *16*, 405–411. [CrossRef]

140. Rykala, J.; Przybylowska, K.; Majsterek, I.; Pasz-Walczak, G.; Sygut, A.; Dziki, A.; Kruk-Jeromin, J. Angiogenesis markers quantification in breast cancer and their correlation with clinicopathological prognostic variables. *Pathol. Oncol. Res.* **2011**, *17*, 809–817. [CrossRef] [PubMed]

141. Barton, D.P.; Cai, A.; Wendt, K.; Young, M.; Gamero, A.; De Cesare, S. Angiogenic protein expression in advanced epithelial ovarian cancer. *Clin. Cancer Res.* **1997**, *3*, 1579–1586. [PubMed]

142. Liu-Chittenden, Y.; Huang, B.; Shim, J.S.; Chen, Q.; Lee, S.-J.; Anders, R.A.; Liu, J.O.; Pan, D. Genetic and pharmacological disruption of the TEAD-YAP complex suppresses the oncogenic activity of YAP. *Genes Dev.* **2012**, *26*, 1300–1305. [CrossRef]

143. Noland, C.L.; Gierke, S.; Schnier, P.D.; Murray, J.; Sandoval, W.N.; Sagolla, M.; Dey, A.; Hannoush, R.N.; Fairbrother, W.J.; Cunningham, C.N. Palmitoylation of TEAD Transcription Factors Is Required for Their Stability and Function in Hippo Pathway Signaling. *Structure* **2016**, *24*, 179–186. [CrossRef] [PubMed]

144. Chan, P.; Han, X.; Zheng, B.; DeRan, M.; Yu, J.; Jarugumilli, G.K.; Deng, H.; Pan, D.; Luo, X.; Wu, X. Autopalmitoylation of TEAD proteins regulates transcriptional output of the Hippo pathway. *Nat. Chem. Biol.* **2016**, *12*, 282–289. [CrossRef] [PubMed]

145. Li, Y.; Liu, S.; Ng, E.Y.; Li, R.; Poulsen, A.; Hill, J.; Pobbati, A.V.; Hung, A.W.; Hong, W.; Keller, T.H.; et al. Structural and ligand-binding analysis of the YAP-binding domain of transcription factor TEAD4. *Biochem. J.* **2018**, *475*, 2043–2055. [CrossRef] [PubMed]

146. Bum-Erdene, K.; Zhou, D.; Gonzalez-Gutierrez, G.; Ghozayel, M.K.; Si, Y.; Xu, D.; Shannon, H.E.; Bailey, B.J.; Corson, T.W.; Pollok, K.E.; et al. Small-Molecule Covalent Modification of Conserved Cysteine Leads to Allosteric Inhibition of the TEADYap Protein-Protein Interaction. *Cell Chem. Biol.* **2019**, *26*, 378–389. [CrossRef] [PubMed]

147. Bao, Y.; Nakagawa, K.; Yang, Z.; Ikeda, M.; Withanage, K.; Ishigami-Yuasa, M.; Okuno, Y.; Hata, S.; Nishina, H.; Hata, Y. A cell-based assay to screen stimulators of the Hippo pathway reveals the inhibitory effect of dobutamine on the YAP-dependent gene transcription. *J. Biochem.* **2011**, *150*, 199–208. [CrossRef]

148. Park, H.W.; Guan, K.-L. Regulation of the Hippo pathway and implications for anticancer drug development. *Trends Pharmacol. Sci.* **2013**, *34*, 581–589. [CrossRef] [PubMed]

149. Borcherding, D.C.; Tong, W.; Hugo, E.R.; Barnard, D.F.; Fox, S.; LaSance, K.; Shaughnessy, E.; Ben-Jonathan, N. Expression and therapeutic targeting of dopamine receptor-1 (D1R) in breast cancer. *Oncogene* **2016**, *35*, 3103–3113. [CrossRef]

150. Sever, R.; Brugge, J.S. Signal transduction in cancer. *Cold Spring Harb. Perspect. Med.* **2015**, *5*. [CrossRef]

151. Onken, M.D.; Makepeace, C.M.; Kaltenbronn, K.M.; Kanai, S.M.; Todd, T.D.; Wang, S.; Broekelmann, T.J.; Rao, P.K.; Cooper, J.A.; Blumer, K.J. Targeting nucleotide exchange to inhibit constitutively active G protein α subunits in cancer cells. *Sci. Signal.* **2018**, *11*, eaao6852. [CrossRef] [PubMed]

152. Annala, S.; Feng, X.; Shridhar, N.; Eryilmaz, F.; Patt, J.; Yang, J.; Pfeil, E.M.; Cervantes-Villagrana, R.D.; Inoue, A.; Häberlein, F.; et al. Direct targeting of $G\alpha_q$ and $G\alpha_{11}$ oncoproteins in cancer cells. *Sci. Signal.* **2019**, *12*, eaau5948. [CrossRef] [PubMed]

153. Sorrentino, G.; Ruggeri, N.; Specchia, V.; Cordenonsi, M.; Mano, M.; Dupont, S.; Manfrin, A.; Ingallina, E.; Sommaggio, R.; Piazza, S.; et al. Metabolic control of YAP and TAZ by the mevalonate pathway. *Nat. Cell Biol.* **2014**, *16*, 357–366. [CrossRef] [PubMed]

154. Oku, Y.; Nishiya, N.; Shito, T.; Yamamoto, R.; Yamamoto, Y.; Oyama, C.; Uehara, Y. Small molecules inhibiting the nuclear localization of YAP/TAZ for chemotherapeutics and chemosensitizers against breast cancers. *FEBS Open Bio* **2015**, *5*, 542–549. [CrossRef] [PubMed]

Review

The Roles of YAP/TAZ and the Hippo Pathway in Healthy and Diseased Skin

Emanuel Rognoni [1] **and Gernot Walko** [2,*]

[1] Centre for Endocrinology, William Harvey Research Institute, Barts and The London School of Medicine, Queen Mary University of London, London EC1M 6BQ, UK; e.rognoni@qmul.ac.uk

[2] Department of Biology and Biochemistry & Centre for Therapeutic Innovation, University of Bath, Claverton Down, Bath BA2 7AY, UK

* Correspondence: g.walko@bath.ac.uk; Tel.: +44-(0)1225-38-6261

Received: 22 March 2019; Accepted: 30 April 2019; Published: 3 May 2019

Abstract: Skin is the largest organ of the human body. Its architecture and physiological functions depend on diverse populations of epidermal cells and dermal fibroblasts. Reciprocal communication between the epidermis and dermis plays a key role in skin development, homeostasis and repair. While several stem cell populations have been identified in the epidermis with distinct locations and functions, there is additional heterogeneity within the mesenchymal cells of the dermis. Here, we discuss the current knowledge of how the Hippo pathway and its downstream effectors Yes-associated protein (YAP) and transcriptional coactivator with PDZ-binding motif (TAZ) contribute to the maintenance, activation and coordination of the epidermal and dermal cell populations during development, homeostasis, wound healing and cancer.

Keywords: Hippo signalling; skin development; stem cells; skin cancer; fibroblasts; fibrosis; wound healing

1. Introduction

The skin is the largest organ of the human body. It forms a protective interface between the body and the external environment. Various skin cell populations act in harmony to provide protection from daily wear and tear, harmful microbes and other assaults from the external environment. In addition, skin forms a barrier against water loss, enables thermoregulation, and relays somatosensory information to the brain to inform reflexes and behaviours [1].

The outermost tissue layer of the skin, the epidermis, is a multilayered (stratified) squamous epithelium that is separated by a basement membrane (BM) from the underlying dermis. The dermis forms the skin scaffold consisting of a dense extracellular matrix (ECM) meshwork and different cell populations, including fibroblasts, sensory neurons and endothelial and immune cells. Maintenance and repair of the epidermis are dependent on stem cells (SCs) that both self-renew and give rise to cells that undergo terminal differentiation. Multiple SC populations have been identified within mammalian epidermis; these are distinguished by their location and the markers that they express [2–5]. However, during tissue regeneration such as wound healing, these SC populations display remarkable plasticity by switching from one SC type to another or by generating a wider range of differentiated lineages than under steady state conditions [1,6]. Likewise, an astonishing heterogeneity and plasticity of dermal cell populations has recently been identified in mouse and human dermis during development and tissue regeneration, which seems to evolve with age [7–11]. Upon tissue damage dermal fibroblasts not only become activated, start proliferating and deposit/ remodel ECM, but are also involved in regenerating HF and the hypodermis [3]. Plasticity in the epidermis and dermis typically resolves as wounds heal. However, in cancer, it can endure [1,6].

Carcinomas originating from the epidermis are by far the most frequently diagnosed human cancers, with cutaneous squamous cell carcinoma being the most dangerous subtype due to its ability to metastasise [12,13]. There is now strong evidence that epidermal cancers are maintained by a subpopulation of epidermal cells termed tumour-initiating cells or cancer SCs, which hijack the homeostatic controls operating in normal epidermal stem cells to endow themselves with the potential to self-renew and fuel the cancer [14,15]. Distinct epithelial SC populations have been associated with distinct tumour types; however, how different fibroblast lineages affect tumour development, type and progression is still unclear [3].

In the epidermis and dermis, the Hippo signalling pathway and its downstream effectors, the transcriptional coactivators Yes-associated protein (YAP) and transcriptional coactivator with PDZ-binding motif (TAZ, also called WW Domain Containing Transcription Regulator 1 (WWTR1)), regulate diverse tissue-specific functions during development, homeostasis and regeneration. While in the epidermis YAP/TAZ are essential to control cell growth and differentiation, in fibroblasts they function predominantly as an intracellular mechanical rheostat to sense the physical cellular environment, promoting ECM remodelling and contractility. The Hippo pathway is a tumour suppressor pathway, since its dysregulation and the resulting YAP/TAZ hyperactivation promote cancer development.

This review is divided into three parts. We first give an overview of the contributions of the different epidermal SC and dermal fibroblast populations to skin tissue development, homeostasis and repair, and the roles of these cells in cancer development (chapters 2–6); in the second part, we review the existing work on the specific roles of YAP/TAZ in controlling epidermal SC and fibroblast functions in healthy and diseased skin, summarise the various mechanisms of YAP/TAZ regulation and discuss YAP/TAZ-targeting therapy-approaches (chapters 7–11). We finally close this review with some future perspectives for YAP/TAZ in skin biology (chapter 12).

2. Skin Architecture

The skin consists of two tissue layers—the epidermis and the underlying dermis—which are separated by a BM (Figure 1). The epidermis is a multilayered (stratified) squamous epithelium consisting of the IFE and associated hair follicles (HFs), sebaceous glands (SGs), and sweat glands. Keratinocytes are the main epidermal cell type. Several other cell types, such as Merkel cells, melanocytes, gamma delta (γ∂) T-cells and Langerhans cells, are also found in mammalian epidermis [1,3,5].

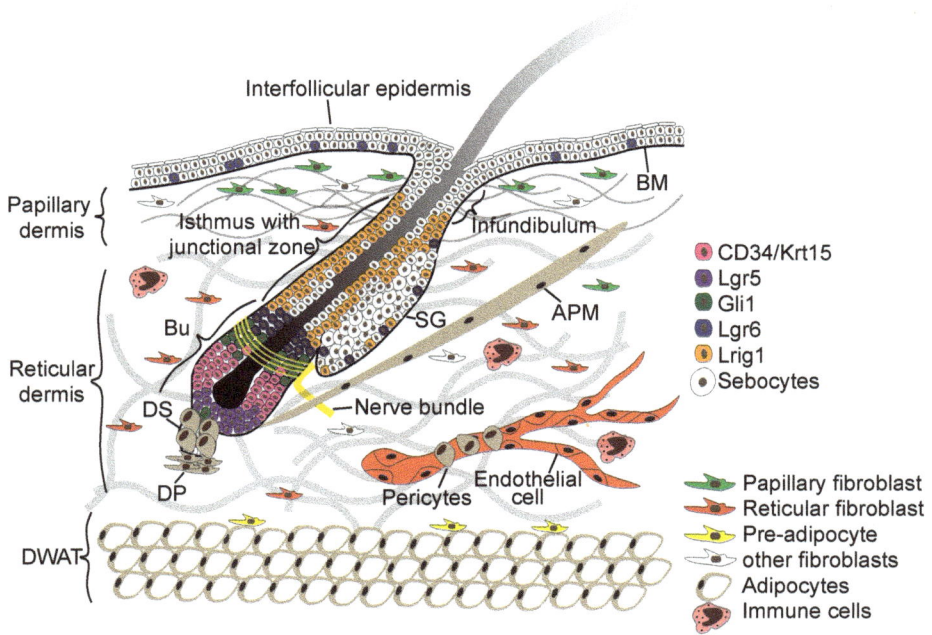

Figure 1. Morphology of the skin. The epidermis and dermis are separated by a BM. In the epidermis, multiple spatially distinct stem cell populations have been identified in the hair follicle bulge, isthmus and sebaceous gland, and their characteristic markers are shown in the colour coded legend. In the dermis, papillary fibroblasts are located in proximity to the BM and are embedded in thin collagen fibres. Reticular fibroblasts populate the central dermis and are surrounded by thick collagen bundles (grey). Preadipocytes are close to the DWAT where the mature adipocytes reside. In addition, specialised fibroblast subpopulations associate with the HF give rise to the DP, DS and APM. Endothelial cells form the blood vessels which are surrounded by pericytes. Sensory neurons are associated with the HF upper bulge SC population and different immune cell types populate different regions of the skin. Abbreviations: APM, arrector pili muscle; BM, basement membrane; Bu, bulge; DP, dermal papilla; DS, dermal sheath; DWAT, dermal white adipose tissue; SG, sebaceous gland.

The interfollicular epidermis (IFE) consists of an inner layer of proliferative basal cells that express keratins Krt5 and Krt14 and are attached to the underlying BM via integrin (ITG) extracellular matrix (ECM) receptors and several layers of suprabasal keratinocytes at various stages of a terminal differentiation programme. Only the innermost (basal) layer is proliferative (Figure 2A). To constantly renew the epidermal barrier, differentiating cells moving outward replenish the terminally differentiated squames that are sloughed from the skin surface [1,5].

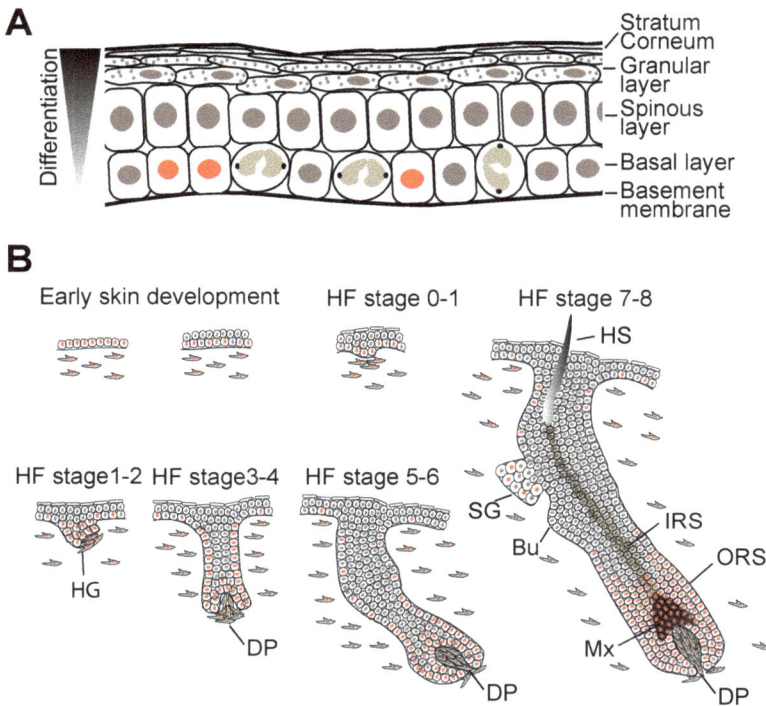

Figure 2. Yes-associated protein (YAP)/transcriptional coactivator with PDZ-binding motif (TAZ) activity in the IFE and HF development. (**A**) The IFE is a stratified squamous epithelium. It is divided into four main layers that are distinguished morphologically according to the differentiation status of the keratinocytes as they cease to proliferate and move upward to produce the skin's barrier. Note that in adult IFE nuclear YAP/TAZ are restricted in cell clusters of the basal layer. (**B**) Early skin and HF development. During early skin development, cells of the epidermis and dermis are highly proliferative and highly positive for nuclear YAP/TAZ. Once the epidermis starts to stratify, only the proliferative cells in the basal layer maintain nuclear YAP/TAZ. HF development is initiated by an epidermal–mesenchymal cross-talk inducing condensation of mesenchymal cells beneath the BM which leads to the formation of a HF placode (HF stages 0–1). The HF placodes further matures into a hair germ (HF stages 1–2), which start to engulf the dermal papilla fibroblasts (HF stages 3–4). At HF stages 5–6 the dermal papilla is fully encapsulated; the HF epithelial cells differentiate into the distinct HF layers and the bulge and SG start to form. All cell compartments are clearly visible at HF stage 7–8 and the HF shaft emerges through the epidermis. YAP is nuclear in the placode and hair germ cells of the epidermis and dermis (HF stage 0–2) and then becomes more restricted to the highly proliferative basal epithelial cells (HF stage 3–8). YAP/TAZ are highly nuclear in the HF matrix, and there are additional cell cluster with nuclear YAP in the IFE and SG. Note that due to its wide-spread expression in skin, only nuclear YAP/TAZ are shown to indicate sites of YAPTAZ activity. Abbreviations: BM, basement membrane; Bu, bulge; DP, dermal papilla; HF, hair follicle; HG, hair germ; HS, hair shaft; IFE, interfollicular epidermis; IRS, inner root sheath; Mx matrix; ORS outer root sheath; SG, sebaceous gland.

Contiguous with the IFE are HFs (Figure 1), which are encased by the BM and frequently cycle between growth and resting phase. Within the HF epithelium multiple SC populations have been identified through the use of lineage tracing and flow cytometry (for a comprehensive review on lineage tracing see [16,17]). These include SCs of the junctional zone between the IFE, HF and SG (Figure 1), which express the receptor tyrosine kinase regulator Lrig1, and cells of the lower hair follicle that express the R-spondin receptor Lgr5, the transmembrane phosphoglycoprotein CD34, keratins Krt15 and Krt19 and Sox9 [2,18]. In addition, Gli1+ and Lgr6+ stem cells are found in the upper hair follicle and with the latter scattered within the IFE [2,18].

Upon growth induction, cells in the HF lower bulge become activated, proliferate and differentiate in the district hair layers through a complex mesenchymal–epithelial cross-talk, while the lower HF portion expands [3,5]. The upper portion of the HF, which includes the infundibulum, junctional zone and sebaceous glands, does not cycle but undergoes frequent turnover governed by multiple resident SC pools, each responsible for maintaining homeostasis of their nearby territory [2,3]. Similar to the IFE, in the SG proliferating cells anchored to the BM support turnover of differentiated cells [2]. In contrast to mouse skin, the epidermis in human skin is proportionately much more interfollicular, is much thicker and the epidermis–dermis junction is undulated [19]. These epidermal invaginations are referred to as rete ridges and their width and depth varies with age and disease [20]. Human IFE SCs have been shown to reside in cluster between rete ridges and decline with age [20–23].

The dermis is composed of different sublayers that are distinguished by cell type, cell density, and extracellular matrix (ECM) composition (Figure 1) [3,24]. The papillary dermis is located closest to the IFE and displays a high fibroblast density. The reticular dermis is the central and largest layer of the dermis, and consists of a thick, highly organised collagen fibre-rich ECM with sparsely embedded fibroblasts. Under the reticular dermis lies the dermal white adipose tissue (DWAT), also referred to as hypodermis, which harbours pre- and mature adipocytes [3]. In addition specialised fibroblast subpopulations are associated with the blood vessels (pericytes) and HFs giving rise to the dermal sheath, dermal papilla and the arrector pili muscle (APM) (Figure 1). The dermis is highly vascularised and innervated, and cells of the immune system traffic through both the dermis and epidermis [24].

3. Skin Morphogenesis and Hair Follicle Development

Shortly after gastrulation, the skin forms as a flat single-layered epithelium from the surface ectoderm and the dermis (arising from the mesoderm) appears homogenous in composition [5]. Initially, unspecified epidermal progenitors divide exclusively parallel to the BM underneath, but within several days, divisions become first oblique and then more perpendicular, leading to asymmetric fates, stratification, and differentiation of the epidermis [5] (Figure 2B). Local induction of WNT signalling in the epidermis and subsequently in the dermis causes epidermal cells with high levels of WNT signalling to cluster into HF placodes, which are characterised by expression of adult SC markers such as the transcription factor (TF) Sox9, and Lrig1 and Lgr6 [5,16]. Production of sonic hedgehog (SHH) by WNT^{hi} cells then induces further HF maturation, during which SC markers begin to segregate into the distinct HF compartments including the HF bulge, junctional zone, isthmus and sebaceous gland [2,18] (Figure 2B). By the time epidermal morphogenesis is complete, the SCs of the different HF compartments then reside in discrete niches whose cellular components and other sources of signalling factors heavily influence their behaviour [1,3,5].

During embryonic development multipotent fibroblasts are highly proliferative and start to differentiate into papillary and reticular lineages [7,9]. While the papillary fibroblasts give rise to the dermal sheath, dermal papilla and APM, reticular cells differentiate to preadipocyte and mature adipocytes [7,9]. Postnatally, fibroblasts stop proliferating and enter a quiescent state for efficient ECM deposition and remodelling. During dermal maturation fibroblast lineages become segregated by increased ECM deposition, start intermixing and lineage marker expression dynamically changes with age. While dermal fibroblast density and the papillary layer decrease with age, there is an increase in adipocyte layer [11,25,26]. Intriguingly, it was shown that the coordinated switch in fibroblast behaviour from being highly proliferative in embryonic development to quiescence postnatally in order to allow efficient ECM deposition/remodelling is balanced by a negative feedback loop which is necessary and sufficient to define dermal architecture during development [27]. So far, the cell intrinsic and extrinsic regulatory mechanisms controlling dermal fibroblast lineage identity, behaviour and fate, are largely unknown.

4. Skin Homeostasis and Tissue Repair

Homeostasis of the IFE is maintained by a delicate balance between basal cell proliferation and suprabasal cell differentiation/stratification (Figure 2A). At the onset of differentiation, basal cells become detached (delaminate) from the BM, stop proliferating, and once located in the suprabasal cell layer, start executing terminal cell differentiation programmes that involve extensive remodelling of intracellular proteins, intercellular junctions, lipid extrusion and nuclear fragmentation. These events culminate in the terminally differentiated cells becoming highly cross-linked scales that are exfoliated from the surface of the skin [28,29]. Intriguingly the temporal and spatial dynamics of differentiation commitment seems to be controlled by autoregulatory network of phosphatases [30].

While several SC populations have been described in the HFs which maintain homeostasis and are able to participate to wound repair, the identity and organisation of SCs within the IFE is still an unresolved matter. Early studies of human and mouse epidermis revealed heterogeneity in the propensity of basal IFE cells to proliferate, and the concept arose that IFE SCs self-renew infrequently, while their progeny undergo a small number of amplifying divisions prior to the onset of terminal differentiation [31,32]. Such so-called transit amplifying cells were also identified in vitro in studies of colony formation by cultured human epidermal cells [33,34]. While it was first believed that the IFE is maintained by a single cell population [35] further studies showed that basal cells in mouse epidermis are heterogeneous and some exhibit SC characteristics [36,37]. Single-cell transcriptomic analysis of cultured human epidermal cells [38] and mouse and human epidermis [39,40] further supports the concept of IFE SC heterogeneity.

In the dermis, clonal lineage tracing revealed that fibroblast turnover is very low lineages and lineages remain in a quiescence state long-term to deposit and remodel their surrounding ECM during homeostasis. If mesenchymal SC actively contribute to dermal homeostasis is still a matter of debate [41].

During wound healing different cell types of the epidermis and dermis are coordinated to regenerate the skin [3,4]. In the initial inflammatory phase, a blood clot forms and immune cells infiltrate the wound bed, which is followed by epidermal and dermal cells starting to proliferate and to migrate into the wound bed in the proliferative phase. During resolution phase, epidermal cells still proliferate, while dermal fibroblasts enter quiescence to allow for efficient ECM deposition and remodelling. Recently it was shown that epidermal cells organise in concentric zones with distinct cellular activity and gene expression [42,43]. In the wound bed central migratory zone, cell migration and differentiation are tightly coordinated to promote epidermal thickening and wound closure. In the second (proliferative) zone cells are highly proliferative, polarised towards the migration direction and control the involvement of surrounding epithelial cells in the unwounded area. Although different epidermal cells from the IFE and HF contribute to wound repair, within the wound healing zones they exhibit similar behaviour in cell proliferation, migration and differentiation. Indeed, during wound healing, SCs exhibit a high degree of plasticity and temporarily lose their lineage restriction [36]. In addition, it was recently shown that differentiated sebaceous duct cells are able to dedifferentiate, proliferate and regenerate the IFE long-term, however the transcriptional and epigenetic mechanisms are still unclear [44].

Upon wounding different fibroblast lineages become activated (α-smooth muscle actin (aSMA)-positive myofibroblasts) at the wound site, quickly resume proliferation and migrate into the wound bed [11,27]. While fibroblast in the reticular and adipocyte layer mediate the initial phase of wound repair, papillary fibroblasts are recruited at a later stage and dermal papilla and APM fibroblasts seem not to contribute at all [9,11,45]. Intriguingly, besides depositing/remodelling ECM in the wound bed, myofibroblasts are able to acquire a dermal papilla or adipocyte fate in response to distinct signals promoting hair follicle and adipocyte regeneration [46–48]. After tissue repair, wound bed fibroblasts re-establish a quiescent state to maintain skin homeostasis. The signals controlling distinct fibroblast lineage recruitment and promoting myofibroblast conversion to other mesenchymal cell populations are largely unclear.

5. Skin Cancer

Skin cancers can be divided into cutaneous melanomas and nonmelanoma skin cancers (NMSCs). NMSCs are by far the most frequently diagnosed human cancers worldwide [49]. NMSCs comprise various types of carcinomas, such as basal cell carcinomas (BCCs), cutaneous squamous cell carcinomas (cSCCs), keratoacanthomas, Merkel cell carcinomas, cutaneous lymphomas, angiosarcomas and various rare adnexal tumours. Approximately 80% of nonmelanoma skin cancers are BCCs and 20% are cSCCs [49,50]. Both types of NMSCs originate from the epidermis. Although BCC is a malignant cancer, it is generally only locally invasive and rarely metastasises [49]. UV radiation is the principal mutagen in BCC pathogenesis, and dysregulation of the SHH/PTCH1/SMO pathway is central to BCC development [49]. Overexpression of the Hedgehog pathway, either through deletion of PTCH1, mutational activation of SMO, or overexpression of GLI1 or GLI2 have been reported in human and mouse BCC [19,49,50]. Compared to BCC, cSCC is a more aggressive tumour that can form lethal metastases and is associated with mutation in RAS GTPases (HRAS and KRAS), cell cycle regulators such as TP53 and CDKN2A and Notch signalling receptors (NOTCH1, NOTCH2, and NOTCH3) [13,14,49,51]. cSCC can be induced in mice through multistage carcinogenesis models employing either UV irradiation or chemical carcinogens or forced expression of oncogenes targeted to the epidermis [14,15,51]. Interestingly, studies using transgenic mice with expression of oncogenic KRAS in different compartments of adult epidermis revealed that both HF and IFE SCs represent the cells of origin of mouse cSCC [14]. However, oncogenic targeting of HF SCs led to formation of more aggressive and less differentiated cSCCs with features of epithelial to mesenchymal transition (EMT) [14]. Like cSCC, murine BCC can also arise from multiple epidermal compartments [51].

Cutaneous melanoma originates from pigment-producing melanocytes [52]. The incidence of cutaneous melanoma is substantially lower than that of NMSC; however, approximately 75% of skin cancer deaths are due to metastatic melanoma [52]. Malignant transformation into melanoma follows a sequential genetic model that results in constitutive activation of oncogenic signal transduction [52,53]. Oncogenic driver mutations in melanoma involve BRAF (~50%), NRAS, KRAS and HRAS (~25%) and NF1 (~15%) [52,53]. Other frequent genetic alterations include activating *TERT* promoter mutations, found in 30–80% of melanomas [52,53].

Communication between tumour cells and their microenvironment, including the tumour stroma (the non-transformed tissue components associated with a tumour), plays an important role in the development and progression of skin cancer [3,54–57]. Besides endothelial and immune cells, a major component of the microenvironment is cancer-associated fibroblasts (CAFs), which play an important role in the evolution of solid tumours. CAFs seem to originate from different mesenchymal populations, ranging from normal fibroblasts and mesenchymal SCs to transdifferentiated epithelial and endothelial cells. In contrast to normal fibroblasts, CAFs either reside within the tumour margin or infiltrate the tumour mass and show increased proliferation, migration, ECM deposition and secretion of growth factors and other ECM modulators [58,59]. To date there have been few studies of how different fibroblast lineages contribute to tumour stroma formation, and whether the tumour stroma differs between different types of skin cancers. Interestingly, one study showed that fibroblasts of the reticular dermis are predisposed to differentiate into CAFs upon cSCC signals, assisting invasion and EMT [60].

6. The Hippo Signalling Pathway

The Hippo pathway is a highly conserved signal transduction pathway that regulates gene expression. The core of the pathway is a kinase cascade that in mammals comprises MST1 (Ste20-like kinase 1; also known as STK4) and MST2 (also known as STK3), the homologues of the *D. melanogaster* Hpo kinase, large tumour suppressor kinase 1 (LATS1) and LATS2 (Warts in *D. melanogaster*), the adaptor proteins Salvador 1 (SAV1) (Sav in *D. melanogaster*), MOB1A and MOB1B (Mats in *D. melanogaster*) and the paralogous transcriptional coactivator proteins YAP and TAZ (Yorkie in *D. melanogaster*) [61–64]. In addition to LATS1/LATS2, NDR1 (STK38) and NDR2 (STK38L) also function as YAP/TAZ kinases [65].

The predominant transcriptional binding partners of YAP/TAZ are TFs of the TEA domain (TEAD) family (TEAD1–TEAD4) (scalloped in *D. melanogaster*) [61–64].

The functions of YAP and TAZ appear to be largely but not entirely redundant [66]. Indeed, there is genetic evidence that—at least in certain tissue contexts—both paralogues might drive distinct transcriptomes [67,68]. Inactivation of YAP usually has stronger consequences on cellular physiology than of TAZ, but this might simply reflect different expression levels of YAP and TAZ [67].

Mechanistically, YAP/TAZ binds to gene enhancer elements in complex with a TEAD TF, interacting with chromatin remodelling factors and modulating RNA polymerase II to drive or repress the expression of target genes, which prominently include cell cycle, cell migration and cell fate regulators [69–76]. Although TEAD factors are their predominant transcriptional interaction partners, YAP/TAZ have been shown to physically interact with other TFs such as p73, RUNX1/2/3 and TBX5 [77–85]. YAP–TEAD complexes alone may likely not be sufficient to execute the different transcriptional programmes. Indeed, bioinformatics analyses of the regulatory regions that are bound by YAP/TAZ-TEAD complexes frequently identified cooperation between YAP/TEAD and other TFs [73,74,85,86]. Therefore, YAP/TAZ cooperates with various TFs and chromatin regulators to regulate target gene expression.

Upstream of YAP/TAZ, activation of MST1/MST2 induces the phosphorylation of SAV1 and MOB1A/MOB1B, which assists MST1/MST2 in the recruitment, phosphorylation and activation of LATS1/LATS2 [61,62,64]. In parallel to MST1/MST2, two groups of MAP4Ks (mitogen-activated protein kinase kinase kinase kinase), MAP4K1/2/3/5 (homologs of Drosophila Happyhour (Hppy)) and MAP4K4/6/7 (homologs of Drosophila Misshapen (Msn)), can also directly phosphorylate and activate LATS1/LATS2 [87–89]. Subsequently, LATS1/LATS2 phosphorylates YAP and TAZ on several serine residues. Of these sites, the most relevant residues that keep YAP/TAZ inhibited are S127 and S381 in human YAP and S89 and S311 in human TAZ [61,62,64]. YAP/TAZ S127/S89 phosphorylation by LATS1/LATS2 creates a binding site for 14-3-3 proteins which contribute to keeping YAP/TAZ in the cytoplasm [90–92]. However, in many cellular contexts this signalling input alone does not appear to be sufficient to inactivate YAP/TAZ, as several studies have documented S127/S89-phosphorylated YAP/TAZ in the nucleus [21,82,93,94]. S381/S311 phosphorylation of YAP/TAZ modulates their protein stability by triggering further phosphorylation by casein kinase 1δ or 1ε (CK1) as well as ubiquitylation by the SKP1-CUL1-F-box protein (SCF) E3 ubiquitin ligase complex and proteasomal degradation [90–92]. It should be noted here that Hippo pathway regulation is not static in either ON or OFF state, but rather it is dynamic. YAP/TAZ is under constant phosphorylation and dephosphorylation and is rapidly trafficked between the cytoplasm and the nucleus [95–98]. Altogether, these observations suggest that additional regulatory inputs need to work together with the core Hippo pathway to fully inhibit YAP/TAZ activity.

The activity of the Hippo pathway is regulated by a multitude of upstream inputs, many of which relay signals from the plasma membrane [62–64]. However, unlike other classical signal transduction pathways, such as the epidermal growth factor (EGF), transforming growth factor-β (TGFβ) or WNT signalling pathways, the Hippo pathway does not appear to have dedicated transmembrane receptors and extracellular ligands. Rather, the Hippo pathway is regulated by a network of upstream signalling components that have roles in other processes such as the establishment of cell adhesion [99–104], cell morphology [105–108] and cell polarity [109–116]. The activity of the Hippo pathway is thus modulated in response to mechanical strains and changes or defects in cell–cell and cell–ECM adhesion [63,64,117] but also nutrient availability and other cellular stresses [118]. Therefore, the Hippo pathway constitutes a sensor for tissue and cellular integrity rather than responding to dedicated extracellular signalling molecules.

The biophysical properties of the extracellular environment and the cell shape are profound regulators of Hippo pathway activity. Mechanical stress, such as that caused when cells are grown on stiff surfaces, triggers YAP/TAZ nuclear translocation, whereas detachment from the ECM causes YAP/TAZ nuclear export [93,102,105–107,119,120]. The effects of the mechanical properties of the ECM on the Hippo pathway are mediated by integrin complexes at cell–ECM adhesion sites (focal adhesions and hemidesmosomes) and changes in the actomyosin cytoskeleton induced by integrin signalling in response to physical ECM properties [100,102,103,121]. Although the mechanisms are not fully understood, mechanical forces may further regulate YAP/TAZ through modulating the structure of nuclear pores, and hence the nuclear translocation of YAP/TAZ [96].

The Hippo pathway is also modulated by extensive crosstalk with other signalling pathways. In particular, these include G protein-coupled receptors (GPCRs) that are activated by lipids (lysophosphatidic acid and sphingosine-1-phosphophate) or hormones (glucagon or adrenaline) and signal through F-actin to regulate YAP/TAZ [122,123]; the WNT pathway, which regulates YAP/TAZ through direct interaction with the β-catenin destruction complex and through destruction complex-independent mechanisms [124–128]; SRC family kinases that either promote YAP nuclear localisation and transcriptional activity directly by phosphorylating tyrosine residues or indirectly by repressing LATS1/LATS2 [94,100,121,129–132]; and the PI3K pathway [100,103,133,134].

7. Expression of YAP and TAZ in Skin during Development, Homeostasis, Regeneration and Cancer

In both mouse and human epidermis, there is a clear correlation between nuclear localisation of YAP and the extent of cycling activity of epidermal SCs. In mouse skin, YAP expression was demonstrated as early as embryonic day 12, when the epidermis exists as a single cell layer of undifferentiated progenitors (Figure 2B). At this time point, YAP was shown to be predominantly nuclear, suggesting it is already active as transcriptional coactivator [135]. On stratification, YAP remains mostly concentrated in the basal epidermal cell layer, but the numbers of basal cells with nuclear YAP localisation wanes as proliferative activity in the IFE gradually diminishes postnatally [100,135]. Similar changes in YAP expression patterns have also been documented in human epidermis [21,100]. The timing and patterning of TAZ expression during epidermal morphogenesis and in the postnatal epidermis appears to coincide with that of YAP [100]. During HF formation, YAP remains nuclear in the proliferative cell populations of the outer root sheet (ORS) and transit-amplifying matrix (Mx) cells, but is cytoplasmic in the terminally differentiated lineages of the inner root sheet (IRS) and hair shaft (HS) (Figure 2B) [135]. In adult epidermis, prominent yet cytoplasmic YAP expression can be detected in the lower HF in telogen when SCs are quiescent, but YAP becomes nuclear in the HF during growth phase (anagen) when SCs are actively cycling [136,137] (Figure 3). YAP is also expressed in basal cells of SGs, where its nuclear localisation correlates with the expression of the proliferation marker Ki67 [138]. In the adult IFE, YAP is cytoplasmic in suprabasal, differentiated cells and nuclear in basal layer cells, but there is ambiguity in the literature about the extent of nuclear YAP abundance in the basal layer [21,99,100,135–137,139,140]. It is worth noting that nuclear YAP abundance in the IFE and HF junctional zone increases as HFs enter anagen [137] (Figure 3), which is consistent with lineage tracing studies showing that basal IFE cells nearest to actively cycling HFs are considerably more proliferative than the ones more distant [141]. Given the variation in HF densities at different body sites, this may explain some of the differences in nuclear YAP abundance in adult mouse IFE reported by different laboratories. In skin fibroblasts YAP is nuclear in proliferative cells and is mainly cytoplasmatic in quiescent cells postnatally [21] (Figure 3).

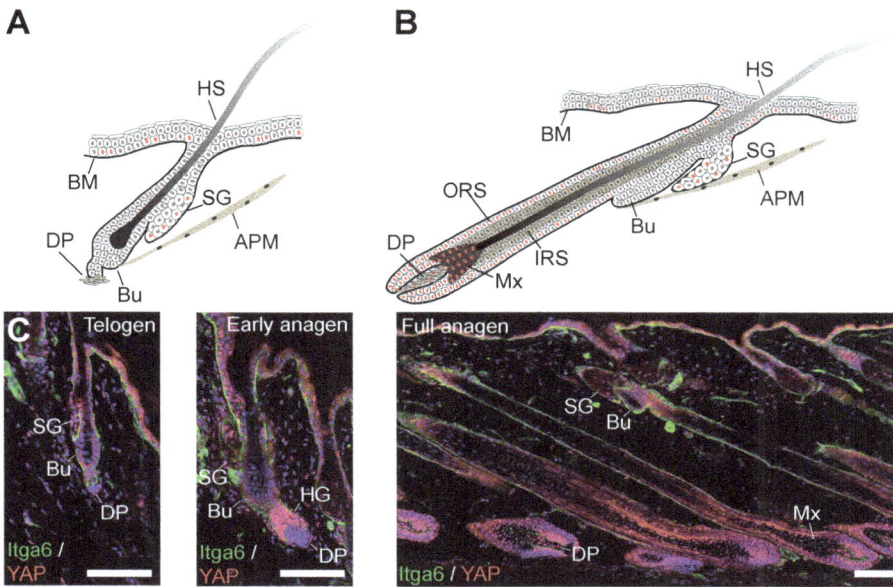

Figure 3. YAP localisation during HF cycling. (**A,B**) Schematic of a HF in the resting phase (telogen) (A) and growth phase (anagen) (B). In telogen phase, YAP is mainly localised to the cytoplasm in bulge and DP cells and there are only scattered cluster of cells with nuclear YAP in the IFE and SG. During anagen the highly proliferative cells of the ORS and HF matrix display strong nuclear YAP localisation, while YAP is cytoplasmatic in differentiating cells of the IRS and HS. Beside cell clusters with nuclear YAP in the IFE and SG during anagen, some DP cells also display YAP in the nucleus. Note that only nuclear YAP is shown. (**C**) Immunostaining for YAP (red) and Itga6 (green) of mouse skin during telogen, early anagen and full anagen. Nuclei are stained with DAPI (blue). Note, the strong increase in nuclear YAP in the HG and infundibulum during anagen induction. Scale bars, 50 µm. Abbreviations: Bu, bulge; DP, dermal papilla; HF, hair follicle; HG, hair germ; HS, hair shaft; IFE, interfollicular epidermis; IRS, inner root sheath; Mx matrix; ORS outer root sheath; SG, sebaceous gland.

Upon skin wounding, increased numbers of cells with nuclear YAP/TAZ can be observed predominantly in the basal cell layer of the migrating epidermal tongue at the wound edge, but also in suprabasal cells [21,100]. This pattern is maintained after wound closure when the IFE is still hyperproliferative, but YAP (and supposedly also TAZ) becomes increasingly cytoplasmic again once tissue homeostasis has been re-established [21]. While YAP is mainly cytoplasmic in fibroblasts in adult dermis, it highly expressed and nuclear in dermal cells in and outside the wound bed in the early wound healing phase [21]. Increased epidermal YAP expression has also been documented in psoriasis, a chronic inflammatory skin disease inducing hyperproliferation and abnormal differentiation of epidermal keratinocytes [142]. Similarly, YAP is highly expressed and nuclear in fibrosis of the skin and other tissues, a disease condition characterised by excessive ECM deposition/remodelling, inflammation and skin cell proliferation [143].

YAP is also overexpressed in most types of human and murine epidermal cancers [101,135,139, 144,145]. In cSCC, increased YAP expression was shown to correlate with disease progression [139,145]. Well-differentiated human cSCC presented nuclear YAP expression at the invasive tumour front, and poorly differentiated cSCCs with mesenchymal features showed a homogeneous stronger nuclear staining [139]. In contrast, TAZ was found to be expressed mainly in the cytoplasm of a subset of well-differentiated and poorly differentiated cSCC, and few cells stained positive for nuclear TAZ [139].

YAP is also strongly expressed and nuclear in spindle cell carcinoma (spSCC), a morphologically distinct type of cSCC with pronounced mesenchymal features [146]. In BCC, YAP is highly expressed and nuclear in superficial, nodular and infiltrative forms, while TAZ is mostly expressed in the tumour-surrounding mesenchyme [139,144,147,148]. In kerathoacanthoma, increased nuclear YAP expression has been observed in subset of tumours with low expression on the adherens junction component α-catenin [101]. YAP is also overexpressed in trichilemmal carcinoma and pilomatrixoma, rare tumours of HFs [149,150].

In a recent study, YAP protein expression was found to be elevated in benign melanocytic nevi and primary cutaneous melanomas, but YAP was present at only very low levels in normal melanocytes in healthy human skin [151]. Interestingly, more melanocytic cells in nevi and in early stage cutaneous melanomas had higher nuclear YAP abundance compared to cells from metastatic disease [151]. This study also found YAP to be ubiquitously expressed across a panel of melanoma cell lines, while TAZ expression was observed in most but not all cell lines [151]. In contrast, a different study found the expression of TAZ to be upregulated in human melanoma cell lines [152].

8. YAP/TAZ Drive Proliferation of Epidermal SCs and Fibroblasts during Development and Tissue Regeneration

In line with the timing and patterning of YAP expression during epidermal development, conditional knockout of YAP in Krt14-expressing epidermal progenitors of mouse skin (*K14-Cre/YAP* knockout mice) resulted in severe epidermal hypoplasia caused by insufficient proliferation of SCs to sustain epidermal morphogenesis, particularly in skin areas with high growth demand [99]. Vice versa, human Krt14 promoter-driven expression during mouse embryogenesis of a mutant YAP transgene (YAP-S127A, hereafter referred to as *K14/YAP-S127A*) with enhanced nuclear localisation led to the formation of a hyperthickened epidermis and impaired differentiation and invagination of HFs as a consequence of increased SC proliferation at the expense of terminal differentiation [99,135]. Accordingly, a genome-wide RNA interference (RNAi) screen identified YAP as an essential regulator of proliferation of human IFE SCs in culture, when they are actively cycling [21]. Expression of the YAP-S127A or NLS-YAP-5SA transgene (bovine Krt5 promoter-driven transgene with all five serine phosphorylation sites mutated and its N-terminus fused to a NLS; hereafter referred to as *K5/NLS-YAP-5SA*) both caused severe tissue dysplasia in adult epidermis that eventually progressed to tumour-like masses resembling cSCCs [99,146]. In stark contrast, adult mice expressing another YAP transgene (YAP-5SA-ΔC, lacking the transactivation domain and expressed under control of the bovine Krt5 promoter; hereafter referred to as *K5/YAP-5SA-ΔC*) displayed enhanced nuclear localisation and developed only a mild skin phenotype with later onset of epidermal hyperthickening and hair loss [136,137]. In fibroblasts, the YAP-5SA-ΔC transgene was found to display reduced transcriptional coactivator activity due to decreased nuclear accumulation [97]. However, YAP-5SA-ΔC showed strong nuclear accumulation in epidermal cells [137]. This suggests that sequences other than the C-terminus are involved in controlling nucleocytoplasmic shuttling in the context of keratinocytes. Interestingly, while *K14/YAP-S127A* transgenic mice displayed loss of terminally differentiated cell types in the IFE [99,135], the hyperthickening of *K5/YAP-5SA-ΔC* IFE was caused by expansion of both the basal and suprabasal cell compartments as well as hyperkeratinisation in the most differentiated cell layers [136]. This suggests that the C-terminus of YAP (including the YAP transactivation domain and PDZ-binding motif) may control the balance between epidermal SC proliferation and differentiation in the IFE. Consistent with the predominant nuclear localisation of YAP in SC-containing compartments during HF growth, *K5/YAP-5SA-ΔC* transgenic mice displayed striking HF abnormalities due to marked expansion of the SC populations in the lower HF [136]. In line with this, two weeks after tamoxifen-induced epidermal depletion of YAP and TAZ (*K5-CreERT/YAP/TAZ*), mice developed progressive hair loss and HF growth was completely blocked in neonates [100]. Apart from a moderate decrease in proliferation the IFE of adult *K5-CreERT/YAP/TAZ* double knockout mice showed no obvious abnormalities consistent with the lower nuclear abundance of YAP (and TAZ) in the basal cell layer of adult compared to foetal and neonatal mice [100]. Surprisingly, two other studies reported no obvious skin phenotypes in epidermis-restricted conditional YAP/TAZ double knockout mice [73,139]. This discrepancy can likely be explained by the different promoters used to drive conditional Cre transgene expression (bovine Krt5 promoter [100] vs. human Krt14 promoter [73,139]), which have different deletion efficiencies and onsets/timings [153–155]. Skin grafting experiments revealed that YAP knockdown significantly impaired SG development, and SGs were found to be grossly enlarged in *K5/YAP-5SA-ΔC* mice, pointing to a role of YAP in controlling SG homeostasis [136,138]. In contrast to the epidermis, the role of YAP/TAZ signalling in dermal fibroblasts during development and maturation remains largely unclear.

Consistent with the increased nuclear localisation of YAP/TAZ upon skin wounding, conditional YAP/TAZ knockout in the adult epidermis or topical application of interfering RNAs onto skin wounds slowed down wound closure due to reduced cell proliferation [100,156]. Similarly, RNAi-mediated YAP knock down in human primary keratinocyte cultures caused impaired regeneration of epidermal tissue in 3D organotypic skin cultures. The hypoplastic epidermis reconstituted by YAP knockdown keratinocytes also displayed premature onset of terminal differentiation, again highlighting the dual role of YAP in balancing SC proliferation and differentiation [21]. Interestingly, nuclear YAP abundance is prominent in basal cells throughout the wound healing zones of the regenerating epidermis [21], including the leading edge closest to the wound where cells are not proliferating but migrate as a sheet [4]. This suggests a role of YAP/TAZ in positively regulating keratinocyte migration, similar to what has been observed in other cell types [85,157,158]. In diabetic wounds with delayed healing, YAP expression is reduced, which can be recapitulated in vitro when dermal fibroblasts are cultured under high glucose condition [159]. In fibroblasts, YAP/TAZ knockdown attenuates key fibroblast functions, including matrix synthesis, contraction, proliferation on stiff matrix, whereas overexpression of activated mutants promotes fibroblast growth on soft matrix and drive fibrosis in vivo [143].

Interestingly, knock-in mice expressing a mutated version of YAP (YAP-S79A) that is unable to interact with TEAD TFs phenocopied the severe skin hypoplasia of YAP conditional knockout mice [99]. Likewise, RNAi-mediated silencing of TEAD expression was shown to significantly impair proliferation of primary mouse and human keratinocytes in culture [21,135]. These findings provided genetic evidence that TEADs are likely to be major transcriptional partners in epidermal SC proliferation. However, it cannot be ruled out that YAP/TAZ might interact also with other TFs and coactivators involved in controlling epidermal SC functions, as there is evidence that YAP/TAZ form a nuclear complex with SMAD2/3 [160] and mediate nuclear accumulation of activated β-catenin [137] in a human keratinocyte cell line. Indeed, the hypoplastic abnormalities in the epidermis of *K5/YAP-5SA-ΔC* transgenic mice were shown to involve activation of WNT16/β-catenin and TGF-β signalling [137,161,162] consistent with the well-established role of WNT and TGF-β signalling in controlling epidermal SC functions [163–167]. Although YAP is by default a transcriptional coactivator, it can also act as a corepressor in a complex with TEAD transcription factors and distinct chromatin-modifying proteins [76]. Thus, the increased expression of terminal differentiation markers observed in YAP knock down keratinocytes in culture [21,135] and in YAP/TAZ double knock out epidermis in vivo [119] could be directly linked to putative corepressor functions of YAP. YAP's coactivator functions to drive TEAD-mediated gene transcription depend on its cofactor WBP2 [21]. While WBP2 knockout mice did not have hair growth abnormalities, they displayed reduced proliferation in the regenerating epidermis after skin wounding, thereby phenocopying some aspects of epidermis-specific loss of YAP/TAZ [21].

RNA sequencing followed by gene set enrichment analysis of human keratinocyte cell lines transfected with YAP-5SA or YAP-specific siRNAs led to identification of YAP-regulated gene sets that included cell cycle reactomes (such as E2F targets or cyclin E-associated genes) and cell growth reactomes (including transcriptional cell cycle regulators such as Myc, and global translation regulators) [100], similar to what has been found in other cell types [73,74]. Interestingly, the epidermal transcriptome of *K5/YAP-5SA-ΔC* transgenic mice was found to be enriched in genes that drive growth of cSCC cells [21,161]; consistent with the idea that aberrant YAP activation induces a premalignant tissue state that can progress further to neoplasia [99,136]. Among the key direct transcriptional targets of YAP/TAZ identified in the epidermis context are *Cyr61* (*CCN1*) [21,100,135], *CTGF* (*CCN2*) [21,119], *ITGB1* [100], *TGFBR3* [161] and the Notch ligands *DLL1* and *JAG2* [119]. Many of these YAP/TAZ target genes appear to function in positive feedback loops that maintain epidermal SC identity. For example, by controlling the expression of Notch ligands, YAP/TAZ appear to keep Notch receptors on the surface of epidermal SCs inactive (a process called cis-inhibition) and unable to receive differentiation-inducing Notch signals coming from neighbouring cells in-trans [119]. YAP/TAZ were previously shown to mediate alternative WNT signalling through the expression of secreted WNT inhibitors to suppress canonical WNT/β-catenin signalling [128]. However, in the epidermis of *K5/YAP-5SA-ΔC* transgenic mice, only WNT16 is induced by YAP, and it behaves as a canonical WNT ligand that activates β-catenin [162]. During skin wound healing, YAP/TAZ modulates the expression of TGF-β1 signalling pathway components, likely by acting in concert with AP-1 TFs and SMAD7 [156,168].

9. YAP/TAZ Activity in Skin is Controlled by Hippo Signalling-Dependent and Independent Signalling Mechanisms

All components of the core Hippo kinase cascade are expressed in both mouse and human keratinocytes [21,99,121,149]. However, there is ambiguity as to what extent Hippo signalling is involved in controlling the activity of YAP/TAZ in the epidermis. Conditional knockout of MST1/MST2 in mouse epidermis proved to be inconsequential for epidermal homeostasis and YAP activity [99]. In contrast, epidermis-restricted conditional knockout of MOB1A and MOB1B in postnatal skin led to gross abnormalities of IFE and HFs caused by marked expansion of the SC populations [149], reminiscent of the skin phenotypes of *K14/YAP-S127A* and *K5/YAP-5SA-ΔC* transgenic mice [99,135,136]. Indeed, dramatic expansion of cells with nuclear YAP was reported in the epidermis of MOB1A/MOB1B double knockout mice, and primary keratinocytes failed to exclude YAP from the nucleus in response to high cell densities, consistent with a defective Hippo pathway [149]. The consequences of conditional LATS1/LATS2 double knockout in the epidermis have not yet been studied. However, reduced expression and activating phosphorylation of LATS1/LATS2 in MOB1A/MOB1B keratinocytes in response to activation of upstream kinases [149], as well as increased YAP transcriptional activity upon RNAi-mediated LATS1/LATS2 ablation of [121], point towards involvement of LATS1/LATS2 in controlling YAP/TAZ in mouse epidermis (Figure 4). Interestingly knockdown of LATS1/LATS2 in a human keratinocyte cell line impacted nuclear YAP transcriptional activity only in confluent but not in dense cultures, indicating that cell compaction and reduced adhesive area are major triggers for YAP/TAZ localisation [21,106].

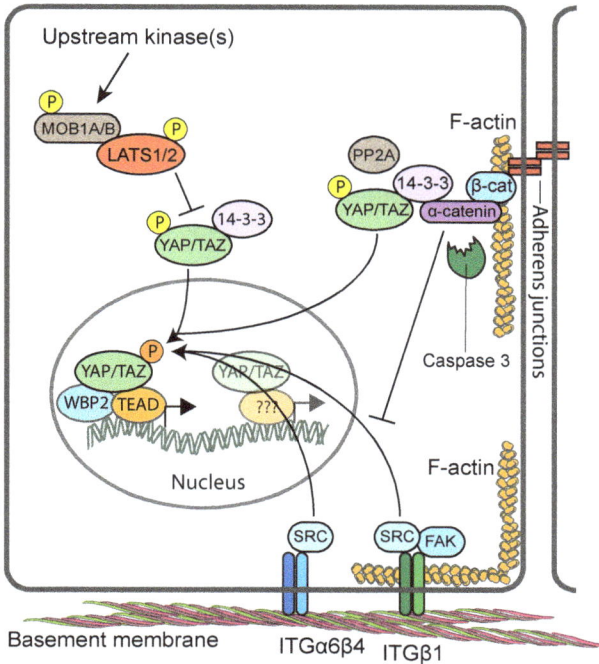

Figure 4. Regulation of YAP/TAZ in epidermal cells. Hippo signalling via MOB1A/MOB1B and LATS1/LATS2 inhibits YAP/TAZ via serine phosphorylation (yellow) to promote cytoplasmic retention. The kinases activating MOB1A/MOB1B and LATS1/LATS2 are not known. Integrin (ITG)–SRC signalling promotes YAP/TAZ nuclear localisation and TEAD binding. SRC can directly phosphorylate YAP/TAZ on tyrosine residues (orange) but may also act indirectly to activate Hippo signalling. A contractile F-actin-myosin cytoskeleton helps stabilise ITGβ1 adhesions and thus may contribute to SRC activation, while ITGβ4 adhesions are part of hemidesmosomal complexes that are stabilised by keratin intermediate filaments (not shown). At adherens junctions, α-catenin controls YAP/TAZ activity and phosphorylation by modulating its interaction with 14-3-3 and the PP2A phosphatase. In proliferating cells of the sebaceous gland, activation of caspase-3 cleaves α-catenin, thus facilitating the activation and nuclear translocation of YAP/TAZ. α-catenin can also inhibit ITGβ4-mediated direct activation of SRC. Putative nuclear interactions of YAP/TAZ with other transcription factors are also indicated.

Indeed, several studies have identified the adherens junction component αE-catenin as a cell density-dependent YAP regulator [99,101,121]. These functions of αE-catenin involve its signalling properties but appear to be independent of cadherin-mediated adhesion [99]. Genetic ablation of αE-catenin in murine epidermis (using *K14-Cre* mice) or more specifically in the HF bulge (using *GFAP-Cre* mice) led to a hyperproliferative phenotype that was caused by increased nuclear abundance of YAP in the basal and suprabasal epidermal cell layers [99,101,121]. Two mechanisms of αE-catenin-mediated regulation of YAP/TAZ have been identified. In one mechanism, αE-catenin was found to promote cytoplasmic YAP localisation and S127 phosphorylation directly by modulating its interaction with 14-3-3 and the PP2Ac phosphatase [99] (Figure 4). This mechanism appears to operate also in SGs, where complex formation between αE-catenin, 14-3-3 and YAP is negatively regulated by Caspase 3-mediated cleavage of αE-catenin [138]. In a second mechanism, αE-catenin was found to supress SRC family kinase (SFK)-mediated tyrosine phosphorylation of YAP which otherwise promotes YAP's nuclear localisation and TEAD binding [121] (Figure 4). Interestingly, both αE-catenin-dependent mechanisms appear to operate independently of LATS1/LATS2. When human keratinocytes were seeded onto ECM-coated polydimethylsiloxane elastomer substrates that mimic the epidermal–dermal interface [20], flattened cells with nuclear YAP where found to cluster at the tips of the substrates, while compacted cells with cytoplasmic YAP populated the sides and troughs of the substrates [21]. This patterning was shown to be dependent on mechanical forces exerted at intercellular junctions and modulated by SFKs and Rho GTPase signalling in response to undulations in the epidermal–dermal interface [21,169]. These findings suggest an additional layer of complexity in the regulation of YAP/TAZ by adherens junctions, namely that their stability and the organization and contractility of the associated actin cytoskeleton control the extent of cell compaction and thereby nucleocytoplasmic shuttling of YAP/TAZ independent of LATS1/LATS2 [21,170].

Integrin and ECM expression not only provide epidermal SC markers, but they also regulate SC fate during homeostasis, tissue repair and cancer progression [171,172]. When human keratinocyte cultures were fractionated by differential adhesion to extracellular matrix, expression of YAP was found to be highest in the rapidly adhering, ITGB1^{hi}, SC-enriched, fraction [21]. In line with this, inhibition of ITGB1 with function-blocking antibodies or RNAi, or inhibition of the downstream effectors SRC and FAK or PI3K profoundly impaired YAP/TAZ nuclear localisation in a human keratinocyte cell line [100]. In vivo, conditional deletion of SRC or FAK in the epidermis, or pharmacological inhibition of SFK activity, led to decreased YAP levels and nuclear localisation in basal keratinocytes in skin [100]. Mechanistically, ITGB1-mediated YAP-activation appears to depend on the integrity and proper organisation of the F-actin cytoskeleton [100,173,174], but less so on actomyosin contractility [21,100]. Interestingly, ITGB1-mediated cell-ECM adhesion is stimulated by CTGF, one of the universal gene targets of YAP/TAZ/TEADs, suggesting a positive feedback loop that maintains epidermal SC identity [175]. Beside ITGB1 the hemidesmosome-associated ITGB4 [176] was also shown to control YAP activity via direct SRC-mediated phosphorylation of YAP [121]. Interestingly, ITGB4-mediated YAP activation is negatively regulated by αE-catenin [121], highlighting cross-talk between cadherins/catenins in adherens junctions and basal integrins as an important mechanism that coordinates SC cycling and terminal differentiation commitment during homeostasis and tissue repair [177].

In fibroblasts, increased ECM stiffness mechano-activates YAP/TAZ, which induces expression of profibrotic mediators such as PAI-1 and ECM proteins that provide a feed-forward loop maintaining fibroblast activation and tissue fibrosis [143]. Using photodegradable hydrogels of tuneable stiffness it was shown that YAP acts as a mechanical rheostat in mesenchymal SCs and promotes specific cell fates in response to past matrix stiffness, suggesting that YAP is involved in the regulation of mechanical cell memory [178]. While it is largely unknown if Hippo signalling differs in different fibroblast subpopulations, it was shown that LATS2 repressed preadipocyte proliferation and promotes adipocyte differentiation by inhibiting Wnt signalling and promoting PPARγ transcriptional activity in the nucleus [179].

Summarised, YAP/TAZ are essential drivers of proliferation of epidermal SCs and fibroblasts during skin development and repair. The activity of epidermal YAP/TAZ is controlled by signalling downstream of adherens junctions and integrins as well as the mechanical forces transduced and imposed by their associated actin cytoskeleton. These mechanisms appear to work in concert with Hippo signalling in a context-dependent manner (Figure 4). Consistent with their solitary nature, mechanical signals seem to act as the predominant signalling cues in the regulation of YAP/TAZ in fibroblasts.

10. YAP and TAZ as Oncogenes in Skin Cancers

In accordance with their increased expression in epidermal cancers, YAP and TAZ were found to play key roles in the development of cSCC and BCC. In a mouse model of cSCC, where activation of oncogenic *Kras*^{G12D} in combination with *Tp53* deletion in the hair follicle lineage (using *Lgr5-CreER/Kras*^{G12D}/*Tp53KO* mice) results in a wide spectrum of cSCC ranging from well-differentiated cSCCs to tumours resembling spSCC, YAP/TAZ were demonstrated to be essential for tumour initiation [139]. Upon conditional deletion of YAP/TAZ in *Lgr5-CreER/Kras*^{G12D}/*Tp53KO* mice, cSCC/spSCC formation was completely abrogated, due to the rapid apoptosis of the oncogene expressing cells [139]. Likewise, conditional deletion of YAP/TAZ in a mouse model of BCC (*K14CreER/SmoM2* mice), which develops invasive BCCs post-expression of mutant SMO, efficiently prevented tumour initiation [139]. A similar study found that conditional deletion of YAP alone significantly reduced the tumour burden of *K14CreER/SmoM2* mice, but did not completely abrogate BCC formation [144]. This suggests that in the context of BCC YAP is the dominant paralogue, but TAZ might provide a compensatory mechanism in YAP-deficient BCC clones. Supporting the critical role of YAP in BCC development, those tumours that did form in *K14CreER/SmoM2* mice with conditional YAP deletion represented SmoM2-expressing clones that had escaped Cre-mediated recombination [144]. Longitudinal tracking of the evolution of YAP-positive versus YAP-null clones further demonstrated that the YAP-null clones were initially outcompeted by YAP-positive clones, and were eventually depleted over time as the tumours progressed to an invasive phenotype [144], potentially due to increased apoptosis [139]. Interestingly, YAP appears to promote the survival of BCC cells independently of WNT and Hedgehog signalling, the major signalling pathways involved in BCC development [144]. However, activation of the Hedgehog effector GLI2 in the epidermis *K5/YAP-5SA-ΔC* was shown to involve β-catenin [148]. YAP was shown to promote BCC initiation and progression via direct interaction with TEAD transcription factors to drive JNK-Jun signalling both at the level of *c-Jun* gene transcription but also upstream of c-Jun by controlling JNK activation [144]. c-Jun is a component of the functionally diverse AP-1 transcription complex, and in several cell types, YAP/TAZ/TEAD and AP-1 where shown to form a complex that synergistically activates target genes directly involved in the cell cycle control of S-phase entry and mitosis [73,74,85]. Indeed, ChIP sequencing analysis revealed co-occupation of chromatin regions by TEADs and AP-1 TFs in BCC [144]. Moreover, in a chemical carcinogen-induced mouse model of epidermal tumorigenesis, YAP/TAZ was also shown to be essential for tumour development [73]. Thus, YAP/TAZ/TEADs and AP-1 complexes appear to interact at multiple levels to promote epidermal tumour initiation and progression.

YAP/TAZ and the TEADs were also found to play key roles in tumour progression of cutaneous melanoma. Several studies found that RNAi-mediated silencing of YAP/TAZ in human melanoma cell lines reduced cell proliferation, survival and anchorage-independent growth in 2D culture, dermal invasion in 3D organotypic skin cultures, and in vivo lung metastasis following tail vein injection [152,180,181]. Importantly, inhibition of YAP/TAZ function in melanoma cells was shown to overcome resistance to BRAF inhibitors [182–184]. Mechanistically, the resistance to BRAF inhibitors appears to involve evasion of the antitumour immune response by YAP/TAZ-driven upregulation of the immune checkpoint ligand PD-L1 [185]. However, a recent study revealed that while targeting of YAP and TAZ provided potent anti-melanoma effects against various human melanoma cell lines as well as uncultured, therapy-naive melanoma cells grown as tumours in patient-derived xenograft (PDX) assays, these effects were not evident in all PDX melanomas and cell lines [151]. Recently, somatic hypermutation of YAP was detected in one patient's melanoma [151]. These first ever described hyperactivating YAP mutations in a human cancer manifested as seven distinct missense point mutations that caused serine to alanine transpositions. Four of these serines are key residues that are phosphorylated by the central Hippo pathway kinases LATS1/LATS2 and NDR1/NDR2 [61,62,64,65]. Consequently, the hypermutant *YAP* allele was shown to code for a highly active YAP protein [151].

Deregulation of the regulatory mechanisms that operate in healthy skin to control YAP/TAZ leads to their hyperactivation in skin cancers: Conditional knockout of αE-catenin in the HF bulge (using *GFAP-Cre* mice) was found to cause development of early onset keratoacanthomas displaying increased nuclear abundance of YAP [101], and there is significant correlation between low αE-catenin abundance and nuclear YAP localisation in human keratoacanthomas and cSCC [99,101]. Interestingly, conditional MOB1A/MOB1B double knockout mice developed trichilemmal carcinomas, but no other epidermal cancer types [149]. 14-3-3σ knockout mice displayed epidermal hyperplasia and increased formation of papillomas and cSCCs in response to chemical carcinogenesis that strongly correlated with enhanced nuclear YAP/TAZ abundance in the basal and suprabasal epidermal cell layers [186,187].

Similar to epidermal cancers, SFKs are also important drivers of YAP/TAZ activity in melanoma cells; however, in this context SFKs were shown to regulate YAP/TAZ indirectly by repressing LATS1/LATS2 [129]. In BRAF inhibitor-resistant melanoma cells, inhibition of actin polymerisation and actomyosin contractility was found to suppress both YAP/TAZ activity and drug-resistance, thus highlighting an important role of the actin cytoskeleton in this context [183].

The cancer microenvironment is characterised by elevated mechanical force at the cell and tissue levels. CAFs play a key role in the aberrant mechano-signalling observed in many tumour types [188]. Consistent with its role as a mechano-sensor, YAP is highly active in CAFs and essential to promote CAF-induced matrix stiffness, cancer cell invasion and angiogenesis. It is believed that similar to fibrosis, YAP establishes feed-forward loop between matrix stiffness and fibroblast activation maintaining CAF phenotype long-term. Mechanistically, matrix stiffness promotes actomyosin-dependent regulation of Src-family kinases to activate YAP by inducing a switch from 14-3-3 protein to TEAD1 and TEAD4 binding. YAP dependent expression of cytoskeletal regulators including ANLA, DIAPH3 and MYL9 then promotes further matrix remodelling and stiffening and thus maintaining YAP activation [120]. Similar to YAP–TEAD signalling the MRTF-SRF signalling axis also responds to extracellular signals and mechanical stimuli. Transcriptional comparison revealed that for the CAF contractile and pro-invasive phenotype both signalling pathways are essential and interdependent on the DNA level, suggesting that MRF-SRF and YAP–TEAD pathway interact indirectly by modulating the cytoskeletal dynamics/contractility [189]. Intriguingly, BCC progression in *K14CreER/SmoM2* mice was shown to be accompanied by activation of RhoA/ROCK signalling, fibroblast activation and ECM remodelling [148]. Therefore, in this context epidermal YAP may also be activated indirectly in response to the increased dermal stiffness due to epidermal Hedgehog signalling activity.

Beside extracellular signals including cell–cell contact, mechanical stress or matrix stiffness, also the metabolic status regulates YAP and LATS activity. Cellular energy starvation induces YAP inactivation via the key energy sensor AMPK through direct phosphorylation at S94 and indirectly through LATS kinase activation [190]. Recently, connections between ECM biophysical properties and metabolic cross-talk between cancer cells and CAF have been elucidated [191]. It was shown that CAF derived-aspartate supports cancer cell proliferation while cancer cell derived glutamate regulates the redox state of CAFs to promote ECM remodelling. This metabolic reprogramming seems to be dependent on the matrix stiffness and to be coordinated by YAP/TAZ through transcriptional regulation of glutaminase and the aspartate/glutamate transporter SLC1A3. In addition to promoting cell proliferation and contractility a role of YAP/TAZ in suppressing cell senescence has been described where YAP directly controls expression of key enzymes involved in deoxynucleotide biosynthesis [192].

In summary, YAP/TAZ signalling seems to be largely dispensable for skin homeostasis in adulthood but is essential for key processes of tumour initiation and progression in both epidermal and dermal cells.

11. Targeting YAP and TAZ for Skin Cancer Treatment

In the clinic, the biggest challenge for BCC, cSCC and cutaneous melanoma remains treatment of patients with advanced or metastatic disease [13,49,50,52,193,194]. For comprehensive reviews on current treatment options see [194–196]. The inconsequentiality of YAP/TAZ inactivation for normal tissue function and their absolute requirement for cancer development and progression in the same tissue makes YAP/TAZ very attractive for cancer therapy, highlighting the possibility that targeting YAP/TAZ may display a large therapeutic window [197–199]. Since all upstream regulators ultimately impact on YAP/TAZ nuclear availability and transcriptional responses, designing compounds able to interfere at these levels may represent a 'universal' anti-YAP/TAZ strategy. One approach that could have potential as a YAP-targeting therapy for cSCC is pharmacological inhibition of SFKs using the drug dasatinib. In orthotopic mouse xenograft models, dasatinib treatment was shown to cause prominent inhibition of tumour growth through interference with SFK-induced YAP nuclear translocation and activation [121]. Of note, topical dasatinib application was recently found to induce regression of murine cSCC with less inflammation, no ulceration and no mortality compared to treatment with

5-fluorouracil, one of the standard chemotherapies for cSCC [200]. SFK inhibition-induced suppression of YAP/TAZ activity might also be beneficial for the treatment of metastatic melanoma [129]. This could be of particular importance in BRAF inhibitor-resistant melanomas, where drug resistance is conferred through YAP/TAZ activation [182–184]. In this context, it worth highlighting that one study found that interfering with the mechano-transducing functions of YAP/TAZ by inhibition of actin remodelling could suppress BRAF inhibitor-resistance [183]. Another strategy to interfere with YAP/TAZ functions that is currently under intensive investigation is to target the TEAD TFs [201–203]. Targeting the YAP/TAZ-TEAD complex should directly diminish the potential side effects expected from targeting the upstream proteins of the pathway which are interconnected with other signalling networks. Current strategies can be categorized into two primary approaches. One is to block the protein–protein interaction between YAP/TAZ and TEADs [204–207]. The other is to target a lipid pocket at the core of the TEADs, occupied by a palmitoyl ligand that is essential for TEAD folding, stability and YAP/TAZ binding [207–209]. The advantages and liabilities of disrupting the YAP/TAZ-TEAD complex through these two distinct mechanisms have yet to be fully elucidated. As initiation and progression of BCC and cSCC appears to critically depend on TEADs [73,144], small molecules able to interfere with YAP/TAZ-TEAD interaction could hold great promise for therapeutic interventions inthese cancers.

12. Summary and Outlook

The research thus far shows that in the epidermis YAP and TAZ promote SC activation and cycling for development and regeneration of HFs and SGs as well as IFE morphogenesis, but they appear to be largely dispensable for IFE homeostasis in adult mice. Because current evidence of YAP/TAZ activation as a cue for SC mobilisation and tissue regeneration comes from studies of mice and cultured cell lines, at the moment, it is not clear whether primary human cells and organs can respond to YAP/TAZ activation in the same way as cells from current murine models. It should be noted here that—at least in culture—mouse and human keratinocytes display distinct growth behaviours [210,211]. It remains thus to be tested if the ambiguous findings related to the involvement of LATS1/LATS2 and other Hippo pathway components in the control of epidermal homeostasis and cancer development reflect such cell-intrinsic differences between mice and humans. Also, the kinase(s) involved in activating MOB1A/MOB1B and LATS1/LATS2 in keratinocytes still remain to be identified.

Beside cell–cell contact and the mechanical microenviroment, the cell metabolism has been recently also identified to play a key role in YAP/TAZ regulation, suggesting that other regulatory mechanisms will be discovered in the future.

The transcriptional programmes executed by YAP/TAZ to promote epidermal SC cycling during development and tissue regeneration are still not fully characterised, as comprehensive YAP/TAZ ChIP sequencing studies have not yet been performed. Likewise, we still know little about the genes and processes under control of YAP/TAZ that drive nonmelanoma skin cancer initiation and progression. Curative therapeutic approaches for BCC and cSCC with distant metastasis are still lacking. Therefore, there is an urgent need to identify and understand the signalling pathways controlling initiation and progression of these epidermal cancers. Their widespread activation in human skin cancers and their essential function as a signalling hub pinpoint YAP/TAZ as prime candidates for effective cancer treatments. Since YAP/TAZ function to suppress terminal differentiation, their inactivation could potentially lead to 'normalisation' of cancer cells by reverting them from a more malignant to a benign differentiated phenotype. To meet the high demands on transcriptional regulators set by cancer cells to fuel their uncontrolled proliferation, YAP/TAZ need to interact with chromatin regulators, transcriptional cofactors, and even the basal transcriptional machinery [72,73,212]. Thus, an in-depth characterisation of the YAP/TAZ-associated machinery that drives this 'transcriptional addiction' in skin cancer cells and enables YAP/TAZ to execute their various downstream transcriptional programmes will help to inform new therapeutic approaches.

In dermal fibroblasts, YAP/TAZ signalling is essential for sensing the physical environment of the cell and influences cell proliferation, ECM deposition and remodelling. Whether YAP/TAZ signalling is differentially regulated in different fibroblast subpopulations is currently unknown. Since YAP signalling is influenced by many other signalling pathways (including WNT and TGF-β1) it is tempting to speculate that YAP/TAZ signalling regulation and activity differs in papillary fibroblasts, characterised by high WNT signalling activity, and in reticular fibroblasts showing a pronounced ECM and immune gene signature [8]. Indeed, reticular fibroblasts have been identified as main drivers for aberrant ECM deposition/remodelling during scar formation and fibrosis, which is associated with increased YAP/TAZ signalling to establish a pathogenic feed-forward loop between matrix stiffening, proliferation and activation. While progress has been made in identifying the signalling pathways that contribute to fibrosis and cancer, we are still lacking a clear understanding of the early pathogenic processes in the dermis inducing and maintaining aberrant fibroblast function. Thus, dissecting the complex YAP/TAZ signalling crosstalk within different fibroblast populations during tissue homeostasis, regeneration and disease will help develop novel treatment strategies targeting aberrant fibroblast behaviour in fibrosis and cancer.

13. Materials and Methods

Figure 4 was created using Servier Medical Art templates, which are licensed under a Creative Commons Attribution 3.0 Unported License.

Author Contributions: Conceptualization, E.R. and G.W.; Writing—Original Draft Preparation, E.R. and G.W.; Writing—Review & Editing, G.W.; Visualization, E.R. and G.W.; Funding Acquisition, G.W.

Acknowledgments: G.W. gratefully acknowledges financial support from the British Skin Foundation (004/SG/18) and the Faculty of Science, University of Bath. E.R. is the recipient of an EMBO Advanced Fellowship (aALTF 523-2017).

Conflicts of Interest: The authors declare no conflicts of interest.

References

1. Belokhvostova, D.; Berzanskyte, I.; Cujba, A.-M.; Jowett, G.; Marshall, L.; Prueller, J.; Watt, F.M. Homeostasis, regeneration and tumour formation in the mammalian epidermis. *Int. J. Dev. Boil.* **2018**, *62*, 571–582. [CrossRef] [PubMed]

2. Schepeler, T.; Page, M.E.; Jensen, K.B. Heterogeneity and plasticity of epidermal stem cells. *Development* **2014**, *141*, 2559–2567. [CrossRef]

3. Rognoni, E.; Watt, F.M. Skin Cell Heterogeneity in Development, Wound Healing, and Cancer. *Trends Cell Biol.* **2018**, *28*, 709–722. [CrossRef]

4. Dekoninck, S.; Blanpain, C. Stem cell dynamics, migration and plasticity during wound healing. *Nat. Cell Boil.* **2019**, *21*, 18–24. [CrossRef] [PubMed]

5. Gonzales, K.A.U.; Fuchs, E. Skin and Its Regenerative Powers: An Alliance between Stem Cells and Their Niche. *Dev. Cell* **2017**, *43*, 387–401. [CrossRef]

6. Ge, Y.; Fuchs, E. Stretching the limits: From homeostasis to stem cell plasticity in wound healing and cancer. *Nat. Rev. Microbiol.* **2018**, *19*, 311–325. [CrossRef]

7. Rinkevich, Y.; Walmsley, G.G.; Hu, M.S.; Maan, Z.N.; Newman, A.M.; Drukker, M.; Januszyk, M.; Krampitz, G.W.; Gurtner, G.C.; Lorenz, H.P. Skin Fibrosis. Identification and Isolation of a Dermal Lineage with Intrinsic Fibrogenic Potential. *Science* **2015**, *348*, aaa2151. [CrossRef]

8. Philippeos, C.; Telerman, S.; Oules, B.; Pisco, A.; Shaw, T.; Elgueta, R.; Lombardi, G.; Driskell, R.; Soldin, M.; Lynch, M.; et al. 1354 Spatial and single-cell transcriptional profiling identifies functionally distinct human dermal fibroblast subpopulations. *J. Invest. Derm.* **2018**, *138*, S230. [CrossRef]

9. Driskell, R.R.; Lichtenberger, B.M.; Hoste, E.; Kretzschmar, K.; Simons, B.D.; Charalambous, M.; Ferrón, S.R.; Herault, Y.; Pavlovic, G.; Ferguson-Smith, A.C.; et al. Distinct fibroblast lineages determine dermal architecture in skin development and repair. *Nat. Cell Boil.* **2013**, *504*, 277–281. [CrossRef]

10. Tabib, T.; Morse, C.; Wang, T.; Chen, W.; Lafyatis, R. Sfrp2/Dpp4 and Fmo1/Lsp1 Define Major Fibroblast Populations in Human Skin. *J. Invest. Derm.* **2018**, *138*, 802–810. [CrossRef] [PubMed]

11. Rognoni, E.; Gomez, C.; Pisco, A.O.; Rawlins, E.L.; Simons, B.D.; Watt, F.M.; Driskell, R.R. Inhibition of Beta-Catenin Signalling in Dermal Fibroblasts Enhances Hair Follicle Regeneration During Wound Healing. *Development* **2016**, *143*, 2522–2535. [CrossRef]

12. Green, A.C.; Olsen, C.M. Cutaneous Squamous Cell Carcinoma: An Epidemiological Review. *Br. J. Derm.* **2017**, *177*, 373–381. [CrossRef] [PubMed]

13. Burton, K.A.; Ashack, K.A.; Khachemoune, A. Cutaneous Squamous Cell Carcinoma: A Review of High-Risk and Metastatic Disease. *Am. J. Clin. Derm.* **2016**, *17*, 491–508. [CrossRef]

14. Sánchez-Danés, A.; Blanpain, C. Deciphering the cells of origin of squamous cell carcinomas. *Nat. Rev. Cancer* **2018**, *18*, 549–561. [CrossRef]

15. De Gruijl, F.R.; Tensen, C.P. Pathogenesis of Skin Carcinomas and a Stem Cell as Focal Origin. *Front. Med.* **2018**, *5*, 165. [CrossRef]

16. Hsu, Y.-C.; Pasolli, H.A.; Fuchs, E. Dynamics Between Stem Cells, Niche and Progeny in the Hair Follicle. *Cell* **2011**, *144*, 92–105. [CrossRef] [PubMed]

17. Kretzschmar, K.; Watt, F.M. Lineage Tracing. *Cell* **2012**, *148*, 33–45. [CrossRef]

18. Kretzschmar, K.; Watt, F.M. Markers of Epidermal Stem Cell Subpopulations in Adult Mammalian Skin. *Cold Spring Harb Perspect Med.* **2014**, *4*, a013631. [CrossRef] [PubMed]

19. Khavari, P.A. Modelling cancer in human skin tissue. *Nat. Rev. Cancer* **2006**, *6*, 270–280. [CrossRef] [PubMed]

20. Viswanathan, P.; Guvendiren, M.; Chua, W.; Telerman, S.B.; Liakath-Ali, K.; Burdick, J.A.; Watt, F.M. Mimicking the topography of the epidermal–dermal interface with elastomer substrates. *Integr. Boil.* **2016**, *8*, 21–29. [CrossRef]

21. Walko, G.; Woodhouse, S.; Pisco, A.O.; Rognoni, E.; Liakath-Ali, K.; Lichtenberger, B.M.; Mishra, A.; Telerman, S.B.; Viswanathan, P.; Logtenberg, M.; et al. A genome-wide screen identifies YAP/WBP2 interplay conferring growth advantage on human epidermal stem cells. *Nat. Commun.* **2017**, *8*, 14744. [CrossRef]

22. Jensen, U.B.; Lowell, S.; Watt, F.M. The spatial relationship between stem cells and their progeny in the basal layer of human epidermis: A new view based on whole-mount labelling and lineage analysis. *Development* **1999**, *126*, 2409–2418.

23. Giangreco, A.; Goldie, S.J.; Failla, V.; Saintigny, G.; Watt, F.M. Human Skin Aging Is Associated with Reduced Expression of the Stem Cell Markers Beta1 Integrin and Mcsp. *J. Invest. Derm.* **2010**, *130*, 604–608. [CrossRef]

24. Lynch, M.D.; Watt, F.M. Fibroblast heterogeneity: Implications for human disease. *J. Clin. Invest.* **2018**, *128*, 26–35. [CrossRef]

25. Donati, G.; Proserpio, V.; Lichtenberger, B.M.; Natsuga, K.; Sinclair, R.; Fujiwara, H.; Watt, F.M. Epidermal Wnt/Beta-Catenin Signaling Regulates Adipocyte Differentiation Via Secretion of Adipogenic Factors. *Proc. Natl. Acad Sci. USA* **2014**, *111*, E1501–E1509. [CrossRef] [PubMed]

26. Salzer, M.C.; Lafzi, A.; Berenguer-Llergo, A.; Youssif, C.; Castellanos, A.; Solanas, G.; Peixoto, F.O.; Attolini, C.S.-O.; Prats, N.; Aguilera, M.; et al. Identity Noise and Adipogenic Traits Characterize Dermal Fibroblast Aging. *Cell* **2018**, *175*, 1575–1590.e22. [CrossRef] [PubMed]

27. Rognoni, E.; Pisco, A.O.; Hiratsuka, T.; Sipilä, K.H.; Belmonte, J.M.; Mobasseri, S.A.; Philippeos, C.; Dilão, R.; Watt, F.M. Fibroblast state switching orchestrates dermal maturation and wound healing. *Mol. Syst. Boil.* **2018**, *14*, e8174. [CrossRef]

28. Blanpain, C.; Fuchs, E. Epidermal homeostasis: A balancing act of stem cells in the skin. *Nat. Rev. Mol. Cell Boil.* **2009**, *10*, 207–217. [CrossRef] [PubMed]

29. Simpson, C.L.; Patel, D.M.; Green, K.J. Deconstructing the skin: Cytoarchitectural determinants of epidermal morphogenesis. *Nat. Rev. Mol. Cell Boil.* **2011**, *12*, 565–580. [CrossRef]

30. Mishra, A.; Oules, B.; Pisco, A.O.; Ly, T.; Liakath-Ali, K.; Walko, G.; Viswanathan, P.; Tihy, M.; Nijjher, J.; Dunn, S.-J.; et al. A protein phosphatase network controls the temporal and spatial dynamics of differentiation commitment in human epidermis. *eLife* **2017**, *6*. [CrossRef]

31. Jones, P.H.; Simons, B.D.; Watt, F.M. Sic Transit Gloria: Farewell to the Epidermal Transit Amplifying Cell? *Cell Stem Cell* **2007**, *1*, 371–381. [CrossRef] [PubMed]

32. Barrandon, Y.; Grasset, N.; Zaffalon, A.; Gorostidi, F.; Claudinot, S.; Droz-Georget, S.L.; Nanba, D.; Rochat, A. Capturing epidermal stemness for regenerative medicine. *Semin. Cell Dev. Boil.* **2012**, *23*, 937–944. [CrossRef]

33. Jones, P.H.; Watt, F.M. Separation of human epidermal stem cells from transit amplifying cells on the basis of differences in integrin function and expression. *Cell* **1993**, *73*, 713–724. [CrossRef]

34. Barrandon, Y.; Green, H. Three clonal types of keratinocyte with different capacities for multiplication. *Proc. Acad. Sci.* **1987**, *84*, 2302–2306. [CrossRef]

35. Doupé, D.P.; Klein, A.M.; Simons, B.D.; Jones, P.H. The Ordered Architecture of Murine Ear Epidermis Is Maintained by Progenitor Cells with Random Fate. *Dev. Cell* **2010**, *18*, 317–323. [CrossRef] [PubMed]

36. Sada, A.; Jacob, F.; Leung, E.; Wang, S.; White, B.S.; Shalloway, D.; Tumbar, T. Defining the cellular lineage hierarchy in the interfollicular epidermis of adult skin. *Nat. Cell Boil.* **2016**, *18*, 619–631. [CrossRef] [PubMed]

37. Mascré, G.; Dekoninck, S.; Drogat, B.; Youssef, K.K.; Brohée, S.; Sotiropoulou, P.A.; Simons, B.D.; Blanpain, C. Distinct contribution of stem and progenitor cells to epidermal maintenance. *Nat. Cell Boil.* **2012**, *489*, 257–262. [CrossRef]

38. Tan, D.W.M.; Jensen, K.B.; Trotter, M.W.B.; Connelly, J.T.; Broad, S.; Watt, F.M. Single-cell gene expression profiling reveals functional heterogeneity of undifferentiated human epidermal cells. *Development* **2013**, *140*, 1433–1444. [CrossRef] [PubMed]

39. Cheng, J.B.; Sedgewick, A.J.; Finnegan, A.I.; Harirchian, P.; Lee, J.; Kwon, S.; Fassett, M.S.; Golovato, J.; Gray, M.; Ghadially, R.; et al. Transcriptional Programming of Normal and Inflamed Human Epidermis at Single-Cell Resolution. *Cell Rep.* **2018**, *25*, 871–883. [CrossRef] [PubMed]

40. Joost, S.; Zeisel, A.; Jacob, T.; Sun, X.; La Manno, G.; Lonnerberg, P.; Linnarsson, S.; Kasper, M. Single-Cell Transcriptomics Reveals that Differentiation and Spatial Signatures Shape Epidermal and Hair Follicle Heterogeneity. *Cell Syst.* **2016**, *3*, 221–237.e9. [CrossRef] [PubMed]

41. Vishnubalaji, R.; Al-Nbaheen, M.; Kadalmani, B.; Aldahmash, A.; Ramesh, T. Skin-derived multipotent stromal cells – an archrival for mesenchymal stem cells. *Cell Tissue Res.* **2012**, *350*, 1–12. [CrossRef] [PubMed]

42. Aragona, M.; Dekoninck, S.; Rulands, S.; Lenglez, S.; Mascré, G.; Simons, B.D.; Blanpain, C. Defining stem cell dynamics and migration during wound healing in mouse skin epidermis. *Nat. Commun.* **2017**, *8*, 14684. [CrossRef]

43. Park, S.; Gonzalez, D.G.; Guirao, B.; Boucher, J.D.; Cockburn, K.; Marsh, E.D.; Mesa, K.R.; Brown, S.; Rompolas, P.; Haberman, A.M.; et al. Tissue-Scale Coordination of Cellular Behaviour Promotes Epidermal Wound Repair in Live Mice. *Nat. Cell Biol.* **2017**, *19*, 155–163. [CrossRef] [PubMed]

44. Donati, G.; Rognoni, E.; Hiratsuka, T.; Liakath-Ali, K.; Hoste, E.; Kar, G.; Kayikci, M.; Russell, R.; Kretzschmar, K.; Mulder, K.W.; et al. Wounding Induces Dedifferentiation of Epidermal Gata6(+) Cells and Acquisition of Stem Cell Properties. *Nat. Cell Biol* **2017**, *19*, 603–613. [CrossRef]

45. Schmidt, B.A.; Horsley, V. Intradermal adipocytes mediate fibroblast recruitment during skin wound healing. *Development* **2013**, *140*, 1517–1527. [CrossRef]

46. Gay, D.; Kwon, O.; Zhang, Z.; Spata, M.; Plikus, M.V.; Holler, P.D.; Ito, M.; Yang, Z.; Treffeisen, E.; Kim, C.D.; et al. Fgf9 from Dermal Gammadelta T Cells Induces Hair Follicle Neogenesis after Wounding. *Nat. Med.* **2013**, *19*, 916–923. [CrossRef]

47. Lim, C.H.; Sun, Q.; Ratti, K.; Lee, S.-H.; Zheng, Y.; Takeo, M.; Lee, W.; Rabbani, P.; Plikus, M.V.; Cain, J.E.; et al. Hedgehog stimulates hair follicle neogenesis by creating inductive dermis during murine skin wound healing. *Nat. Commun.* **2018**, *9*, 4903. [CrossRef] [PubMed]

48. Plikus, M.V.; Guerrero-Juarez, C.F.; Ito, M.; Li, Y.R.; Dedhia, P.H.; Zheng, Y.; Shao, M.; Gay, D.L.; Ramos, R.; Hsi, T.-C.; et al. Regeneration of fat cells from myofibroblasts during wound healing. *Science* **2017**, *355*, 748–752. [CrossRef] [PubMed]

49. Didona, D.; Paolino, G.; Bottoni, U.; Cantisani, C. Non Melanoma Skin Cancer Pathogenesis Overview. *Biomedicines* **2018**, *6*, 6. [CrossRef] [PubMed]

50. Fernandes, A.R.; Santos, A.C.; Sanchez-Lopez, E.; Kovacevic, A.B.; Espina, M.; Calpena, A.C.; Veiga, F.J.; Garcia, M.L.; Souto, E.B. Neoplastic Multifocal Skin Lesions: Biology, Etiology, and Targeted Therapies for Nonmelanoma Skin Cancers. *Skin Pharm. Physiol.* **2018**, *31*, 59–73. [CrossRef] [PubMed]

51. Thieu, K.; Ruiz, M.E.; Owens, D.M. Cells of Origin and Tumor-Initiating Cells for Nonmelanoma Skin Cancers. *Cancer Lett.* **2013**, *338*, 82–88. [CrossRef]

52. Schadendorf, D.; van Akkooi, A.C.J.; Berking, C.; Griewank, K.G.; Gutzmer, R.; Hauschild, A.; Stang, A.; Roesch, A.; Ugurel, S. Melanoma. *Lancet* **2018**, *392*, 971–984. [CrossRef]

53. Testa, U.; Castelli, G.; Pelosi, E. Melanoma: Genetic Abnormalities, Tumor Progression, Clonal Evolution and Tumor Initiating Cells. *Med Sci.* **2017**, *5*, 28.

54. Nissinen, L.; Farshchian, M.; Riihilä, P.; Kähäri, V.-M. New perspectives on role of tumor microenvironment in progression of cutaneous squamous cell carcinoma. *Cell Tissue Res.* **2016**, *365*, 691–702. [CrossRef] [PubMed]

55. Guerra, L.; Odorisio, T.; Zambruno, G.; Castiglia, D. Stromal microenvironment in type VII collagen-deficient skin: The ground for squamous cell carcinoma development. *Matrix Biol.* **2017**, *63*, 1–10. [CrossRef]

56. Arwert, E.N.; Hoste, E.; Watt, F.M. Epithelial stem cells, wound healing and cancer. *Nat. Rev. Cancer* **2012**, *12*, 170–180. [CrossRef] [PubMed]

57. Lim, Y.Z.; South, A.P. Tumour–stroma crosstalk in the development of squamous cell carcinoma. *Int. J. Biochem. Cell Boil.* **2014**, *53*, 450–458. [CrossRef] [PubMed]

58. Kalluri, R.; Zeisberg, M. Fibroblasts in Cancer. *Nat. Rev. Cancer* **2006**, *6*, 392–401. [CrossRef] [PubMed]

59. LeBleu, V.S.; Kalluri, R. A peek into cancer-associated fibroblasts: Origins, functions and translational impact. *Model. Mech.* **2018**, *11*, dmm029447. [CrossRef]

60. Hogervorst, M.; Rietveld, M.; De Gruijl, F.; El Ghalbzouri, A. A shift from papillary to reticular fibroblasts enables tumour–stroma interaction and invasion. *Br. J. Cancer* **2018**, *118*, 1089–1097. [CrossRef] [PubMed]

61. Meng, Z.; Moroishi, T.; Guan, K.-L. Mechanisms of Hippo pathway regulation. *Genes Dev.* **2016**, *30*, 1–17. [CrossRef]

62. Moya, I.M.; Halder, G. Hippo–YAP/TAZ signalling in organ regeneration and regenerative medicine. *Nat. Rev. Mol. Cell Boil.* **2018**, 1. [CrossRef] [PubMed]

63. Ma, S.; Meng, Z.; Chen, R.; Guan, K.-L. The Hippo Pathway: Biology and Pathophysiology. *Annu. Biochem.* **2018**, *88*. [CrossRef] [PubMed]

64. Piccolo, S.; Dupont, S.; Cordenonsi, M. The Biology of YAP/TAZ: Hippo Signaling and Beyond. *Physiol. Rev.* **2014**, *94*, 1287–1312. [CrossRef]

65. Sharif, A.A.; Hergovich, A. The NDR/LATS protein kinases in immunology and cancer biology. *Semin. Cancer Biol.* **2018**, *48*, 104–114. [CrossRef] [PubMed]

66. Callus, B.A.; Finch-Edmondson, M.L.; Fletcher, S.; Wilton, S.D. YAPping about and not forgetting TAZ. *Febs Lett.* **2019**, *593*, 253–276. [CrossRef] [PubMed]

67. Plouffe, S.W.; Lin, K.C.; Moore, J.L., 3rd; Tan, F.E.; Ma, S.; Ye, Z.; Qiu, Y.; Ren, B.; Guan, K.L. The Hippo Pathway Effector Proteins Yap and Taz Have Both Distinct and Overlapping Functions in the Cell. *J. Biol. Chem.* **2018**, *293*, 11230–11240. [CrossRef] [PubMed]

68. Sun, C.; De Mello, V.; Mohamed, A.; Quiroga, H.P.O.; Al Bloshi, A.; Tremblay, A.M.; Von Kriegsheim, A.; Vargesson, N.; Matallanas, D.; Wackerhage, H.; et al. Common and Distinctive Functions of the Hippo Effectors Taz and Yap in Skeletal Muscle Stem Cell Function. *Stem Cells* **2017**, *35*, 1958–1972. [CrossRef] [PubMed]

69. Oh, H.; Slattery, M.; Ma, L.; Crofts, A.; White, K.P.; Mann, R.S.; Irvine, K.D. Genome-wide association of Yorkie with chromatin and chromatin remodeling complexes. *Cell Rep.* **2013**, *3*, 309–318. [CrossRef] [PubMed]

70. Oh, H.; Slattery, M.; Ma, L.; White, K.P.; Mann, R.S.; Irvine, K.D. Yorkie promotes transcription by recruiting a Histone methyltransferase complex. *Cell Rep.* **2014**, *8*, 449–459. [CrossRef]

71. Ikmi, A.; Gaertner, B.; Seidel, C.; Srivastava, M.; Zeitlinger, J.; Gibson, M.C. Molecular Evolution of the Yap/Yorkie Proto-Oncogene and Elucidation of Its Core Transcriptional Program. *Mol. Boil. Evol.* **2014**, *31*, 1375–1390. [CrossRef] [PubMed]

72. Galli, G.G.; Carrara, M.; Yuan, W.-C.; Valdes-Quezada, C.; Gurung, B.; Pepe-Mooney, B.; Zhang, T.; Geeven, G.; Gray, N.S.; De Laat, W.; et al. YAP drives growth by controlling transcriptional pause release from dynamic enhancers. *Mol. Cell* **2015**, *60*, 328–337. [CrossRef]

73. Zanconato, F.; Forcato, M.; Battilana, G.; Azzolin, L.; Quaranta, E.; Bodega, B.; Rosato, A.; Bicciato, S.; Cordenonsi, M.; Piccolo, S. Genome-wide association between YAP/TAZ/TEAD and AP-1 at enhancers drives oncogenic growth. *Nat. Cell Boil.* **2015**, *17*, 1218–1227. [CrossRef] [PubMed]

74. Stein, C.; Bardet, A.F.; Roma, G.; Bergling, S.; Clay, I.; Ruchti, A.; Agarinis, C.; Schmelzle, T.; Bouwmeester, T.; Schübeler, D.; et al. YAP1 Exerts Its Transcriptional Control via TEAD-Mediated Activation of Enhancers. *PLoS Genet.* **2015**, *11*, e1005465. [CrossRef]

75. Zhao, B.; Ye, X.; Yu, J.; Li, L.; Li, W.; Li, S.; Yu, J.; Lin, J.D.; Wang, C.-Y.; Chinnaiyan, A.M.; et al. TEAD mediates YAP-dependent gene induction and growth control. *Genome Res.* **2008**, *22*, 1962–1971. [CrossRef]

76. Kim, M.; Kim, T.; Johnson, R.L.; Lim, D.-S. Transcriptional Co-repressor Function of the Hippo Pathway Transducers YAP and TAZ. *Cell Rep.* **2015**, *11*, 270–282. [CrossRef] [PubMed]

77. Kim, M.K.; Jang, J.W.; Bae, S.C. DNA Binding Partners of Yap/Taz. *BMB Rep.* **2018**, *51*, 126–133. [CrossRef] [PubMed]

78. Murakami, M.; Nakagawa, M.; Olson, E.N.; Nakagawa, O. A WW domain protein TAZ is a critical coactivator for TBX5, a transcription factor implicated in Holt–Oram syndrome. *Proc. Acad. Sci.* **2005**, *102*, 18034–18039. [CrossRef]

79. Rosenbluh, J.; Nijhawan, D.; Cox, A.G.; Li, X.; Neal, J.T.; Schafer, E.J.; Zack, T.I.; Wang, X.; Tsherniak, A.; Schinzel, A.C.; et al. Beta-Catenin-Driven Cancers Require a Yap1 Transcriptional Complex for Survival and Tumorigenesis. *Cell* **2012**, *151*, 1457–1473. [CrossRef] [PubMed]

80. Jang, J.W.; Kim, M.K.; Lee, Y.S.; Lee, J.W.; Kim, D.M.; Song, S.H.; Lee, J.Y.; Choi, B.Y.; Min, B.; Chi, X.Z.; et al. Rac-Lats1/2 Signaling Regulates Yap Activity by Switching between the Yap-Binding Partners Tead4 and Runx3. *Oncogene* **2017**, *36*, 999–1011. [CrossRef]

81. Vitolo, M.I.; Anglin, I.E.; Mahoney, W.M.; Renoud, K.J.; Gartenhaus, R.B.; Bachman, K.E.; Passaniti, A. The RUNX2 transcription factor cooperates with the YES-associated protein, YAP65, to promote cell transformation. *Cancer Biol. Ther.* **2007**, *6*, 856–863. [CrossRef]

82. Papaspyropoulos, A.; Bradley, L.; Thapa, A.; Leung, C.Y.; Toskas, K.; Koennig, D.; Pefani, D.-E.; Raso, C.; Grou, C.; Hamilton, G.; et al. RASSF1A uncouples Wnt from Hippo signalling and promotes YAP mediated differentiation via p73. *Nat. Commun.* **2018**, *9*, 424. [CrossRef]

83. Strano, S.; Monti, O.; Pediconi, N.; Baccarini, A.; Fontemaggi, G.; Lapi, E.; Mantovani, F.; Damalas, A.; Citro, G.; Sacchi, A.; et al. The Transcriptional Coactivator Yes-Associated Protein Drives p73 Gene-Target Specificity in Response to DNA Damage. *Mol. Cell* **2005**, *19*, 429. [CrossRef]

84. Strano, S.; Monti, O.; Baccarini, A.; Sudol, M.; Sacchi, A.; Blandino, G. Physical interaction with yes-associated protein enhances p73 transcriptional activity. *J. Biol. Chem.* **2001**, *37*, S279.

85. Liu, X.; Li, H.; Rajurkar, M.; Li, Q.; Cotton, J.L.; Ou, J.; Zhu, L.J.; Goel, H.L.; Mercurio, A.M.; Park, J.-S.; et al. Tead and AP1 coordinate transcription and motility. *Cell Rep.* **2016**, *14*, 1169–1180. [CrossRef]

86. Croci, O.; De Fazio, S.; Biagioni, F.; Donato, E.; Caganova, M.; Curti, L.; Doni, M.; Sberna, S.; Aldeghi, D.; Biancotto, C.; et al. Transcriptional integration of mitogenic and mechanical signals by Myc and YAP. *Genome Res.* **2017**, *31*, 2017–2022. [CrossRef]

87. Meng, Z.; Moroishi, T.; Mottier-Pavie, V.; Plouffe, S.W.; Hansen, C.G.; Hong, A.W.; Park, H.W.; Mo, J.-S.; Lu, W.; Lu, S.; et al. MAP4K family kinases act in parallel to MST1/2 to activate LATS1/2 in the Hippo pathway. *Nat. Commun.* **2015**, *6*, 8357. [CrossRef]

88. Zheng, Y.; Wang, W.; Liu, B.; Deng, H.; Uster, E.; Pan, D. Identification of Happyhour/MAP4K as alternative Hpo/Mst-like kinases in the Hippo kinase cascade. *Dev. Cell* **2015**, *34*, 642–655. [CrossRef]

89. Li, Q.; Li, S.; Mana-Capelli, S.; Flach, R.J.R.; Danai, L.V.; Amcheslavsky, A.; Nie, Y.; Kaneko, S.; Yao, X.; Chen, X.; et al. The conserved Misshapen-Warts-Yorkie pathway acts in enteroblasts to regulate intestinal stem cells in Drosophila. *Dev. Cell* **2014**, *31*, 291–304. [CrossRef]

90. Zhao, B.; Wei, X.; Li, W.; Udan, R.S.; Yang, Q.; Kim, J.; Xie, J.; Ikenoue, T.; Yu, J.; Li, L.; et al. Inactivation of YAP oncoprotein by the Hippo pathway is involved in cell contact inhibition and tissue growth control. *Genome Res.* **2007**, *21*, 2747–2761. [CrossRef]

91. Zhao, B.; Li, L.; Tumaneng, K.; Wang, C.Y.; Guan, K.L. A Coordinated Phosphorylation by Lats and Ck1 Regulates Yap Stability through Scf(Beta-Trcp). *Genes Dev.* **2010**, *24*, 72–85. [CrossRef]

92. Liu, C.Y.; Zha, Z.Y.; Zhou, X.; Zhang, H.; Huang, W.; Zhao, D.; Li, T.; Chan, S.W.; Lim, C.J.; Hong, W.; et al. The Hippo Tumor Pathway Promotes Taz Degradation by Phosphorylating a Phosphodegron and Recruiting the Scf{Beta}-Trcp E3 Ligase. *J. Biol. Chem.* **2010**, *285*, 37159–37169. [CrossRef]

93. Das, A.; Fischer, R.S.; Pan, D.; Waterman, C.M. YAP Nuclear Localization in the Absence of Cell-Cell Contact Is Mediated by a Filamentous Actin-dependent, Myosin II- and Phospho-YAP-independent Pathway during Extracellular Matrix Mechanosensing*. *J. Boil. Chem.* **2016**, *291*, 6096–6110. [CrossRef]

94. Elbediwy, A.; Vanyai, H.; Diaz-De-La-Loza, M.-D.-C.; Frith, D.; Snijders, A.P.; Thompson, B.J. Enigma proteins regulate YAP mechanotransduction. *J. Cell Sci.* **2018**, *131*, jcs.221788. [CrossRef] [PubMed]

95. Kofler, M.; Speight, P.; Little, D.; Di Ciano-Oliveira, C.; Szászi, K.; Kapus, A. Mediated nuclear import and export of TAZ and the underlying molecular requirements. *Nat. Commun.* **2018**, *9*, 4966. [CrossRef] [PubMed]

96. Elosegui-Artola, A.; Andreu, I.; Beedle, A.E.; Lezamiz, A.; Uroz, M.; Kosmalska, A.J.; Oria, R.; Kechagia, J.Z.; Rico-Lastres, P.; Le Roux, A.-L.; et al. Force Triggers YAP Nuclear Entry by Regulating Transport across Nuclear Pores. *Cell* **2017**, *171*, 1397–1410.e14. [CrossRef] [PubMed]

97. Ege, N.; Dowbaj, A.M.; Jiang, M.; Howell, M.; Hooper, S.; Foster, C.; Jenkins, R.P.; Sahai, E. Quantitative Analysis Reveals that Actin and Src-Family Kinases Regulate Nuclear YAP1 and Its Export. *Cell Syst.* **2018**, *6*, 692–708.e13. [CrossRef]

98. Manning, S.A.; Dent, L.G.; Kondo, S.; Zhao, Z.W.; Plachta, N.; Harvey, K.F. Dynamic Fluctuations in Subcellular Localization of the Hippo Pathway Effector Yorkie In Vivo. *Curr. Boil.* **2018**, *28*, 1651–1660.e4. [CrossRef] [PubMed]

99. Schlegelmilch, K.; Mohseni, M.; Kirak, O.; Pruszak, J.; Rodriguez, J.R.; Zhou, D.; Kreger, B.T.; Vasioukhin, V.; Avruch, J.; Brummelkamp, T.R.; et al. Yap1 Acts Downstream of Alpha-Catenin to Control Epidermal Proliferation. *Cell* **2011**, *144*, 782–795. [CrossRef] [PubMed]

100. Elbediwy, A.; Vincent-Mistiaen, Z.I.; Spencer-Dene, B.; Stone, R.K.; Boeing, S.; Wculek, S.K.; Cordero, J.; Tan, E.H.; Ridgway, R.; Brunton, V.G.; et al. Integrin signalling regulates YAP and TAZ to control skin homeostasis. *Development* **2016**, *143*, 1674–1687. [CrossRef]

101. Silvis, M.R.; Kreger, B.T.; Lien, W.H.; Klezovitch, O.; Rudakova, G.M.; Camargo, F.D.; Lantz, D.M.; Seykora, J.T.; Vasioukhin, V. Alpha-Catenin Is a Tumor Suppressor That Controls Cell Accumulation by Regulating the Localization and Activity of the Transcriptional Coactivator Yap1. *Sci Signal.* **2011**, *4*, ra33. [CrossRef]

102. Zhao, B.; Li, L.; Wang, L.; Wang, C.-Y.; Yu, J.; Guan, K.-L. Cell detachment activates the Hippo pathway via cytoskeleton reorganization to induce anoikis. *Genes Dev.* **2012**, *26*, 54–68. [CrossRef]

103. Kim, N.-G.; Gumbiner, B.M. Adhesion to fibronectin regulates Hippo signaling via the FAK–Src–PI3K pathway. *J. Cell Boil.* **2015**, *210*, 503–515. [CrossRef]

104. Kim, N.-G.; Koh, E.; Chen, X.; Gumbiner, B.M. E-cadherin mediates contact inhibition of proliferation through Hippo signaling-pathway components. *Proc. Acad. Sci.* **2011**, *108*, 11930–11935. [CrossRef]

105. Dupont, S.; Morsut, L.; Aragona, M.; Enzo, E.; Giulitti, S.; Cordenonsi, M.; Zanconato, F.; Le Digabel, J.; Forcato, M.; Bicciato, S.; et al. Role of YAP/TAZ in mechanotransduction. *Nat. Cell Boil.* **2011**, *474*, 179–183. [CrossRef]

106. Aragona, M.; Panciera, T.; Manfrin, A.; Giulitti, S.; Michielin, F.; Elvassore, N.; Dupont, S.; Piccolo, S. A Mechanical Checkpoint Controls Multicellular Growth through YAP/TAZ Regulation by Actin-Processing Factors. *Cell* **2013**, *154*, 1047–1059. [CrossRef]

107. Wada, K.-I.; Itoga, K.; Okano, T.; Yonemura, S.; Sasaki, H. Hippo pathway regulation by cell morphology and stress fibers. *Development* **2011**, *138*, 3907–3914. [CrossRef]

108. Bao, M.; Xie, J.; Piruska, A.; Huck, W.T.S. 3D microniches reveal the importance of cell size and shape. *Nat. Commun.* **2017**, *8*, 1962. [CrossRef]

109. Chen, C.-L.; Gajewski, K.M.; Hamaratoglu, F.; Bossuyt, W.; Sansores-Garcia, L.; Tao, C.; Halder, G. The apical-basal cell polarity determinant Crumbs regulates Hippo signaling in Drosophila. *Proc. Acad. Sci.* **2010**, *107*, 15810–15815. [CrossRef]

110. Sun, G.; Irvine, K.D. Regulation of Hippo Signaling by Jun Kinase Signaling During Compensatory Cell Proliferation and Regeneration, and in Neoplastic Tumors. *Dev. Biol.* **2011**, *350*, 139–151. [CrossRef]

111. Zhou, B.; Flodby, P.; Luo, J.; Castillo, D.R.; Liu, Y.; Yu, F.X.; McConnell, A.; Varghese, B.; Li, G.; Chimge, N.O.; et al. Claudin-18-Mediated Yap Activity Regulates Lung Stem and Progenitor Cell Homeostasis and Tumorigenesis. *J. Clin Invest.* **2018**, *128*, 970–984. [CrossRef]

112. Szymaniak, A.D.; Mahoney, J.E.; Cardoso, W.V.; Varelas, X. Crumbs3-mediated polarity directs airway epithelial cell fate through the Hippo pathway effector Yap. *Dev. Cell* **2015**, *34*, 283–296. [CrossRef] [PubMed]

113. Narimatsu, M.; Samavarchi-Tehrani, P.; Varelas, X.; Wrana, J.L. Distinct Polarity Cues Direct Taz/Yap and Tgfbeta Receptor Localization to Differentially Control Tgfbeta-Induced Smad Signaling. *Dev. Cell* **2015**, *32*, 652–656. [CrossRef] [PubMed]

114. Yang, C.-C.; Graves, H.K.; Moya, I.M.; Tao, C.; Hamaratoglu, F.; Gladden, A.B.; Halder, G.; Hamaratoğlu, F. Differential regulation of the Hippo pathway by adherens junctions and apical–basal cell polarity modules. *Proc. Acad. Sci.* **2015**, *112*, 1785–1790. [CrossRef]

115. Yu, J.; Zheng, Y.; Dong, J.; Klusza, S.; Deng, W.-M.; Pan, D. Kibra functions as a tumor suppressor protein that regulates Hippo signaling in conjunction with Merlin and Expanded. *Dev. Cell* **2010**, *18*, 288–299. [CrossRef] [PubMed]

116. Hamaratoglu, F.; Willecke, M.; Kango-Singh, M.; Nolo, R.; Hyun, E.; Tao, C.; Jafar-Nejad, H.; Halder, G. The Tumour-Suppressor Genes Nf2/Merlin and Expanded Act through Hippo Signalling to Regulate Cell Proliferation and Apoptosis. *Nat. Cell Biol.* **2006**, *8*, 27–36. [CrossRef]

117. Elbediwy, A.; Thompson, B.J.; Vincent-Mistiaen, Z.I.; Vincent-Mistiaen, Z.I. YAP and TAZ in epithelial stem cells: A sensor for cell polarity, mechanical forces and tissue damage. *BioEssays* **2016**, *38*, 644–653. [CrossRef]

118. Koo, J.H.; Guan, K.-L. Interplay between YAP/TAZ and Metabolism. *Cell Metab.* **2018**, *28*, 196–206. [CrossRef]

119. Totaro, A.; Castellan, M.; Battilana, G.; Zanconato, F.; Azzolin, L.; Giulitti, S.; Cordenonsi, M.; Piccolo, S. YAP/TAZ link cell mechanics to Notch signalling to control epidermal stem cell fate. *Nat. Commun.* **2017**, *8*, 15206. [CrossRef]

120. Calvo, F.; Ege, N.; Grande-García, A.; Hooper, S.; Jenkins, R.P.; Chaudhry, S.I.; Harrington, K.; Williamson, P.; Moeendarbary, E.; Charras, G.; et al. Mechanotransduction and YAP-dependent matrix remodelling is required for the generation and maintenance of cancer-associated fibroblasts. *Nat. Cell Boil.* **2013**, *15*, 637–646. [CrossRef] [PubMed]

121. Li, P.; Silvis, M.R.; Honaker, Y.; Lien, W.H.; Arron, S.T.; Vasioukhin, V. Alphae-Catenin Inhibits a Src-Yap1 Oncogenic Module That Couples Tyrosine Kinases and the Effector of Hippo Signaling Pathway. *Genes Dev.* **2016**, *30*, 798–811. [CrossRef]

122. Yu, F.-X.; Zhao, B.; Panupinthu, N.; Jewell, J.L.; Lian, I.; Wang, L.H.; Zhao, J.; Yuan, H.; Tumaneng, K.; Li, H.; et al. Regulation of the Hippo-YAP pathway by G-protein coupled receptor signaling. *Cell* **2012**, *150*, 780–791. [CrossRef]

123. Miller, E.; Yang, J.; DeRan, M.; Wu, C.; Su, A.I.; Bonamy, G.M.; Liu, J.; Peters, E.C.; Wu, X. Identification of Serum-Derived Sphingosine-1-Phosphate as a Small Molecule Regulator of YAP. *Chem. Boil.* **2012**, *19*, 955–962. [CrossRef]

124. Lim, S.K.; Lu, S.Y.; Kang, S.-A.; Tan, H.J.; Li, Z.; Wee, Z.N.A.; Guan, J.S.; Chichili, V.P.R.; Sivaraman, J.; Putti, T.; et al. Wnt Signaling Promotes Breast Cancer by Blocking ITCH-Mediated Degradation of YAP/TAZ Transcriptional Coactivator WBP2. *Cancer Res* **2016**, *76*, 6278–6289. [CrossRef]

125. Cai, J.; Maitra, A.; Anders, R.A.; Taketo, M.M.; Pan, D. Beta-Catenin Destruction Complex-Independent Regulation of Hippo-Yap Signaling by Apc in Intestinal Tumorigenesis. *Genes Dev.* **2015**, *29*, 1493–1506. [CrossRef] [PubMed]

126. Azzolin, L.; Panciera, T.; Soligo, S.; Enzo, E.; Bicciato, S.; Dupont, S.; Bresolin, S.; Frasson, C.; Basso, G.; Guzzardo, V.; et al. Yap/Taz Incorporation in the Beta-Catenin Destruction Complex Orchestrates the Wnt Response. *Cell* **2014**, *158*, 157–170. [CrossRef] [PubMed]

127. Lee, Y.; Kim, N.H.; Cho, E.S.; Yang, J.H.; Cha, Y.H.; Kang, H.E.; Yun, J.S.; Cho, S.B.; Lee, S.-H.; Paclíková, P.; et al. Dishevelled has a YAP nuclear export function in a tumor suppressor context-dependent manner. *Nat. Commun.* **2018**, *9*, 2301. [CrossRef]

128. Park, H.W.; Kim, Y.C.; Yu, B.; Moroishi, T.; Mo, J.-S.; Plouffe, S.W.; Meng, Z.; Lin, K.C.; Yu, F.-X.; Alexander, C.M.; et al. Alternative Wnt Signaling Activates YAP/TAZ. *Cell* **2015**, *162*, 780–794. [CrossRef]

129. Lamar, J.M.; Xiao, Y.; Norton, E.; Jiang, Z.G.; Gerhard, G.M.; Kooner, S.; Warren, J.S.A.; Hynes, R.O. Src Tyrosine Kinase Activates the Yap/Taz Axis and Thereby Drives Tumor Growth and Metastasis. *J. Biol. Chem.* **2019**, *294*, 2302–2317. [CrossRef]

130. Yui, S.; Azzolin, L.; Maimets, M.; Pedersen, M.T.; Fordham, R.P.; Hansen, S.L.; Larsen, H.L.; Guiu, J.; Alves, M.R.; Rundsten, C.F.; et al. YAP/TAZ-Dependent Reprogramming of Colonic Epithelium Links ECM Remodeling to Tissue Regeneration. *Cell Stem Cell* **2018**, *22*, 35–49.e7. [CrossRef] [PubMed]

131. Si, Y.; Cao, X.; Dai, X.; Xu, L.; Guo, X.; Yan, H.; Zhu, C.; Zhou, Q.; Tang, M.; Xia, Z.; et al. Src Inhibits the Hippo Tumor Suppressor Pathway through Tyrosine Phosphorylation of Lats1. *Cancer Res* **2017**, *77*, 4868–4880. [CrossRef]

132. Taniguchi, K.; Wu, L.-W.; Grivennikov, S.I.; De Jong, P.R.; Lian, I.; Yu, F.-X.; Wang, K.; Ho, S.B.; Boland, B.S.; Chang, J.T.; et al. A gp130-Src-YAP Module Links Inflammation to Epithelial Regeneration. *Nat. Cell Boil.* **2015**, *519*, 57–62. [CrossRef]

133. Fan, R.; Kim, N.-G.; Gumbiner, B.M. Regulation of Hippo pathway by mitogenic growth factors via phosphoinositide 3-kinase and phosphoinositide-dependent kinase-1. *Proc. Acad. Sci.* **2013**, *110*, 2569–2574. [CrossRef]

134. Basu, S.; Totty, N.F.; Irwin, M.S.; Sudol, M.; Downward, J. Akt Phosphorylates the Yes-Associated Protein, YAP, to Induce Interaction with 14-3-3 and Attenuation of p73-Mediated Apoptosis. *Mol. Cell* **2003**, *11*, 11–23. [CrossRef]

135. Zhang, H.; Pasolli, H.A.; Fuchs, E. Yes-associated protein (YAP) transcriptional coactivator functions in balancing growth and differentiation in skin. *Proc. Acad. Sci.* **2011**, *108*, 2270–2275. [CrossRef] [PubMed]

136. Beverdam, A.; Claxton, C.; Zhang, X.; James, G.; Harvey, K.F.; Key, B. Yap Controls Stem/Progenitor Cell Proliferation in the Mouse Postnatal Epidermis. *J. Invest. Derm.* **2013**, *133*, 1497–1505. [CrossRef]

137. Akladios, B.; Mendoza-Reinoso, V.; Samuel, M.S.; Hardeman, E.C.; Khosrotehrani, K.; Key, B.; Beverdam, A. Epidermal Yap2-5sa-Deltac Drives Beta-Catenin Activation to Promote Keratinocyte Proliferation in Mouse Skin in Vivo. *J. Invest. Derm.* **2017**, *137*, 716–726. [CrossRef] [PubMed]

138. Yosefzon, Y.; Soteriou, D.; Feldman, A.; Kostić, L.; Koren, E.; Brown, S.; Ankawa, R.; Sedov, E.; Glaser, F.; Fuchs, Y. Caspase-3 Regulates YAP-Dependent Cell Proliferation and Organ Size. *Mol. Cell* **2018**, *70*, 573–587.e4. [CrossRef]

139. Debaugnies, M.; Sánchez-Danés, A.; Rorive, S.; Raphaël, M.; Liagre, M.; Parent, M.; Brisebarre, A.; Salmon, I.; Blanpain, C. YAP and TAZ are essential for basal and squamous cell carcinoma initiation. *EMBO Rep.* **2018**, *19*, e45809. [CrossRef]

140. Iglesias-Bartolome, R.; Torres, D.; Marone, R.; Feng, X.; Martin, D.; Simaan, M.; Chen, M.; Weinstein, L.S.; Taylor, S.S.; Molinolo, A.A.; et al. Inactivation of a Galpha(S)-Pka Tumour Suppressor Pathway in Skin Stem Cells Initiates Basal-Cell Carcinogenesis. *Nat. Cell Biol.* **2015**, *17*, 793–803. [CrossRef]

141. Roy, E.; Neufeld, Z.; Cerone, L.; Wong, H.Y.; Hodgson, S.; Livet, J.; Khosrotehrani, K. Bimodal behaviour of interfollicular epidermal progenitors regulated by hair follicle position and cycling. *EMBO J.* **2016**, *35*, 2658–2670. [CrossRef]

142. Jia, J.; Li, C.; Yang, J.; Wang, X.; Li, R.; Luo, S.; Li, Z.; Liu, J.; Liu, Z.; Zheng, Y. Yes-associated protein promotes the abnormal proliferation of psoriatic keratinocytes via an amphiregulin dependent pathway. *Sci. Rep.* **2018**, *8*, 14513. [CrossRef] [PubMed]

143. Liu, F.; Lagares, D.; Choi, K.M.; Stopfer, L.; Marinkovic, A.; Vrbanac, V.; Probst, C.K.; Hiemer, S.E.; Sisson, T.H.; Horowitz, J.C.; et al. Mechanosignaling through Yap and Taz Drives Fibroblast Activation and Fibrosis. *Am. J. Physiol. Lung Cell Mol. Physiol.* **2015**, *308*, L344–L357. [CrossRef]

144. Maglic, D.; Schlegelmilch, K.; Dost, A.F.; Panero, R.; Dill, M.T.; A Calogero, R.; Camargo, F.D. YAP-TEAD signaling promotes basal cell carcinoma development via a c-JUN/AP1 axis. *EMBO J.* **2018**, *37*, e98642. [CrossRef]

145. Jia, J.; Li, C.; Luo, S.; Liu-Smith, F.; Yang, J.; Wang, X.; Wang, N.; Lai, B.; Lei, T.; Wang, Q.; et al. Yes-Associated Protein Contributes to the Development of Human Cutaneous Squamous Cell Carcinoma via Activation of RAS. *J. Invest. Derm.* **2016**, *136*, 1267–1277. [CrossRef] [PubMed]

146. Vincent-Mistiaen, Z.; Elbediwy, A.; Vanyai, H.; Cotton, J.; Stamp, G.; Nye, E.; Spencer-Dene, B.; Thomas, G.J.; Mao, J.; Thompson, B.; et al. YAP drives cutaneous squamous cell carcinoma formation and progression. *eLife* **2018**, *7*. [CrossRef] [PubMed]

147. Quan, T.; Xu, Y.; Qin, Z.; Robichaud, P.; Betcher, S.; Calderone, K.; He, T.; Johnson, T.M.; Voorhees, J.J.; Fisher, G.J. Elevated Yap and Its Downstream Targets Ccn1 and Ccn2 in Basal Cell Carcinoma: Impact on Keratinocyte Proliferation and Stromal Cell Activation. *Am. J. Pathol.* **2014**, *184*, 937–943. [CrossRef]

148. Akladios, B.; Reinoso, V.M.; Cain, J.E.; Wang, T.; Lambie, D.L.; Watkins, D.N.; Beverdam, A. Positive regulatory interactions between YAP and Hedgehog signalling in skin homeostasis and BCC development in mouse skin in vivo. *PLoS ONE* **2017**, *12*, e0183178. [CrossRef]

149. Nishio, M.; Hamada, K.; Kawahara, K.; Sasaki, M.; Noguchi, F.; Chiba, S.; Mizuno, K.; Suzuki, S.O.; Dong, Y.; Tokuda, M.; et al. Cancer susceptibility and embryonic lethality in Mob1a/1b double-mutant mice. *J. Clin. Invest.* **2012**, *122*, 4505–4518. [CrossRef] [PubMed]

150. Cappellesso, R.; Bellan, A.; Saraggi, D.; Salmaso, R.; Ventura, L.; Fassina, A. Yap Immunoreactivity Is Directly Related to Pilomatrixoma Size and Proliferation Rate. *Arch. Derm. Res.* **2015**, *307*, 379–383. [CrossRef]

151. Zhang, X.; Tang, J.Z.; Vergara, I.A.; Zhang, Y.; Szeto, P.; Yang, L.; Mintoff, C.; Colebatch, A.; McIntosh, L.; Mitchell, K.A.; et al. Somatic hypermutation of the YAP oncogene in a human cutaneous melanoma. *Mol. Cancer Res.* **2019**. [CrossRef] [PubMed]

152. Nallet-Staub, F.; Marsaud, V.; Li, L.; Gilbert, C.; Dodier, S.; Bataille, V.; Sudol, M.; Herlyn, M.; Mauviel, A. Pro-Invasive Activity of the Hippo Pathway Effectors Yap and Taz in Cutaneous Melanoma. *J. Invest. Derm.* **2014**, *134*, 123–132. [CrossRef]

153. Wong, S.Y.; Reiter, J.F. Wounding mobilizes hair follicle stem cells to form tumors. *Proc. Acad. Sci.* **2011**, *108*, 4093–4098. [CrossRef]

154. Peterson, S.C.; Eberl, M.; Vagnozzi, A.N.; Belkadi, A.; Veniaminova, N.A.; Verhaegen, M.E.; Bichakjian, C.K.; Ward, N.L.; Dlugosz, A.A.; Wong, S.Y. Basal cell carcinoma preferentially arises from stem cells within hair follicle and mechanosensory niches. *Cell Stem Cell* **2015**, *16*, 400–412. [CrossRef]

155. Brakebusch, C.; Grose, R.; Quondamatteo, F.; Ramirez, A.; Jorcano, J.L.; Pirro, A.; Svensson, M.; Herken, R.; Sasaki, T.; Timpl, R.; et al. Skin and Hair Follicle Integrity Is Crucially Dependent on Beta 1 Integrin Expression on Keratinocytes. *EMBO J.* **2000**, *19*, 3990–4003. [CrossRef] [PubMed]

156. Lee, M.-J.; Byun, M.R.; Furutani-Seiki, M.; Hong, J.-H.; Jung, H.-S. YAP and TAZ Regulate Skin Wound Healing. *J. Inves. Derm.* **2014**, *134*, 518–525. [CrossRef] [PubMed]

157. Qiao, Y.; Chen, J.; Lim, Y.B.; Finch-Edmondson, M.L.; Seshachalam, V.P.; Qin, L.; Jiang, T.; Low, B.C.; Singh, H.; Lim, C.T.; et al. YAP Regulates Actin Dynamics through ARHGAP29 and Promotes Metastasis. *Cell Rep.* **2017**, *19*, 1495–1502. [CrossRef] [PubMed]

158. Nardone, G.; La Cruz, J.O.-D.; Vrbsky, J.; Martini, C.; Přibyl, J.; Skládal, P.; Pešl, M.; Caluori, G.; Pagliari, S.; Martino, F.; et al. YAP regulates cell mechanics by controlling focal adhesion assembly. *Nat. Commun.* **2017**, *8*, 15321. [CrossRef]

159. Um, J.; Yu, J.; Park, K.-S. Substance P accelerates wound healing in type 2 diabetic mice through endothelial progenitor cell mobilization and Yes-associated protein activation. *Mol. Med. Rep.* **2017**, *15*, 3035–3040. [CrossRef] [PubMed]

160. Grannas, K.; Arngarden, L.; Lonn, P.; Mazurkiewicz, M.; Blokzijl, A.; Zieba, A.; Soderberg, O. Crosstalk between Hippo and Tgfbeta: Subcellular Localization of Yap/Taz/Smad Complexes. *J. Mol Biol* **2015**, *427*, 3407–3415. [CrossRef]

161. Corley, S.M.; Mendoza-Reinoso, V.; Giles, N.; Singer, E.S.; Common, J.E.; Wilkins, M.R.; Beverdam, A. Plau and Tgfbr3 are YAP-regulated genes that promote keratinocyte proliferation. *Cell Death* **2018**, *9*, 1106. [CrossRef]

162. Mendoza-Reinoso, V.; Beverdam, A. Epidermal Yap Activity Drives Canonical Wnt16/Beta-Catenin Signaling to Promote Keratinocyte Proliferation in Vitro and in the Murine Skin. *Stem Cell Res.* **2018**, *29*, 15–23. [CrossRef]

163. Lim, X.; Tan, S.H.; Koh, W.L.C.; Chau, R.M.W.; Yan, K.S.; Kuo, C.J.; Van Amerongen, R.; Klein, A.M.; Nusse, R.; Koh, W.L.C.; et al. Interfollicular epidermal stem cells self-renew via autocrine Wnt signaling. *Science* **2013**, *342*, 1226–1230. [CrossRef] [PubMed]

164. Choi, Y.S.; Zhang, Y.; Xu, M.; Yang, Y.; Ito, M.; Peng, T.; Cui, Z.; Nagy, A.; Hadjantonakis, A.K.; Lang, R.A.; et al. Distinct Functions for Wnt/Beta-Catenin in Hair Follicle Stem Cell Proliferation and Survival and Interfollicular Epidermal Homeostasis. *Cell Stem Cell* **2013**, *13*, 720–733. [CrossRef]

165. Lin, H.Y.; Yang, L.T. Differential Response of Epithelial Stem Cell Populations in Hair Follicles to Tgf-Beta Signaling. *Dev. Biol.* **2013**, *373*, 394–406. [CrossRef]

166. Oshimori, N.; Fuchs, E. Paracrine Tgf-Beta Signaling Counterbalances Bmp-Mediated Repression in Hair Follicle Stem Cell Activation. *Cell Stem Cell* **2012**, *10*, 63–75. [CrossRef]

167. Rognoni, E.; Widmaier, M.; Jakobson, M.; Ruppert, R.; Ussar, S.; Katsougkri, D.; Bottcher, R.T.; Lai-Cheong, J.E.; Rifkin, D.B.; McGrath, J.A.; et al. Kindlin-1 Controls Wnt and Tgf-Beta Availability to Regulate Cutaneous Stem Cell Proliferation. *Nat. Med.* **2014**, *20*, 350–359. [CrossRef] [PubMed]

168. Qin, Z.; Xia, W.; Fisher, G.J.; Voorhees, J.J.; Quan, T. Yap/Taz Regulates Tgf-Beta/Smad3 Signaling by Induction of Smad7 Via Ap-1 in Human Skin Dermal Fibroblasts. *Cell Commun. Signal.* **2018**, *16*, 18. [CrossRef]

169. Mobasseri, S.A.; Zijl, S.; Salameti, V.; Walko, G.; Stannard, A.; Garcia-Manyes, S.; Watt, F.M. Patterning of human epidermal stem cells on undulating elastomer substrates reflects differences in cell stiffness. *Acta Biomater.* **2019**, *87*, 256–264. [CrossRef] [PubMed]

170. Hirata, H.; Samsonov, M.; Sokabe, M. Actomyosin contractility provokes contact inhibition in E-cadherin-ligated keratinocytes. *Sci. Rep.* **2017**, *7*, 46326. [CrossRef]

171. Fujiwara, H.; Ferreira, M.; Donati, G.; Marciano, D.K.; Linton, J.M.; Sato, Y.; Hartner, A.; Sekiguchi, K.; Reichardt, L.F.; Watt, F.M. The basement membrane of hair follicle stem cells is a muscle cell niche. *Cell* **2011**, *144*, 577–589. [CrossRef]

172. Margadant, C.; Charafeddine, R.A.; Sonnenberg, A. Unique and redundant functions of integrins in the epidermis. *Faseb J.* **2010**, *24*, 4133–4152. [CrossRef]

173. Totaro, A.; Castellan, M.; Di Biagio, D.; Piccolo, S. Crosstalk between YAP/TAZ and Notch Signaling. *Trends Cell Biol.* **2018**, *28*, 560–573. [CrossRef] [PubMed]

174. Zhou, K.; Muroyama, A.; Underwood, J.; Leylek, R.; Ray, S.; Soderling, S.H.; Lechler, T. Actin-related protein2/3 complex regulates tight junctions and terminal differentiation to promote epidermal barrier formation. *Proc. Acad. Sci.* **2013**, *110*, E3820–E3829. [CrossRef] [PubMed]

175. Kiwanuka, E.; Andersson, L.; Caterson, E.J.; Junker, J.P.; Gerdin, B.; Eriksson, E. Ccn2 Promotes Keratinocyte Adhesion and Migration Via Integrin Alpha5beta1. *Exp. Cell Res.* **2013**, *319*, 2938–2946. [CrossRef]

176. Walko, G.; Castañón, M.J.; Wiche, G. Molecular architecture and function of the hemidesmosome. *Cell Tissue Res.* **2015**, *360*, 529–544. [CrossRef]

177. Livshits, G.; Kobielak, A.; Fuchs, E. Governing epidermal homeostasis by coupling cell–cell adhesion to integrin and growth factor signaling, proliferation, and apoptosis. *Proc. Acad. Sci.* **2012**, *109*, 4886–4891. [CrossRef]

178. Yang, C.; Tibbitt, M.W.; Basta, L.; Anseth, K.S. Mechanical memory and dosing influence stem cell fate. *Nat. Mater.* **2014**, *13*, 645–652. [CrossRef] [PubMed]

179. An, Y.; Kang, Q.; Zhao, Y.; Hu, X.; Li, N. Lats2 Modulates Adipocyte Proliferation and Differentiation via Hippo Signaling. *PLoS ONE* **2013**, *8*, e72042. [CrossRef] [PubMed]

180. Xiong, H.; Yu, Q.; Gong, Y.; Chen, W.; Tong, Y.; Wang, Y.; Xu, H.; Shi, Y. Yes-Associated Protein (YAP) Promotes Tumorigenesis in Melanoma Cells Through Stimulation of Low-Density Lipoprotein Receptor-Related Protein 1 (LRP1). *Sci. Rep.* **2017**, *7*, 15528. [CrossRef] [PubMed]

181. Stern, P.; Liu, H.; Lamar, J.M.; Schindler, J.W.; Jiang, Z.-G.; O Hynes, R. The Hippo pathway target, YAP, promotes metastasis through its TEAD-interaction domain. *Proc. Acad. Sci.* **2012**, *109*, E2441–E2450.

182. Fisher, M.L.; Grun, D.; Adhikary, G.; Xu, W.; Eckert, R.L. Inhibition of YAP function overcomes BRAF inhibitor resistance in melanoma cancer stem cells. *Oncotarget* **2017**, *8*, 110257–110272. [CrossRef] [PubMed]

183. Kim, M.H.; Kim, J.; Hong, H.; Lee, S.H.; Lee, J.K.; Jung, E.; Kim, J. Actin Remodeling Confers Braf Inhibitor Resistance to Melanoma Cells through Yap/Taz Activation. *EMBO J.* **2016**, *35*, 462–478. [CrossRef]

184. Lin, L.; Sabnis, A.J.; Chan, E.; Olivas, V.; Cade, L.; Pazarentzos, E.; Asthana, S.; Neel, D.; Yan, J.J.; Lu, X.; et al. The Hippo effector YAP promotes resistance to RAF- and MEK-targeted cancer therapies. *Nat. Genet.* **2015**, *47*, 250–256. [CrossRef]

185. Kim, M.H.; Kim, C.G.; Kim, S.-K.; Shin, S.J.; Choe, E.A.; Park, S.-H.; Shin, E.-C.; Kim, J. YAP-Induced PD-L1 Expression Drives Immune Evasion in BRAFi-Resistant Melanoma. *Cancer Immunol. Res.* **2018**, *6*, 255–266. [CrossRef] [PubMed]

186. Li, Q.; Sambandam, S.A.; Lu, H.J.; Thomson, A.; Kim, S.H.; Lu, H.; Xin, Y.; Lu, Q. 14-3-3sigma and P63 Play Opposing Roles in Epidermal Tumorigenesis. *Carcinogenesis* **2011**, *32*, 1782–1788. [CrossRef] [PubMed]

187. Sambandam, S.A.T.; Kasetti, R.B.; Xue, L.; Dean, D.C.; Lu, Q.; Li, Q. 14-3-3sigma Regulates Keratinocyte Proliferation and Differentiation by Modulating Yap1 Cellular Localization. *J. Invest. Derm.* **2015**, *135*, 1621–1628. [CrossRef] [PubMed]

188. Dufort, C.C.; Paszek, M.J.; Weaver, V.M. Balancing forces: Architectural control of mechanotransduction. *Nat. Rev. Mol. Cell Boil.* **2011**, *12*, 308–319. [CrossRef]

189. Foster, C.T.; Gualdrini, F.; Treisman, R. Mutual dependence of the MRTF–SRF and YAP–TEAD pathways in cancer-associated fibroblasts is indirect and mediated by cytoskeletal dynamics. *Genome Res.* **2017**, *31*, 2361–2375. [CrossRef]

190. Mo, J.-S.; Meng, Z.; Kim, Y.C.; Park, H.W.; Hansen, C.G.; Kim, S.; Lim, D.-S.; Guan, K.-L. Cellular energy stress induces AMPK-mediated regulation of YAP and the Hippo pathway. *Nat. Cell Boil.* **2015**, *17*, 500–510. [CrossRef]

191. Bertero, T.; Oldham, W.M.; Grasset, E.M.; Bourget, I.; Boulter, E.; Pisano, S.; Hofman, P.; Bellvert, F.; Meneguzzi, G.; Bulavin, D.V.; et al. Tumor-Stroma Mechanics Coordinate Amino Acid Availability to Sustain Tumor Growth and Malignancy. *Cell Metab.* **2019**, *29*, 124–140.e10. [CrossRef]

192. Santinon, G.; Brian, I.; Pocaterra, A.; Romani, P.; Franzolin, E.; Rampazzo, C.; Bicciato, S.; Dupont, S. dNTP metabolism links mechanical cues and YAP/TAZ to cell growth and oncogene-induced senescence. *EMBO J.* **2018**, *37*, e97780. [CrossRef]

193. Puig, S.; Berrocal, A. Management of high-risk and advanced basal cell carcinoma. *Clin. Transl. Oncol.* **2015**, *17*, 497–503. [CrossRef]

194. Henriques, V.; Martins, T.; Link, W.; Ferreira, B.I. The Emerging Therapeutic Landscape of Advanced Melanoma. *Curr. Pharm. Des.* **2018**, *24*, 249–558. [CrossRef] [PubMed]

195. Chong, K.; Daud, A.; Ortiz-Urda, S.; Arron, S.T. Cutting Edge in Medical Management of Cutaneous Oncology. *Semin. Cutan. Med. Surg.* **2012**, *31*, 140–149. [CrossRef] [PubMed]

196. Spallone, G.; Botti, E.; Costanzo, A. Targeted Therapy in Nonmelanoma Skin Cancers. *Mol. Cell. Basis Metastasis: Road Ther.* **2011**, *3*, 2255–2273. [CrossRef]

197. Zanconato, F.; Battilana, G.; Cordenonsi, M.; Piccolo, S. YAP/TAZ as therapeutic targets in cancer. *Curr. Opin. Pharmacol.* **2016**, *29*, 26–33. [CrossRef] [PubMed]

198. Zanconato, F.; Cordenonsi, M.; Piccolo, S. Yap/Taz at the Roots of Cancer. *Cancer Cell* **2016**, *29*, 783–803. [CrossRef] [PubMed]

199. Johnson, R.; Halder, G. The Two Faces of Hippo: Targeting the Hippo Pathway for Regenerative Medicine and Cancer Treatment. *Nat. Rev. Drug Discov.* **2014**, *13*, 63–79. [CrossRef]

200. Yang, X.; Daifallah, A.E.M.; Shankar, S.; Beer, J.; Marshall, C.; Dentchev, T.; Seykora, F.; D'Armas, S.; Hahn, J.; Lee, V.; et al. Topical kinase inhibitors induce regression of cutaneous squamous cell carcinoma. *Exp. Dermatol.* **2019**. [CrossRef]

201. Gibault, F.; Sturbaut, M.; Bailly, F.; Melnyk, P.; Cotelle, P. Targeting Transcriptional Enhanced Associate Domains (Teads). *J. Med. Chem.* **2018**, *61*, 5057–5072. [CrossRef] [PubMed]

202. Santucci, M.; Vignudelli, T.; Ferrari, S.; Mor, M.; Scalvini, L.; Bolognesi, M.L.; Uliassi, E.; Costi, M.P. The Hippo Pathway and YAP/TAZ–TEAD Protein–Protein Interaction as Targets for Regenerative Medicine and Cancer Treatment. *J. Med. Chem.* **2015**, *58*, 4857–4873. [CrossRef]

203. Holden, J.K.; Cunningham, C.N. Targeting the Hippo Pathway and Cancer through the TEAD Family of Transcription Factors. *Cancers* **2018**, *10*, 81. [CrossRef] [PubMed]

204. Bum-Erdene, K.; Zhou, D.; Gonzalez-Gutierrez, G.; Ghozayel, M.K.; Si, Y.; Xu, D.; Shannon, H.E.; Bailey, B.J.; Corson, T.W.; Pollok, K.E.; et al. Small-Molecule Covalent Modification of Conserved Cysteine Leads to Allosteric Inhibition of the TEAD·Yap Protein-Protein Interaction. *Cell Chem. Boil.* **2019**, *26*, 378–389.e13. [CrossRef]

205. Zhou, Z.; Hu, T.; Xu, Z.; Lin, Z.; Zhang, Z.; Feng, T.; Zhu, L.; Rong, Y.; Shen, H.; Luk, J.M.; et al. Targeting Hippo pathway by specific interruption of YAP-TEAD interaction using cyclic YAP-like peptides. *Faseb J.* **2015**, *29*, 724–732. [CrossRef] [PubMed]

206. Smith, S.A.; Sessions, R.B.; Shoemark, D.K.; Williams, C.; Ebrahimighaei, R.; McNeill, M.C.; Crump, M.P.; McKay, T.R.; Harris, G.; Newby, A.C.; et al. Antiproliferative and Antimigratory Effects of a Novel YAP–TEAD Interaction Inhibitor Identified Using in Silico Molecular Docking. *J. Med. Chem.* **2019**, *62*, 1291–1305. [CrossRef]

207. Crawford, J.J.; Bronner, S.M.; Zbieg, J.R. Hippo Pathway Inhibition by Blocking the Yap/Taz-Tead Interface: A Patent Review. *Expert Opin. Ther. Pat.* **2018**, *28*, 868–873. [CrossRef]

208. Noland, C.L.; Gierke, S.; Schnier, P.D.; Murray, J.; Sandoval, W.N.; Sagolla, M.; Dey, A.; Hannoush, R.N.; Fairbrother, W.J.; Cunningham, C.N. Palmitoylation of TEAD Transcription Factors Is Required for Their Stability and Function in Hippo Pathway Signaling. *Structure* **2016**, *24*, 179–186. [CrossRef] [PubMed]

209. Chan, P.; Han, X.; Zheng, B.; DeRan, M.; Yu, J.; Jarugumilli, G.K.; Deng, H.; Pan, D.; Luo, X.; Wu, X. Autopalmitoylation of Tead Proteins Regulates Transcriptional Output of the Hippo Pathway. *Nat. Chem. Biol.* **2016**, *12*, 282–289. [CrossRef] [PubMed]

210. Watt, F.M. Selective migration of terminally differentiating cells from the basal layer of cultured human epidermis. *J. Cell Boil.* **1984**, *98*, 16–21. [CrossRef]

211. Watt, F.M.; Green, H. Stratification and terminal differentiation of cultured epidermal cells. *Nat. Cell Boil.* **1982**, *295*, 434–436. [CrossRef]

212. Zanconato, F.; Battilana, G.; Forcato, M.; Filippi, L.; Azzolin, L.; Manfrin, A.; Quaranta, E.; Di Biagio, D.; Sigismondo, G.; Guzzardo, V.; et al. Transcriptional addiction in cancer cells is mediated by YAP/TAZ through BRD4. *Nat. Med.* **2018**, *24*, 1599–1610. [CrossRef] [PubMed]

cells

MDPI

Review

Role of the Hippo Pathway in Fibrosis and Cancer

Cho-Long Kim [1], Sue-Hee Choi [1] and Jung-Soon Mo [2,*]

[1] Department of Biomedical Sciences, Cancer Biology Graduate Program, Ajou University Graduate School of Medicine, Suwon 16499, Korea; cholong92@ajou.ac.kr (C.-L.K.); tngml1984@ajou.ac.kr (S.-H.C.)

[2] Genomic Instability Research Center (GIRC), Ajou University School of Medicine, Suwon 16499, Korea

* Correspondence: j5mo@ajou.ac.kr; Tel.: +82-31-219-7803

Received: 17 April 2019; Accepted: 14 May 2019; Published: 16 May 2019

Abstract: The Hippo pathway is the key player in various signaling processes, including organ development and maintenance of tissue homeostasis. This pathway comprises a core kinases module and transcriptional activation module, representing a highly conserved mechanism from *Drosophila* to vertebrates. The central MST1/2-LATS1/2 kinase cascade in this pathway negatively regulates YAP/TAZ transcription co-activators in a phosphorylation-dependent manner. Nuclear YAP/TAZ bind to transcription factors to stimulate gene expression, contributing to the regenerative potential and regulation of cell growth and death. Recent studies have also highlighted the potential role of Hippo pathway dysfunctions in the pathology of several diseases. Here, we review the functional characteristics of the Hippo pathway in organ fibrosis and tumorigenesis, and discuss its potential as new therapeutic targets.

Keywords: YAP; TAZ; MST1/2; LATS1/2; myofibroblast; ECM; fibrosis; EMT; cancer

1. Introduction

The Hippo pathway is an evolutionarily conserved signaling transduction nexus that influences a wide range of biological processes during the growth and development of tissues and organs. This signaling pathway has recently been recognized as a master regulator of the malignant progression of many cancers by regulating cell proliferation and stem/progenitor-cell expansion [1]. Genetic studies in *Drosophila* provided the first identification of the Warts (Wts) as tumor suppressors [2,3].

Subsequent molecular studies further defined the details of core components of the Hippo pathway and their cellular functions, including cell growth, proliferation, survival, and organ-size control, demonstrating high levels of conservation from *Drosophila* to mammals. The Hippo pathway consists of a core kinases module and a transcriptional activation module. In *Drosophila*, the Hippo pathway includes a kinase cascade of Ste20-like kinase Hippo (Hpo) and NDR family kinase Wts [2–8]. Hpo forms a complex with the scaffolding protein Sav to phosphorylate and activate Wts, which physically associates with its regulatory protein Mob as Tumor Suppressor (Mats) [9–11]. Phosphorylated and activated Wts then inactivates Yorkie (Yki) via phosphorylation at serine residues [12], allowing it to interact with 14-3-3 and thereby accumulate in the cytosol [12–15]. The core components of the mammalian Hippo pathway include a kinase cascade of mammalian sterile 20-like kinase 1/2 (MST1/2) and large tumor suppressor kinase 1/2 (LATS1/2) [8,16–18]. MST1/2 and its regulatory protein SAV1 physically interact with and phosphorylate LATS1/2 kinase. Active LATS1/2 kinase in complex with Mob1 phosphorylates and inactivates the transcriptional co-activator Yes-associated protein (YAP) and transcriptional co-activator with a PDZ-binding motif (TAZ) to influence their subcellular localization in a 14-3-3-dependent manner [12–15,19]. YAP has also been identified as YES1-interacting protein, a tyrosine kinase, which is a key downstream effector of the Hippo pathway to function as a transcriptional co-activator [20]. TAZ (encoded by *WWTR*) is composed of a WW domain and PDZ-binding motif was demonstrated to play a key role as another mediator of the Hippo pathway similar to its paralog YAP [21]. Based on studies in *Drosophila* and mammals, the Hippo pathway kinase LATS1/2 regulates YAP/TAZ activity and their subcellular localization by direct phosphorylation of five consensus HX(R/H/K)XX(S/T) motifs [22–25]. Phosphorylated YAP/TAZ associate with 14-3-3 protein, leading to their cytoplasmic accumulation and the consequent loss of transcriptional co-activator function. Cytosolic YAP/TAZ are further phosphorylated by CK1 and LATS1/2, and are then rapidly destroyed by the ubiquitin-proteasome and lysosome system [26]. Conversely, unphosphorylated YAP/TAZ translocate to the nucleus and bind to the TEAD family transcription factors to activate the transcription of target genes that promote cell proliferation and inhibit apoptosis [27,28]. Besides TEADs, YAP/TAZ can interact with several other proteins, including Smad, Runx family, p73, ErbB4, Pax3, TTF1, PPARγ, and TBX5, which are responsible for transcriptional activation, and VGLL4, which is responsible for transcriptional repression [29].

Despite these functional similarities of YAP and TAZ, they nevertheless differ in their physiological and developmental functions. Much progress has been made in identifying the multiple new regulators of the Hippo pathway. Moreover, accumulating evidence, both physiological and pathological, points to a role of the central upstream Hippo kinase cascade in regulating the downstream effectors YAP/TAZ that sense extracellular environment signals, and integrate intracellular signaling to maintain a variety of biological functions, including tissue homeostasis and cancer [30–32]. Here, we review this research progress of the specific roles of YAP/TAZ to provide new insight into these important mediators of the Hippo pathway as effectors of fibrosis and cancer progression.

2. The Hippo Pathway in Organ Fibrosis

Myofibroblasts are responsible for fibrogenesis and produce a series of extracellular matrix (ECM) components such as collagens, laminins, and fibronectins to promote their contractile force. Recently, the Hippo pathway was shown to contribute to the pathogenesis of fibrosis, in which hyperactive YAP/TAZ accumulate in both the epithelial and stromal tissue compartments of fibrotic tissues.

2.1. Lung Fibrosis

The basic unit of the lungs is the pulmonary alveoli comprised of epithelial and ECM layers surrounded by capillaries [33]. During lung development and repair, respiratory epithelial and mesenchymal progenitors show dramatic changes in their cellular behavior to maintain the appropriate cell type composition and structural organization of the lung epithelium through a re-epithelialization process [34]. The alveolar epithelial cells are the main source of fibroblasts and myofibroblasts, which convert to a mesenchymal cell phenotype following injury, and could trigger the development of fibrosis through a process known as epithelial-mesenchymal transition (EMT). The Hippo pathway controls epithelial progenitor cell proliferation, migration, and differentiation in the developing and mature lungs [35,36].

TAZ is proposed to act as a co-activator of various transcriptional factors through binding of the WW domain to the (L/P)PXY motif of transcriptional factors. During fetal lung development, TAZ and thyroid transcription factor 1 (TTF-1) are co-expressed in respiratory epithelial cells, and TAZ directly interacts with the (L/P)PXY motif of TTF-1 to activate the transcription of target genes, including surfactant protein C (SP-C) that maintains lung morphogenesis [37]. Moreover, Mitani et al. [38] showed that TAZ knockout mice exhibited abnormal lung alveolarization. Microarray analysis of wild-type and TAZ-deficient mouse lungs revealed that connective tissue growth factor (*Ctgf*), a known direct TAZ-TEAD target gene, is the main player in peripheral epithelial cell differentiation and lung development. Previous studies have also suggested that CTGF promotes EMT, lung fibrosis, and lung development, implying a correlation between TAZ and CTGF in lung physiological and pathological functions [39–41].

Moreover, the matrix mechanical environment influences the subcellular localization of YAP/TAZ in lung fibroblasts [42]. An active YAP/TAZ positive feed-forward mechanism was shown to amplify a profibrotic response to induce fibrosis in the lung tissue of mice following tail vein injection of

YAP 5SA/TAZ 4SA-overexpressing fibroblasts. In a similar fashion, YAP and TAZ were found to be highly expressed in the fibroblastic foci of the lungs from patients with idiopathic pulmonary fibrosis (IPF) with obvious nuclear expression of YAP/TAZ, indicating that YAP/TAZ play a significant role in fibrosis [43]. When cultured on a stiff matrix, YAP/TAZ accumulated in the nucleus, and the lung fibroblasts exhibited increased proliferation, contraction, and ECM production, comparable to the results for cells cultured on a soft matrix. However, silencing of YAP and TAZ by small interfering RNAs (siRNAs) reduced the fibroblast responses induced by a stiff matrix, suggesting that YAP and TAZ activation are indispensable for these phenotypic features of myofibroblasts. In further support of the association of the function of YAP/TAZ with lung fibrosis, Liu et al. showed that the forced expression of a constitutively active form of YAP and TAZ in lung fibroblasts increased the expression of CTGF and PAI-1, also known as SERPINE1, along with that of ECM-related proteins such as collagens and fibronectin on a stiff matrix. However, an active TAZ mutant could not induce the expression of these ECM proteins in cells cultured on a soft matrix. These findings suggested that both mechanical stress and fully activated TAZ are necessary to induce the expression of profibrotic genes in the development of tissue fibrosis [44]. RNA sequencing identified that the genes that are positively regulated by TAZ in lung fibroblasts are linked to cell migration and motility, and are closely related to genes regulated by transforming growth factor (TGF)-β, a central mediator in lung fibrosis [43]. In addition, the profibrotic effect of YAP/TAZ was demonstrated in murine models of pulmonary fibrosis. Deletion of a single copy of *Taz* (WWTR1) attenuated bleomycin-induced pulmonary fibrosis and diminished collagen deposition and lung elastance, indicating that TAZ is a key molecule in fibrogenic events [38]. More recently, emerging evidence suggests the involvement of microRNA (miRNA) in IPF progression. For example, YAP1/Twist-induced fibroblast activation and fibrogenesis were blocked by miR15a [45]. IPF is a progressive chronic interstitial lung disease and is usually a fatal condition with no available cure [46]. IPF is characterized by fibroblast proliferation, remodeling of the ECM, and the contraction of fibroblasts, which contribute to tissue tension or stiffness that results in irreversible distortion of the lung architecture [47]. Single-cell RNA profiles from both normal and IPF lung epithelial cells demonstrated that gene expression patterns of IPF cells were in a multilineage-like state, which was not detected during normal lung development [48]. Transcriptomic analysis of IPF cells identified abnormal activation of several pathways that regulate the multilineage differentiation program, including TGF-β, Hippo, PI3K/AKT, p53, and Wnt signaling cascades. Nuclear YAP localization was shown to be increased in IPF lung epithelial cells [49]. YAP cooperates with mTOR-PI3K-AKT signaling to activate aberrant cell proliferation and migration, and inhibits the epithelial cell differentiation that may be involved in the development of IPF. Thus, mechanistically, YAP/TAZ activity contributes to the pathogenesis of IPF in both lung fibroblasts and epithelial cells. Indeed, Speight et al. showed that alpha-smooth muscle actin (α-SMA) expression is regulated by TAZ in the injured epithelium during phenotypic changes such as the epithelial–myofibroblast transition [50]. Similarly, recent reports demonstrated that stiff matrices activated fibroblasts and hepatic stellate cells (HSCs) to promote matrix synthesis through YAP/TAZ-mediated gene induction, including *CTGF* and *PAI-1* [35,43]. Taken together, these observations suggest that YAP and TAZ are important mechanosensors of TGF-β-dependent and -independent lung fibrogenesis (Figure 1a).

2.2. Kidney Fibrosis

The kidney comprises numerous distinct cell types that are organized into a renal filter, including glomerular endothelial cells, glomerular basement membrane, and podocytes [51]. In mice, YAP/TAZ are essential for epithelial cell growth and establishing the proper kidney architecture [52–54]. Several studies have demonstrated the involvement of YAP/TAZ in the progression of cystic kidney disease (CKD) in patients and in a mouse model based on their enhanced nuclear localization in epithelial cells [55–57]. Various renal injuries resulting from IgA nephropathy or membranous nephropathy lead to renal tubulointerstitial fibrosis (TIF), which is the end stage of kidney failure known as progressive fibrotic chronic kidney disease. TAZ protein levels were shown to be markedly increased in both mouse *Sav1*-depleted kidneys after unilateral ureteral obstruction (UUO) and in patients with TIF/CKD [58]. High-level expression and nuclear accumulation of TAZ in the renal tubulointerstitium were demonstrated in three mouse models of renal injury [59]. Consistent with these findings, the YAP protein levels increased in both the cytoplasm and nucleus in regenerative and poorly differentiated renal tubules during the transition from acute kidney injury to CKD [60]. In addition, activation of YAP/TAZ after UUO led to TGF-β-induced EMT-like features in renal TIF [58]. As expected, deletion of *Sav1* in renal tubular epithelial cells induced YAP/TAZ activation and increased the EMT phenotype via induction of SNAI2 and a-SMA, which may improve renal fibrosis.

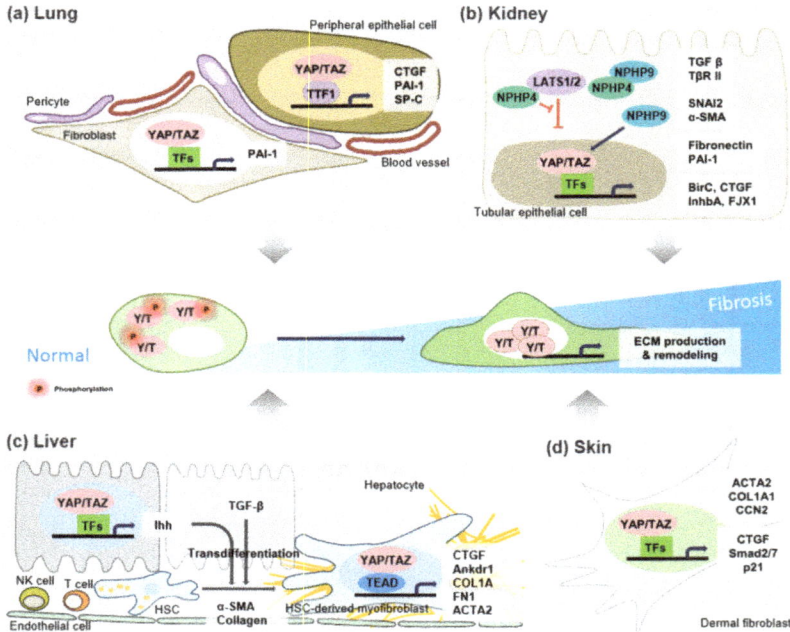

Figure 1. The role of the Hippo pathway in the pathogenesis of tissue fibrosis. In normal cells, YAP/TAZ are phosphorylated and localized in the cytoplasm, leading to their inactivation. However, in many pathological situations, YAP/TAZ localize to the nucleus, and up-regulate their target genes to remodel the ECM. This action leads to tissue fibrosis. (**a**) In respiratory epithelial cells, binding YAP/TAZ with TTF-1 synergistically activates the expression of target genes, including SP-C, CTGF, and PAI-1. Also, active fibroblasts promote matrix synthesis via YAP/TAZ that are critical to the induction of profibrotic genes, such as CTGF and PAI-1, in the development of lung fibrosis. (**b**) NPHP4, by interacting with LATS1/2, promotes YAP/TAZ transcriptional activity. In contrast, NPHP9, another NPH proteins (NPHPs), directly interacts with YAP/TAZ and induces its nuclear localization to regulate a subset of target genes involved in renal fibrosis. (**c**) YAP/TAZ-mediated Indian hedgehog (*Ihh*) gene induction, which increases transcriptional reprogramming by modulating profibrotic gene expression in HSCs. In response to Ihh and TGF-β, HSCs transdifferentiate into the ECM-producing myofibroblasts. (**d**) YAP/TAZ are mainly localized with the nucleus, resulting in the profibrotic gene expression in dermal fibrolasts. See text for further details. The red spheres indicate the phosphorylation of YAP/TAZ by kinase.

Other studies have demonstrated that the Hippo pathway plays a crucial role in tissue repair after renal injury and is involved in maintaining the kidney filtration barrier (Figure 1b) [53,61,62]. Podocyte-specific deletion of *Yap1* in mice induced cell death and the progression of significant proteinuria and glomerular lesions [63]. In kidney fibrosis, TGF-β-induced profibrotic Smad signaling by controlling Smad2/3 localization is regulated by stiffening of the ECM, and this mechanosensing is mediated by YAP/TAZ in renal fibroblasts [64]. The abundance and activation of YAP/TAZ were detected in podocytes of a puromycin aminonucleoside-induced injury model of glomerular disease [62]. Gene expression analysis in the podocytes further indicated that overexpressed YAP upregulated ECM constituents (i.e., COL6A1, BCAM, and ADAMTS1) and growth factor-related proteins (i.e., PLOD2, PDAP1, and PRMT5), which changed the mechanical properties of the glomerular basement membrane, a prognostic marker of glomerular disease progression.

2.3. Liver Fibrosis

The Hippo pathway also plays an important role in liver development [65]. *Yap* overexpression or *Mst1/2*-knockout in the liver and other upstream regulators causing hyperactivation of YAP was shown to contribute to hepatomegaly and liver tumorigenesis [14,66–68]. Large livers comprise activated progenitor cells, hyperproliferating hepatocytes, and bile duct cells, and are more likely to develop liver cancer at later stages [14,66,68]. Thus, the Hippo signaling pathway may control the activation of progenitor cells and their differentiation towards the hepatocyte lineage. HSCs are the predominant mediators of liver fibrosis/cirrhosis and are rapidly activated and transdifferentiate into fibrogenic hepatic myofibroblasts during liver injury. HSCs from a CCl4-induced mouse liver fibrosis model and from the livers of patients with hepatitis C virus infection showed predominant nuclear YAP localization, which could be used for staging fibrosis [69]. High abundance of nuclear YAP denotes the activation of YAP in myofibroblasts. YAP inhibition by specific siRNA or treatment with verteporfin (VP), which disrupts the YAP/TAZ-TEAD interaction, decreased target gene expression and prevented the transdifferentiation of quiescent HSCs into myofibroblasts. YAP has been confirmed as a critical driver of HSC activation based on studies with in vivo and in vitro models of liver fibrosis. Consistent with this concept, Hedgehog signaling-mediated YAP activation was found to be necessary for HSC transdifferentiation and proliferation by a glutaminolytic process [70]. TAZ expression is elevated in hepatocytes of the liver of patients with non-alcoholic steatohepatitis (NASH) and in NASH model mice [71].

The key mechanism driving the transition from steatosis to NASH is TAZ-mediated Indian hedgehog (*Ihh*) gene induction, which enhances transcriptional reprogramming by modulating profibrotic gene expression in HSCs, the fibrogenic precursor cells of ECM-producing myofibroblasts. In the same study, treatment of TAZ-specific siRNA to the hepatocytes isolated from NASH mouse model suppressed hepatic inflammation, fibrosis, and cell death, whereas reconstituted TAZ expression promoted the progression from steatosis to NASH (Figure 1c).

2.4. Heart Fibrosis

The adult mammalian heart has restricted regenerative capacity and repair potential. Thus, after injury or disease, a large number of cardiomyocytes are lost and a fibrotic scar is formed, which can result in heart failure [72]. Conversely, the neonatal heart has potential for enhanced cardiomyocyte proliferation. Therefore, elucidation of the regulatory mechanisms underlying cardiomyocyte proliferation may provide new insight for regenerative repair strategies. Several studies have also demonstrated a role of the Hippo pathway in regulating cardiac development. During mouse embryogenesis, loss of *Mst1/2* or *Lats*, parts of the central kinase cascade, induced cardiomyocyte proliferation [73]. Similarly, in damaged hearts, inactivated *Salv* and *Lats* led to re-entry of the mitotic cell cycle and subsequent cytokinesis, indicating that Hippo signaling is a major regulator of cardiomyocyte renewal and regeneration [74]. A constitutively active S112A mutant form of YAP enhanced cardiomyocyte proliferation and heart growth, whereas cardiac-specific deletion of *Yap* decreased cardiomyocyte proliferation in the embryonic and postnatal mouse heart [75,76]. Consistent with these findings, deletion of *Yap* and *Taz* was shown to result in lethal cardiomyopathy, whereas activated *Yap* stimulated cardiac regeneration and enhanced contractility after myocardial infarction in the adult mouse heart [77]. Insulin-like growth factor signaling and the Wnt pathway cooperatively contribute to the beneficial effects of YAP on fetal cardiac growth, homeostasis, and normal function of the postnatal heart [73,75,78]. In addition, Angiotensin II (ANG II) induced cardiovascular fibrosis in a YAP/TAZ-dependent manner. Lovastatin mediated YAP/TAZ suppression decreased ANG II-induced fibrogenic gene expression as well as cardiovascular fibrosis [79].

2.5. Skin Fibrosis

YAP also plays an important role in maintaining normal skin homeostasis in response to extracellular cues by promoting cell proliferation in epidermal stem and progenitor cells, and tissue expansion [80,81] Genetic analysis showed that YAP is highly expressed and shows nuclear localization in epidermal progenitor and stem cells, revealing a requirement for TEAD to mediate YAP's functions in controlling their self-renewal and maintaining their undifferentiated state [82]. Moreover, enhanced nuclear YAP/TAZ levels have been detected in the dermis during the skin wound-healing process. Downregulation of YAP/TAZ expression delayed healing and reduced the expression levels of TGF-β and its target genes such as p21 and Smad family members [83]. The expression and activation of YAZ/TAZ are increased in Dupuytren disease tissues, which are characterized by the formation of myofibroblast-rich cords and nodules in the hands [84]. This study further showed that YAP1 is involved in the differentiation of human dermal fibroblasts to myofibroblasts, and is necessary for maintenance of a smooth muscle cell-like contractile phenotype in primary Dupuytren myofibroblasts. In addition, YAP/TAZ levels were found to be elevated in systemic sclerosis (SSc) biopsy sections, mainly localized within the nucleus, and they mediated the profibrotic responses in dermal fibroblasts [85]. In mice, depletion of YAP/TAZ was effective in inhibiting bleomycin-induced fibrosis, a model of SSc dermal fibrosis, via inhibition of the PI3K-AKT-GSK3-β pathway. Mechanistically, dimethyl fumarate significantly blocked nuclear YAP localization in SSc dermal fibroblasts and prevented bleomycin-induced skin fibrosis (Figure 1d).

2.6. Tumor Fibrosis

In normal conditions, myofibroblasts modulate the wound-healing process and participate in tissue remodeling via producing a large amount of ECM, matrix metalloproteinases (MMPs), and tissue inhibitors of metalloproteinases (TIMPs) [86]. Myofibroblast activation ultimately occurs via MMPs and TIMPs-mediated ECM reorganization and contraction, conferring the contractile properties of myofibroblasts containing contractile actin filament bundles. In solid tumors, cancer cells reciprocally interact with non-cancerous stromal cells and the ECM to create a permissive tumor microenvironment for proliferation and maintenance of cancer cell-like features [87–90]. In particular, fibroblasts that generally reside in the tumor microenvironment are denoted as cancer-associated fibroblasts (CAFs). During tumorigenesis, both cancer cells and infiltrated immune cells release various cytokines, growth factors, and mediators to stimulate the transdifferentiation and activation of CAFs [91]. CAFs were also suggested to facilitate cancer cell invasion and metastasis through the production of soluble factors and remodeling of the ECM. A co-culture model of cancer cells and stromal fibroblasts revealed that both protease and force-mediated ECM remodeling enabled the invasion of cancer cells via a mechanism requiring CAFs [92]. CAFs mainly promote cancer cell invasion by altering the organization and physical properties of the basement membrane, leading to the formation of gaps that allow for migration [93]. For instance, mRNA expression profiling analyses showed that YAP activation is critical for the establishment and maintenance of CAFs, and that its function is necessary for ECM stiffness modulation in the progression of mammary tumors [94]. YAP/TAZ upregulation is a representative marker of high-grade breast cancer, which is required for CAFs to induce ECM stiffening, angiogenesis, and invasion of cancer cells [95]. Remarkably, increased ECM stiffness promoted cellular YAP activity to induce its target genes, and therefore support their maintenance in the tumor microenvironment by regulating matrix stiffening. Furthermore, ECM stiffness and YAP/TAZ activity function via a reinforcing feed-forward loop to preserve the CAF phenotype. Strikingly, the MRTF-SRF pathway is activated in CAFs and is required for its contractility and pro-invasive properties [96]. Activation of the MRTF-SRF pathway, which is also mechano-responsive and mandatory for maintenance of the CAF phenotype, is dependent on the transcription co-activator YAP. MRTF and YAP can indirectly activate each other through their ability to modulate actin cytoskeletal dynamics and tumor invasion under control of Rho GTPase signaling.

2.7. The Links between Fibrosis and Cancer via the Hippo Pathway

The wound-healing process in injured tissue results in the infiltration of stromal cells, including immune cells, endothelial cells, pericytes, adipocytes, and fibroblasts, to the local tissue. This results in the release of cytokines and growth factor from infiltrated immune cells, which triggers the activation and transdifferentiation of fibroblasts into myofibroblasts. Upon injury, changes to the microenvironment activate myofibroblasts to secrete several growth factors, chemokines, and ECM components, including fibronectin, collagens, laminin, fibronectin, and tenascin [86,97]. In the case of a chronic injury that causes long-term inflammation, myofibroblasts rapidly increase the production of ECM, resulting in excessive deposition and resultant fibrosis. Stressed myoblasts are key actors in fibrosis development, and act as drivers in metastasis and invasion during tumor progression, thus representing the main link between fibrosis and cancer development [91,98–100]. Fibroblast-mediated ECM changes in the stroma may trigger a dramatic alteration in the Hippo pathway in both fibrosis and tumor progression (Figure 2). Recent findings imply that ECM cues and YAP/TAZ collaborate to operate a positive feedback circuit in cancer cells that contributes to matrix stiffening, resulting in further activation of YAP/TAZ oncogenic activity.

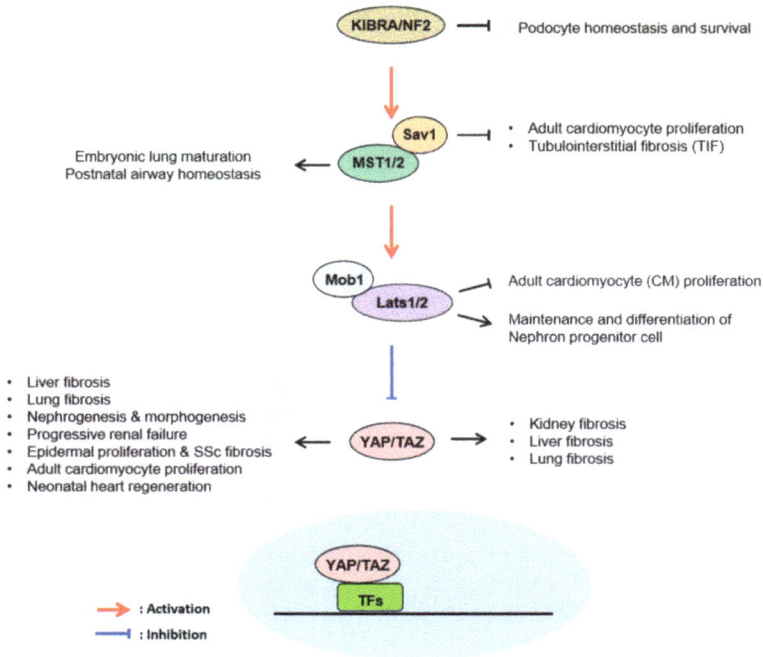

Figure 2. The Hippo pathway drives fibrosis across the organs.

3. The Hippo Pathway in Cancer Progression

3.1. Genetic and Epigenetic Altherations in the Hippo Pathway Genes

Actively proliferating cells or cancer cells typically show a common feature of highly enriched YAP/TAZ protein in the nucleus, which exists in a dephosphorylated form and can enhance fundamental cellular functions, including cell proliferation and tumorigenesis [101]. Extensive research implicates dysregulation of the Hippo pathway in a wide spectrum of human cancers across the body [7,102]. However, only limited cancers have been associated with somatic mutations in components of the Hippo pathway compared with those of key growth and oncogenic signaling pathways. For instance, NF2, an upstream regulator of the Hippo pathway, is an extensively studied tumor suppressor, and a high frequency of NF2-inactivating mutations is a characteristic of several human cancers, including acoustic neuromas, meningiomas, schwannomas, and mesothelioma [103–105]. Interestingly, the core MST1/2-LATS1/2 kinase module functions as a tumor suppressor in transgenic and knockout mouse models, and no somatic mutation in these genes has been discovered in human cancer to date. However, methylation-mediated epigenetic mechanisms contribute to downregulation of LATS1/2, MST1/2, and its regulator RASSF1 in some cancer types [106–108]. This suggests that loss-of-function mutations in tumor suppressor genes may be mediated by non-mutational mechanisms [67,68,108–112]. Mutation of the *YAP* gene at the chromosome 11q22 amplicon has been reported in various human cancers [113,114]. Recently, a YAP and transcription factor E3 (TFE3) (*YAP-TFE3*) fusion gene, and a TAZ and calmodulin-binding transcription activator 1 (CAMTA1) (*TAZ-CAMTA1*) fusion gene were reported in cases of epithelioid hemangioendothelioma, a rare type of soft tissue sarcoma [115,116]. Although the effects of these fusions on the functions of the translated proteins in cancer development and progression remain to be determined, they may be functionally related to the transcriptional regulatory properties ascribed to both TAZ-CAMTA1 and YAP-TFE3.

3.2. YAP/TAZ as Mediators of the Mechanical Cues Shaping the Tumor Microenvironment

To respond to changes in the extracellular environment, cells harmonize an intracellular signaling network to achieve the desired cellular behavior during physiological growth and regeneration. ECM mechanical cues are powerful determinants of cell proliferation, differentiation, and cell death, acting as potent regulators of YAP/TAZ. For instance, the cell-cell contact resulting from a high cell density activates the growth inhibition signaling pathway that is predominantly mediated by the Hippo pathway [22,117,118]. As the cell density increases, LATS1/2 are phosphorylated and activated by cell-to-cell contact, leading to the inactivation of YAP/TAZ, which constitutes a central mechanism for contact inhibition. This contact-mediated inhibition in turn regulates YAP/TAZ-TEAD-mediated transcription, which is also important for embryo development [119]. Dupont et al. [120] first reported that the intracellular localization and activity of YAP/TAZ correlate with ECM rigidity and cell spreading. For example, YAP/TAZ translocate to the nucleus and function as transcriptional co-activators in response to exposure to a stiff ECM, and can stimulate the proliferation and survival of cancer cells. However, on soft matrices, cells are rounded and minimally adhesive by limiting the adhesive area. As expected, the main subcellular localization of YAP/TAZ is the cytoplasm. YAP/TAZ are downstream effectors of RhoA-ROCK-mediated actin cytoskeleton rearrangements, and thus regulate development and tumorigenic processes. In particular, a comprehensive study suggested that LATS1/2-dependent and -independent mechanisms synergize mechanical cues mediated by YAP/TAZ activation. When a cell detaches from the ECM, it undergoes the apoptotic process of anoikis owing to a dysregulated cytoskeleton and increased LATS1/2 kinase activity, which in turn inhibits YAP/TAZ activity [121].

Furthermore, Hippo pathway activity and YAP/TAZ localization depend on cytoskeletal dynamics in response to a mechanical input, including cell spreading, stretching, size, and shape [120,122,123]. Integrin-mediated cell adhesion to the ECM and association with intracellular actin filaments is required for many physiological functions [124]. Focal adhesions comprising focal adhesion kinase (FAK), Src family kinases, and integrins act as a sensor of the physical properties of the environment surrounding the ECM through forces exerted on the adhesions and via intracellular signals in response to extracellular physiochemical stimuli. FAK activates the RhoA-ROCK signaling pathway that controls remodeling of the actin cytoskeleton, which increases the stress fiber assembly, large microfilament bundles, and cell contractility (Figure 3c). In addition to FAK, ILK has been implicated in integrin-mediated signaling and promotes the inhibitory phosphorylation of MYPT1, leading to inactivation of Merlin, which is a key upstream regulator of the Hippo pathway (Figure 3a) [125]. This integrin-mediated FAK–Src signaling triggers YAP/TAZ activation. Moreover, integrin-dependent cell adhesion induces p21-activated kinase 1 (PAK1) activation, which is required for YAP/TAZ nuclear localization in a Merlin-dependent manner (Figure 3b). The small GTPase RAC1 and its downstream effector PAK1 promote the direct phosphorylation of NF2, which potentiates the physical interaction between NF2 and YAP, resulting in the nuclear translocation of activated YAP/TAZ [126]. In addition, fibronectin adhesion-induced stimulation of the FAK-Src-PI3K pathway acts as an upstream negative regulator of the Hippo pathway, which leads to activation of YAP/TAZ in a LATS1/2 kinase-dependent manner [127]. Interestingly, the surrounding ECM activates YAP/TAZ directly to promote the transcription of genes involved in cytoskeleton remodeling through the RhoA-ROCK pathway [128].

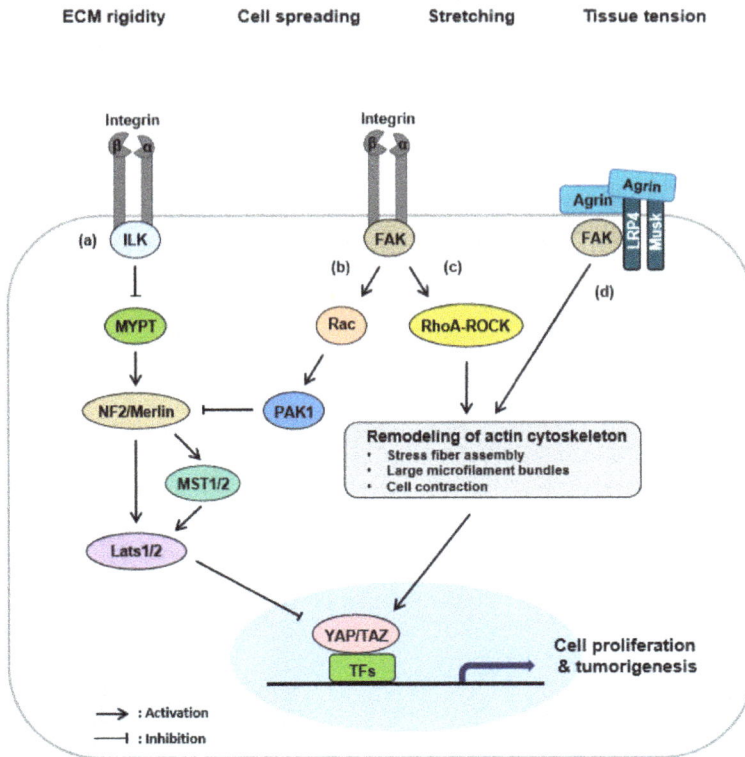

Figure 3. Schematic model of regulation of the Hippo pathway by numerous signaling pathways in cancer. (**a**) Integrin-ILK signaling inhibits MYPT1, leading to inactivation of the MST1/2-LATS1/2 kinase cascade, which triggers YAP/TAZ activation. (**b**) RAC1 and PAK1 phosphorylate NF2 that inhibits the interaction between YAP and NF2, resulting in YAP/TAZ activation. (**c**) Integrin-FAK signaling activates the RhoA-ROCK pathway to control remodeling of the actin cytoskeleton. (**d**) In response to ECM, Agrin enhances RhoA-ROCK-dependent actin cytoskeletal remodeling. As a later event, YAP/TAZ are translocated to the nucleus to interact with transcription factors (TFs), which drive profibrotic and tumorigenic effects.

Morphological manipulation without cell–cell contact and changes in the number of stress fibers according to the adhesion area enhances the nuclear localization of YAP through activation of the Hippo pathway [129]. Moreover, deficiency of destrin, an actin-depolymerizing factor, leads to the induction of F-actin polymerization and the consequent hyperproliferation of corneal epithelial cells [130]. In contrast, reduction of the F-actin binding protein Cofilin, CapZ, and gesolin, which are essential elements of the actin cytoskeleton, aberrantly activates F-actin polymerization and therefore restricts activation of the Hippo pathway to in turn positively regulate YAP/TAZ activity [123]. Agrin, a proteoglycan, promotes liver carcinogenesis as an ECM sensor, leading to enhancement of YAP stability and activity through the combinatory activation of integrin-FAK- and Lrp4/MuSK receptor-mediated signaling pathways (Figure 3d). In this context, Agrin transduces stiffness signals via RhoA-dependent actin cytoskeletal rearrangements. Stimulated ILK-PAK1 signaling antagonizes Merlin and LATS1/2 activation [131–133]. Glypican-3, a member of the glypican family of heparan sulfate proteoglycans, is highly and specifically expressed in hepatocellular carcinoma (HCC) [134]. Glypican-3 targeting HN3 was shown to significantly inhibit the growth of HCC cells and xenograft tumors in nude mice by inhibiting YAP function. Disturbed fluid flow is widely recognized to induce the expression of YAP/TAZ target genes through the coordination of Rho GTPase activities, highlighting YAP/TAZ as a key node in mediating mechanical cues and maintaining vascular homeostasis [135–137]. Moreover, the mechanical stability of the nucleus dictates YAP nuclear entry [138], and a mechano-sensing mechanism has been identified to be directly mediated by nuclear pore complexes. On a stiff substrate, focal adhesions and stress fibers lead to nuclear flattening, which stretches the nuclear pores by transducing mechanical forces. These stretched and curved nuclear pores induce YAP import due to the relaxation of mechanical restriction to nuclear pore complexes.

Overall, Rho GTPase and cytoskeletal tension are responsible for the transduction of mechanical stress in cells and have substantial effects on YAP/TAZ activity. For example, an F-actin disruptor and Rho GTPase inhibitor considerably affect YAP nuclear translocation and activity [120,121,129]. However, it remains unclear precisely how the actin cytoskeleton and mechanical stress converge on YAP/TAZ regulation. Detachment of cells from the ECM triggers LATS1/2, lead to inactivation of YAP/TAZ in a cytoskeleton- or JNK-dependent manner [129,139]. Accordingly, the detailed mechanisms underlying involvement of the Hippo pathway kinases MST1/2 and LATS1/2 as mediators to connect mechanical cue remain to be fully elucidated. Beyond ECM remodeling, CAFs are also capable of secreting a wide range of mediators, including growth factors, chemokines, MMPs, TIMPs, and ECM components [100]. These secreted mediators can act as ligands for receptors expressed by other cell types in the tumor microenvironment, which initiate the signaling cascades that activate YAP/TAZ and lead to tumor progression. For example, in response to epidermal growth factor, its receptor regulates the Hippo pathway through the PI3K-PDK1 pathway, resulting in inactivation of LATS1/2 and activation of YAP [140]. Vascular endothelial growth factor, a major driver of blood vessel formation, activates YAP/TAZ via actin cytoskeleton dynamics to facilitate a transcriptional program to further control cytoskeleton dynamics and initiate a proper angiogenic response [141]. Fibroblast growth factor 5-driven fibroblast growth factor receptor (FGFR) signaling promotes YAP activation, setting off a feed-forward loop whereby YAP forms a complex with TBX5 and induces the expression of FGFR1, 2, and 4 in human cholangiocarcinoma cell lines (Figure 4d) [142]. YAP and KRAS coordinately regulate the cell cycle and DNA replication program by forming a complex with the transcription factor E2F in pancreatic ductal adenocarcinoma (Figure 4g) [143]. Jang et al. (Figure 4a) [144] showed that mechanical stress regulates the cell cycle via the Hippo pathway. In breast cancer cells, YAP and TEAD directly bind to the *Skp2* promoter to enhance its transcription upon receiving a mechanical cue, which is necessary for cell cycle progression. Interaction of YAP with mutant p53 together form a complex with NF-Y, a nuclear transcription factor, which upregulates the expression of cell cycle genes, including cyclin A (*CCNA*), cyclin B (*CCNB*), and *CDK1*, in breast cancer cells (Figure 4b) [145]. YAP also enhances malignant mesothelioma cell growth and survival via the upregulation of cyclin D1 (CCND1) and FOXM1 in cooperation with TEAD (Figure 4a) [146]. Furthermore, YAP/TAZ/TEAD cooperate with AP-1 in the transcriptional amplification of the c-Myc oncogene to promote cell cycle progression in skin cancer (Figure 4c) [147].

3.3. YAP/TAZ Signaling in the EMT

The EMT is a complex transdifferentiation process that allows for epithelial cells to acquire the stem-like properties of mesenchymal cells, and is an important process in both fibrosis and cancer metastasis. The EMT is executed through a series of transcription factor networks, including TWIST1/2, ZEB1/2, Snail, and Slug. During cancer progression, cancer cells acquire aggressive features such as migration, invasion, chemoresistance, and apoptosis resistance that are all triggered by the EMT [148]. YAP/TAZ signaling has also been shown to play a role in the EMT in mammary epithelial cells. Constitutively active TAZ was shown to promote cell proliferation and EMT [25]. Moreover, YAP levels were found to be elevated in the stem cell-like cells derived from non-small lung cancer cells, and contributed to their self-renewal through Sox2. These effects required a physical interaction between YAP1 and Oct4 (Figure 4e) [149]. YAP and KRAS act cooperatively with the transcription factor FOS to activate Slug and vimentin-mediated EMT (Figure 4f) [150]. TAZ activity is required for the self-renewal and tumorigenesis of breast cancer cells. Twist and Snail-mediated EMT processes induce Scribble delocalization so that TAZ is released from Scribble, which stimulates the accumulation of stem-like progenitor cells in breast cancer [151]. These findings suggest that TAZ-induced EMT functions as a feed-forward mechanism of TAZ activation. The direct interaction between ZEB1 and YAP, but not TAZ, stimulated the transcription of a ZEB1/YAP target gene set that is known to promote cancer aggressiveness (Figure 4h) [152]. Indeed, this target gene set was found to be an excellent predictor of poor clinical outcomes in patients with particularly hormone receptor-negative cancer. Similarly, Snail/Slug-YAP/TAZ complexes control YAP/TAZ protein levels and the expression of target genes that modulate self-renewal, differentiation, and bone formation (Figure 4i) [153]. Therefore, crosstalk between YAP/TAZ and EMT transcription factors such as ZEB1/2, Snail/Slug, and Twist modulates the EMT programs, thereby acquiring the characteristics of malignant cancer cells. A genome-wide expression study revealed that TAZ protein levels were higher in breast cancer stem cells compared with those of differentiated breast cancer cells. A high level of TAZ was also associated with the chemoresistance and migratory potential of breast cancer stem cells [154]. The YAP-TEAD complex can directly bind to the *SOX* promoter region to upregulate *SOX9* levels and induced a cancer stem cell phenotype in esophageal cancer cells (Figure 4a) [155]. YAP and the COX2 signaling pathway coordinately regulate urothelial cancer cells with stem cell-like properties via SOX2 induction [156]. Conversely, SOX2 maintains cancer stem cells in osteoblastoma by direct inhibition of NF2 and WWC1, upstream activators of the Hippo pathway, leading to activation of YAP [157] (Figure 4).

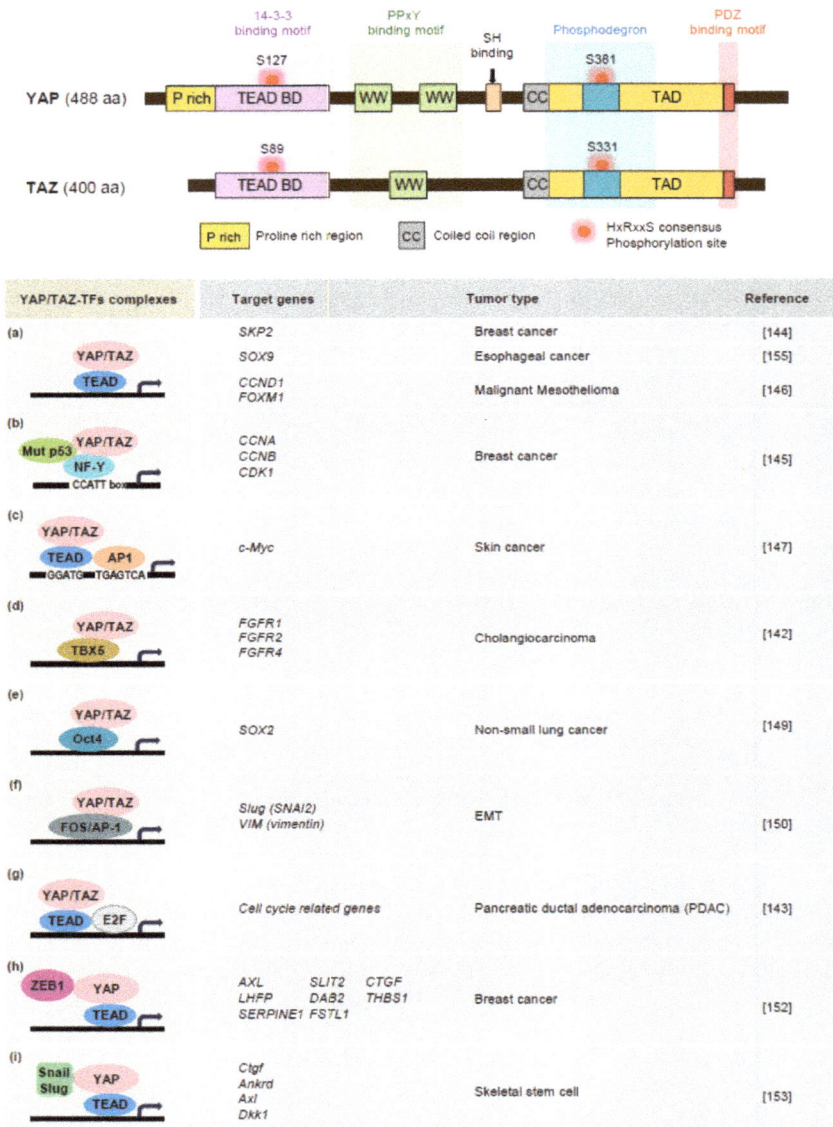

Figure 4. YAP/TAZ-mediated regulatory networks in cancer. YAP/TAZ are the major downstream mediators of the Hippo pathway and are composed of several domains (see Figure 1). The nuclear YAP/TAZ interacts with multiple transcription factors and signaling pathway that collectively contribute to tumorigenesis.

4. Conclusions

Accumulating evidence has highlighted the pivotal role of the Hippo pathway and its contribution to organ and tissue homeostasis under various chemical and physical signals. The MST1/2-LATS1/2 kinase cascade incorporates various upstream inputs to regulate YAP/TAZ activity and subcellular localization so as to maintain balanced processes, including physiological growth, development, and regeneration. In recent years, increasing research efforts have focused on unmasking the physiological/pathological functions of the Hippo pathway using animal models and patients, revealing a critical role in both fibrosis and cancer, which are inextricably linked. YAP/TAZ acts as a cellular mechanosensor that induces the transdifferentiation of fibroblasts to myofibroblasts in tissue fibrosis. In response to matrix stiffness, YAP/TAZ activation drives profibrotic and tumorigenic effects. YAP/TAZ signaling is also capable of initiating the EMT via EMT-related transcription factors, and this YAP/TAZ-dependent EMT program during tumorigenesis may further stimulate stemness and tumorigenicity.

These new insights into the diverse functions of the Hippo pathway will help in discovering novel therapeutic targets for the treatment of fibrosis and cancer. For example, VP is a small molecule that modulates the YAP and TEAD interaction and could reverse fibrosis. [49,64,69]. Given the clear profibrotic effect of YAP/TAZ, such pharmacological inhibitors targeting the Hippo pathway may be beneficial for treating both fibrosis and cancer (Table 1) [69,79,158–165]. However, targeting the Hippo pathway may also cause undesired harmful effects because of its central roles in organ development, homeostasis, and regeneration. Therefore, further studies to uncover the other mechanisms or unknown molecules regulating YAP/TAZ activity are expected to offer promising therapeutic targets in fibrosis and cancer progression.

Table 1. Potential targets for effective Hippo pathway regulation.

Name	Mode of Action	Tested Application	Reference
Verteporfin	Disruption of YAP/TAZ- TEAD complex	• HSCs isolated from C57BL/6J mice or *Smo* flox/flox mice • IPF lung epithelial cells • UUO-induced renal fibrosis • CCl4-induced liver fibrosis mouse model	[49,63,68]
Melatonin	Inhibiting of the expression and activation of YAP1 via binding to MT1&MT2 melatonin receptors	• Bleomycin induced experimental lung fibrosis in mice	[158]
Morin	Increased expression of MST1 and Lats1 Decreased expression of YAP/TAZ	• Diethylnitrosamine (Den)-induced liver fibrosis rat model • Hepatic stellate cells derived form human	[159]
ω-3 PUFA	YAP/TAZ degradation in a proteasome-dependent manner	• CCl4-induced liver fibrosis mouse model • Hepatic stellate cells derived from human and rat	[160]
Fasudil	Inhibitory effect on Rho/ROCK signaling	• Bleomycin-induced mouse lung fibrosis	[161]
Dobutamine	Induced the cytoplasmic translocation of YAP	• U2OS cell line	[162]
Statin	YAP/TAZ nuclear localization and activity	• Breast cancer cell lines	[163]
Lovastatin	YAP/TAZ nuclear localization and activity	• CCl4-induced liver fibrosis mouse model • Human aortic vascular smooth muscle cells • Human kidney cells	[79]
JQ-1	Inhibitor of bromodomain-containing protein 4 (BRD4) mediated profibrotic transcription	• CCl4-induced liver fibrosis mouse model • Hepatic stellate cells derived from mouse	[164]

Author Contributions: Conceptualization, J.M. and C.K.; Writing—Original Draft Preparation, J.M., C.K. and S.C.; Writing—Review & Editing, J.M. and C.K.; Visualization, S.C.; Funding Acquisition, J.M.

Acknowledgments: We apologize for many of most important research contributions to the Hippo pathway, fibrosis, and cancer fields that we could not cite in this review. This work is supported by Basic Science Research Program through the National Research Foundation of Korea (NRF) funded by the Ministry of Science, Ict and Future Planning [NRF-2016R1C1B2016135 (Young Researcher Program)].

Conflicts of Interest: The authors declare that there is no conflict of interest.

References

1. Yu, F.X.; Zhao, B.; Guan, K.L. Hippo Pathway in Organ Size Control, Tissue Homeostasis, and Cancer. *Cell* **2015**, *163*, 811–828. [CrossRef] [PubMed]
2. Harvey, K.F.; Pfleger, C.M.; Hariharan, I.K. The Drosophila Mst ortholog, hippo, restricts growth and cell proliferation and promotes apoptosis. *Cell* **2003**, *114*, 457–467. [CrossRef]
3. Jia, J.; Zhang, W.; Wang, B.; Trinko, R.; Jiang, J. The Drosophila Ste20 family kinase dMST functions as a tumor suppressor by restricting cell proliferation and promoting apoptosis. *Genes Dev.* **2003**, *17*, 2514–2519. [CrossRef]
4. Justice, R.W.; Zilian, O.; Woods, D.F.; Noll, M.; Bryant, P.J. The Drosophila tumor suppressor gene warts encodes a homolog of human myotonic dystrophy kinase and is required for the control of cell shape and proliferation. *Genes Dev.* **1995**, *9*, 534–546. [CrossRef]
5. Pantalacci, S.; Tapon, N.; Leopold, P. The Salvador partner Hippo promotes apoptosis and cell-cycle exit in Drosophila. *Nat. Cell Biol.* **2003**, *5*, 921–927. [CrossRef] [PubMed]
6. Xu, T.; Wang, W.; Zhang, S.; Stewart, R.A.; Yu, W. Identifying tumor suppressors in genetic mosaics: The Drosophila lats gene encodes a putative protein kinase. *Development* **1995**, *121*, 1053–1063. [PubMed]
7. Udan, R.S.; Kango-Singh, M.; Nolo, R.; Tao, C.; Halder, G. Hippo promotes proliferation arrest and apoptosis in the Salvador/Warts pathway. *Nat. Cell Biol.* **2003**, *5*, 914–920. [CrossRef] [PubMed]
8. Wu, S.; Huang, J.; Dong, J.; Pan, D. hippo encodes a Ste-20 family protein kinase that restricts cell proliferation and promotes apoptosis in conjunction with salvador and warts. *Cell* **2003**, *114*, 445–456. [CrossRef]
9. Kango-Singh, M.; Nolo, R.; Tao, C.; Verstreken, P.; Hiesinger, P.R.; Bellen, H.J.; Halder, G. Shar-pei mediates cell proliferation arrest during imaginal disc growth in Drosophila. *Development* **2002**, *129*, 5719–5730. [CrossRef] [PubMed]
10. Tapon, N.; Harvey, K.F.; Bell, D.W.; Wahrer, D.C.; Schiripo, T.A.; Haber, D.; Hariharan, I.K. salvador Promotes both cell cycle exit and apoptosis in Drosophila and is mutated in human cancer cell lines. *Cell* **2002**, *110*, 467–478. [CrossRef]
11. Lai, Z.C.; Wei, X.; Shimizu, T.; Ramos, E.; Rohrbaugh, M.; Nikolaidis, N.; Ho, L.L.; Li, Y. Control of cell proliferation and apoptosis by mob as tumor suppressor, mats. *Cell* **2005**, *120*, 675–685. [CrossRef] [PubMed]
12. Ren, F.; Zhang, L.; Jiang, J. Hippo signaling regulates Yorkie nuclear localization and activity through 14-3-3 dependent and independent mechanisms. *Dev. Biol.* **2010**, *337*, 303–312. [CrossRef]
13. Huang, J.; Wu, S.; Barrera, J.; Matthews, K.; Pan, D. The Hippo signaling pathway coordinately regulates cell proliferation and apoptosis by inactivating Yorkie, the Drosophila Homolog of YAP. *Cell* **2005**, *122*, 421–434. [CrossRef] [PubMed]
14. Dong, J.; Feldmann, G.; Huang, J.; Wu, S.; Zhang, N.; Comerford, S.A.; Gayyed, M.F.; Anders, R.A.; Maitra, A.; Pan, D. Elucidation of a universal size-control mechanism in Drosophila and mammals. *Cell* **2007**, *130*, 1120–1133. [CrossRef] [PubMed]
15. Oh, H.; Irvine, K.D. In vivo regulation of Yorkie phosphorylation and localization. *Development* **2008**, *135*, 1081–1088. [CrossRef] [PubMed]
16. Chan, E.H.; Nousiainen, M.; Chalamalasetty, R.B.; Schafer, A.; Nigg, E.A.; Sillje, H.H. The Ste20-like kinase Mst2 activates the human large tumor suppressor kinase Lats1. *Oncogene* **2005**, *24*, 2076–2086. [CrossRef]
17. Praskova, M.; Xia, F.; Avruch, J. MOBKL1A/MOBKL1B phosphorylation by MST1 and MST2 inhibits cell proliferation. *Curr. Biol.* **2008**, *18*, 311–321. [CrossRef]
18. Callus, B.A.; Verhagen, A.M.; Vaux, D.L. Association of mammalian sterile twenty kinases, Mst1 and Mst2, with hSalvador via C-terminal coiled-coil domains, leads to its stabilization and phosphorylation. *FEBS J.* **2006**, *273*, 4264–4276. [CrossRef] [PubMed]

19. Staley, B.K.; Irvine, K.D. Hippo signaling in Drosophila: Recent advances and insights. *Dev. Dyn.* **2012**, *241*, 3–15. [CrossRef]

20. Sudol, M. Yes-associated protein (YAP65) is a proline-rich phosphoprotein that binds to the SH3 domain of the Yes proto-oncogene product. *Oncogene* **1994**, *9*, 2145–2152.

21. Kanai, F.; Marignani, P.A.; Sarbassova, D.; Yagi, R.; Hall, R.A.; Donowitz, M.; Hisaminato, A.; Fujiwara, T.; Ito, Y.; Cantley, L.C.; et al. TAZ: A novel transcriptional co-activator regulated by interactions with 14-3-3 and PDZ domain proteins. *EMBO J.* **2000**, *19*, 6778–6791. [CrossRef]

22. Zhao, B.; Wei, X.; Li, W.; Udan, R.S.; Yang, Q.; Kim, J.; Xie, J.; Ikenoue, T.; Yu, J.; Li, L.; et al. Inactivation of YAP oncoprotein by the Hippo pathway is involved in cell contact inhibition and tissue growth control. *Genes Dev.* **2007**, *21*, 2747–2761. [CrossRef]

23. Hao, Y.; Chun, A.; Cheung, K.; Rashidi, B.; Yang, X. Tumor suppressor LATS1 is a negative regulator of oncogene YAP. *J. Biol. Chem.* **2008**, *283*, 5496–5509. [CrossRef] [PubMed]

24. Oka, T.; Mazack, V.; Sudol, M. Mst2 and Lats kinases regulate apoptotic function of Yes kinase-associated protein (YAP). *J. Biol. Chem.* **2008**, *283*, 27534–27546. [CrossRef] [PubMed]

25. Lei, Q.Y.; Zhang, H.; Zhao, B.; Zha, Z.Y.; Bai, F.; Pei, X.H.; Zhao, S.; Xiong, Y.; Guan, K.L. TAZ promotes cell proliferation and epithelial-mesenchymal transition and is inhibited by the hippo pathway. *Mol. Cell. Biol.* **2008**, *28*, 2426–2436. [CrossRef]

26. Liu, C.Y.; Zha, Z.Y.; Zhou, X.; Zhang, H.; Huang, W.; Zhao, D.; Li, T.; Chan, S.W.; Lim, C.J.; Hong, W.; et al. The hippo tumor pathway promotes TAZ degradation by phosphorylating a phosphodegron and recruiting the SCF{beta}-TrCP E3 ligase. *J. Biol. Chem.* **2010**, *285*, 37159–37169. [CrossRef] [PubMed]

27. Zhao, B.; Ye, X.; Yu, J.; Li, L.; Li, W.; Li, S.; Yu, J.; Lin, J.D.; Wang, C.Y.; Chinnaiyan, A.M.; et al. TEAD mediates YAP-dependent gene induction and growth control. *Genes Dev.* **2008**, *22*, 1962–1971. [CrossRef]

28. Zhang, H.; Liu, C.Y.; Zha, Z.Y.; Zhao, B.; Yao, J.; Zhao, S.; Xiong, Y.; Lei, Q.Y.; Guan, K.L. TEAD transcription factors mediate the function of TAZ in cell growth and epithelial-mesenchymal transition. *J. Biol. Chem.* **2009**, *284*, 13355–13362. [CrossRef] [PubMed]

29. Meng, Z.; Moroishi, T.; Guan, K.L. Mechanisms of Hippo pathway regulation. *Genes Dev.* **2016**, *30*, 1–17. [CrossRef] [PubMed]

30. Zanconato, F.; Cordenonsi, M.; Piccolo, S. YAP/TAZ at the Roots of Cancer. *Cancer Cell* **2016**, *29*, 783–803. [CrossRef]

31. Mo, J.S.; Park, H.W.; Guan, K.L. The Hippo signaling pathway in stem cell biology and cancer. *EMBO Rep.* **2014**, *15*, 642–656. [CrossRef]

32. Mo, J.S. The role of extracellular biophysical cues in modulating the Hippo-YAP pathway. *BMB Rep.* **2017**, *50*, 71–78. [CrossRef]

33. Corvol, H.; Flamein, F.; Epaud, R.; Clement, A.; Guillot, L. Lung alveolar epithelium and interstitial lung disease. *Int. J. Biochem. Cell Biol.* **2009**, *41*, 1643–1651. [CrossRef]

34. Herriges, M.; Morrisey, E.E. Lung development: Orchestrating the generation and regeneration of a complex organ. *Development* **2014**, *141*, 502–513. [CrossRef] [PubMed]

35. Lange, A.W.; Sridharan, A.; Xu, Y.; Stripp, B.R.; Perl, A.K.; Whitsett, J.A. Hippo/Yap signaling controls epithelial progenitor cell proliferation and differentiation in the embryonic and adult lung. *J. Mol. Cell Biol.* **2015**, *7*, 35–47. [CrossRef]

36. Zhao, R.; Fallon, T.R.; Saladi, S.V.; Pardo-Saganta, A.; Villoria, J.; Mou, H.; Vinarsky, V.; Gonzalez-Celeiro, M.; Nunna, N.; Hariri, L.P.; et al. Yap tunes airway epithelial size and architecture by regulating the identity, maintenance, and self-renewal of stem cells. *Dev. Cell* **2014**, *30*, 151–165. [CrossRef] [PubMed]

37. Park, K.S.; Whitsett, J.A.; Di Palma, T.; Hong, J.H.; Yaffe, M.B.; Zannini, M. TAZ interacts with TTF-1 and regulates expression of surfactant protein-C. *J. Biol. Chem.* **2004**, *279*, 17384–17390. [CrossRef]

38. Mitani, A.; Nagase, T.; Fukuchi, K.; Aburatani, H.; Makita, R.; Kurihara, H. Transcriptional coactivator with PDZ-binding motif is essential for normal alveolarization in mice. *Am. J. Respir. Crit. Care Med.* **2009**, *180*, 326–338. [CrossRef] [PubMed]

39. Kasai, H.; Allen, J.T.; Mason, R.M.; Kamimura, T.; Zhang, Z. TGF-beta1 induces human alveolar epithelial to mesenchymal cell transition (EMT). *Respir. Res.* **2005**, *6*, 56. [CrossRef]

40. Lasky, J.A.; Ortiz, L.A.; Tonthat, B.; Hoyle, G.W.; Corti, M.; Athas, G.; Lungarella, G.; Brody, A.; Friedman, M. Connective tissue growth factor mRNA expression is upregulated in bleomycin-induced lung fibrosis. *Am. J. Physiol.* **1998**, *275*, L365–L371. [CrossRef]

41. Pan, L.H.; Yamauchi, K.; Uzuki, M.; Nakanishi, T.; Takigawa, M.; Inoue, H.; Sawai, T. Type II alveolar epithelial cells and interstitial fibroblasts express connective tissue growth factor in IPF. *Eur. Respir. J.* **2001**, *17*, 1220–1227. [CrossRef]

42. Liu, F.; Lagares, D.; Choi, K.M.; Stopfer, L.; Marinkovic, A.; Vrbanac, V.; Probst, C.K.; Hiemer, S.E.; Sisson, T.H.; Horowitz, J.C.; et al. Mechanosignaling through YAP and TAZ drives fibroblast activation and fibrosis. *Am. J. Physiol. Lung Cell. Mol. Physiol.* **2015**, *308*, L344–L357. [CrossRef]

43. Noguchi, S.; Saito, A.; Mikami, Y.; Urushiyama, H.; Horie, M.; Matsuzaki, H.; Takeshima, H.; Makita, K.; Miyashita, N.; Mitani, A.; et al. TAZ contributes to pulmonary fibrosis by activating profibrotic functions of lung fibroblasts. *Sci. Rep.* **2017**, *7*, 42595. [CrossRef]

44. Jorgenson, A.J.; Choi, K.M.; Sicard, D.; Smith, K.M.; Hiemer, S.E.; Varelas, X.; Tschumperlin, D.J. TAZ activation drives fibroblast spheroid growth, expression of profibrotic paracrine signals, and context-dependent ECM gene expression. *Am. J. Physiol. Cell Physiol.* **2017**, *312*, C277–C285. [CrossRef]

45. Chen, Y.; Zhao, X.; Sun, J.; Su, W.; Zhang, L.; Li, Y.; Liu, Y.; Zhang, L.; Lu, Y.; Shan, H.; et al. YAP1/Twist promotes fibroblast activation and lung fibrosis that conferred by miR-15a loss in IPF. *Cell Death Differ.* **2019**. [CrossRef]

46. Swigris, J.J.; Gould, M.K.; Wilson, S.R. Health-related quality of life among patients with idiopathic pulmonary fibrosis. *Chest* **2005**, *127*, 284–294. [CrossRef] [PubMed]

47. Katzenstein, A.L.; Myers, J.L. Idiopathic pulmonary fibrosis: Clinical relevance of pathologic classification. *Am. J. Respir. Crit. Care Med.* **1998**, *157*, 1301–1315. [CrossRef] [PubMed]

48. Xu, Y.; Mizuno, T.; Sridharan, A.; Du, Y.; Guo, M.; Tang, J.; Wikenheiser-Brokamp, K.A.; Perl, A.T.; Funari, V.A.; Gokey, J.J.; et al. Single-cell RNA sequencing identifies diverse roles of epithelial cells in idiopathic pulmonary fibrosis. *JCI Insight* **2016**, *1*, e90558. [CrossRef]

49. Gokey, J.J.; Sridharan, A.; Xu, Y.; Green, J.; Carraro, G.; Stripp, B.R.; Perl, A.T.; Whitsett, J.A. Active epithelial Hippo signaling in idiopathic pulmonary fibrosis. *JCI Insight* **2018**, *3*. [CrossRef] [PubMed]

50. Speight, P.; Kofler, M.; Szaszi, K.; Kapus, A. Context-dependent switch in chemo/mechanotransduction via multilevel crosstalk among cytoskeleton-regulated MRTF and TAZ and TGFbeta-regulated Smad3. *Nat. Commun.* **2016**, *7*, 11642. [CrossRef]

51. Miner, J.H. The glomerular basement membrane. *Exp. Cell Res.* **2012**, *318*, 973–978. [CrossRef] [PubMed]

52. Hossain, Z.; Ali, S.M.; Ko, H.L.; Xu, J.; Ng, C.P.; Guo, K.; Qi, Z.; Ponniah, S.; Hong, W.; Hunziker, W. Glomerulocystic kidney disease in mice with a targeted inactivation of Wwtr1. *Proc. Natl. Acad. Sci. USA* **2007**, *104*, 1631–1636. [CrossRef]

53. Reginensi, A.; Scott, R.P.; Gregorieff, A.; Bagherie-Lachidan, M.; Chung, C.; Lim, D.S.; Pawson, T.; Wrana, J.; McNeill, H. Yap- and Cdc42-dependent nephrogenesis and morphogenesis during mouse kidney development. *PLoS Genet.* **2013**, *9*, e1003380. [CrossRef] [PubMed]

54. Tian, Y.; Kolb, R.; Hong, J.H.; Carroll, J.; Li, D.; You, J.; Bronson, R.; Yaffe, M.B.; Zhou, J.; Benjamin, T. TAZ promotes PC2 degradation through a SCFbeta-Trcp E3 ligase complex. *Mol. Cell. Biol.* **2007**, *27*, 6383–6395. [CrossRef]

55. Habbig, S.; Bartram, M.P.; Sagmuller, J.G.; Griessmann, A.; Franke, M.; Muller, R.U.; Schwarz, R.; Hoehne, M.; Bergmann, C.; Tessmer, C.; et al. The ciliopathy disease protein NPHP9 promotes nuclear delivery and activation of the oncogenic transcriptional regulator TAZ. *Hum. Mol. Genet.* **2012**, *21*, 5528–5538. [CrossRef] [PubMed]

56. Habbig, S.; Bartram, M.P.; Muller, R.U.; Schwarz, R.; Andriopoulos, N.; Chen, S.; Sagmuller, J.G.; Hoehne, M.; Burst, V.; Liebau, M.C.; et al. NPHP4, a cilia-associated protein, negatively regulates the Hippo pathway. *J. Cell Biol.* **2011**, *193*, 633–642. [CrossRef]

57. Happe, H.; van der Wal, A.M.; Leonhard, W.N.; Kunnen, S.J.; Breuning, M.H.; de Heer, E.; Peters, D.J. Altered Hippo signalling in polycystic kidney disease. *J. Pathol.* **2011**, *224*, 133–142. [CrossRef]

58. Seo, E.; Kim, W.Y.; Hur, J.; Kim, H.; Nam, S.A.; Choi, A.; Kim, Y.M.; Park, S.H.; Chung, C.; Kim, J.; et al. The Hippo-Salvador signaling pathway regulates renal tubulointerstitial fibrosis. *Sci. Rep.* **2016**, *6*, 31931. [CrossRef]

59. Anorga, S.; Overstreet, J.M.; Falke, L.L.; Tang, J.; Goldschmeding, R.G.; Higgins, P.J.; Samarakoon, R. Deregulation of Hippo-TAZ pathway during renal injury confers a fibrotic maladaptive phenotype. *FASEB J. Off. Publ. Fed. Am. Soc. Exp. Biol.* **2018**, *32*, 2644–2657. [CrossRef]

60. Xu, J.; Li, P.X.; Wu, J.; Gao, Y.J.; Yin, M.X.; Lin, Y.; Yang, M.; Chen, D.P.; Sun, H.P.; Liu, Z.B.; et al. Involvement of the Hippo pathway in regeneration and fibrogenesis after ischaemic acute kidney injury: YAP is the key effector. *Clin. Sci.* **2016**, *130*, 349–363. [CrossRef] [PubMed]

61. Wennmann, D.O.; Vollenbroker, B.; Eckart, A.K.; Bonse, J.; Erdmann, F.; Wolters, D.A.; Schenk, L.K.; Schulze, U.; Kremerskothen, J.; Weide, T.; et al. The Hippo pathway is controlled by Angiotensin II signaling and its reactivation induces apoptosis in podocytes. *Cell Death Dis.* **2014**, *5*, e1519. [CrossRef]

62. Rinschen, M.M.; Grahammer, F.; Hoppe, A.K.; Kohli, P.; Hagmann, H.; Kretz, O.; Bertsch, S.; Hohne, M.; Gobel, H.; Bartram, M.P.; et al. YAP-mediated mechanotransduction determines the podocyte's response to damage. *Sci. Signal.* **2017**, *10*, eaaf8165. [CrossRef]

63. Schwartzman, M.; Reginensi, A.; Wong, J.S.; Basgen, J.M.; Meliambro, K.; Nicholas, S.B.; D'Agati, V.; McNeill, H.; Campbell, K.N. Podocyte-Specific Deletion of Yes-Associated Protein Causes FSGS and Progressive Renal Failure. *J. Am. Soc. Nephrol. JASN* **2016**, *27*, 216–226. [CrossRef] [PubMed]

64. Szeto, S.G.; Narimatsu, M.; Lu, M.; He, X.; Sidiqi, A.M.; Tolosa, M.F.; Chan, L.; De Freitas, K.; Bialik, J.F.; Majumder, S.; et al. YAP/TAZ Are Mechanoregulators of TGF-beta-Smad Signaling and Renal Fibrogenesis. *J. Am. Soc. Nephrol. JASN* **2016**, *27*, 3117–3128. [CrossRef] [PubMed]

65. Wang, C.; Zhang, L.; He, Q.; Feng, X.; Zhu, J.; Xu, Z.; Wang, X.; Chen, F.; Li, X.; Dong, J. Differences in Yes-associated protein and mRNA levels in regenerating liver and hepatocellular carcinoma. *Mol. Med. Rep.* **2012**, *5*, 410–414. [CrossRef]

66. Camargo, F.D.; Gokhale, S.; Johnnidis, J.B.; Fu, D.; Bell, G.W.; Jaenisch, R.; Brummelkamp, T.R. YAP1 increases organ size and expands undifferentiated progenitor cells. *Curr. Biol.* **2007**, *17*, 2054–2060. [CrossRef]

67. Zhou, D.; Conrad, C.; Xia, F.; Park, J.S.; Payer, B.; Yin, Y.; Lauwers, G.Y.; Thasler, W.; Lee, J.T.; Avruch, J.; et al. Mst1 and Mst2 maintain hepatocyte quiescence and suppress hepatocellular carcinoma development through inactivation of the Yap1 oncogene. *Cancer Cell* **2009**, *16*, 425–438. [CrossRef]

68. Lu, L.; Li, Y.; Kim, S.M.; Bossuyt, W.; Liu, P.; Qiu, Q.; Wang, Y.; Halder, G.; Finegold, M.J.; Lee, J.S.; et al. Hippo signaling is a potent in vivo growth and tumor suppressor pathway in the mammalian liver. *Proc. Natl. Acad. Sci. USA* **2010**, *107*, 1437–1442. [CrossRef]

69. Mannaerts, I.; Leite, S.B.; Verhulst, S.; Claerhout, S.; Eysackers, N.; Thoen, L.F.; Hoorens, A.; Reynaert, H.; Halder, G.; van Grunsven, L.A. The Hippo pathway effector YAP controls mouse hepatic stellate cell activation. *J. Hepatol.* **2015**, *63*, 679–688. [CrossRef]

70. Du, K.; Hyun, J.; Premont, R.T.; Choi, S.S.; Michelotti, G.A.; Swiderska-Syn, M.; Dalton, G.D.; Thelen, E.; Rizi, B.S.; Jung, Y.; et al. Hedgehog-YAP Signaling Pathway Regulates Glutaminolysis to Control Activation of Hepatic Stellate Cells. *Gastroenterology* **2018**, *154*, 1465.e13–1479.e13. [CrossRef] [PubMed]

71. Wang, X.; Zheng, Z.; Caviglia, J.M.; Corey, K.E.; Herfel, T.M.; Cai, B.; Masia, R.; Chung, R.T.; Lefkowitch, J.H.; Schwabe, R.F.; et al. Hepatocyte TAZ/WWTR1 Promotes Inflammation and Fibrosis in Nonalcoholic Steatohepatitis. *Cell Metab.* **2016**, *24*, 848–862. [CrossRef] [PubMed]

72. Laflamme, M.A.; Murry, C.E. Heart regeneration. *Nature* **2011**, *473*, 326–335. [CrossRef] [PubMed]

73. Heallen, T.; Zhang, M.; Wang, J.; Bonilla-Claudio, M.; Klysik, E.; Johnson, R.L.; Martin, J.F. Hippo pathway inhibits Wnt signaling to restrain cardiomyocyte proliferation and heart size. *Science* **2011**, *332*, 458–461. [CrossRef] [PubMed]

74. Heallen, T.; Morikawa, Y.; Leach, J.; Tao, G.; Willerson, J.T.; Johnson, R.L.; Martin, J.F. Hippo signaling impedes adult heart regeneration. *Development* **2013**, *140*, 4683–4690. [CrossRef] [PubMed]

75. Xin, M.; Kim, Y.; Sutherland, L.B.; Qi, X.; McAnally, J.; Schwartz, R.J.; Richardson, J.A.; Bassel-Duby, R.; Olson, E.N. Regulation of insulin-like growth factor signaling by Yap governs cardiomyocyte proliferation and embryonic heart size. *Sci. Signal.* **2011**, *4*, ra70. [CrossRef]

76. Von Gise, A.; Lin, Z.; Schlegelmilch, K.; Honor, L.B.; Pan, G.M.; Buck, J.N.; Ma, Q.; Ishiwata, T.; Zhou, B.; Camargo, F.D.; et al. YAP1, the nuclear target of Hippo signaling, stimulates heart growth through cardiomyocyte proliferation but not hypertrophy. *Proc. Natl. Acad. Sci. USA* **2012**, *109*, 2394–2399. [CrossRef]

77. Xin, M.; Kim, Y.; Sutherland, L.B.; Murakami, M.; Qi, X.; McAnally, J.; Porrello, E.R.; Mahmoud, A.I.; Tan, W.; Shelton, J.M.; et al. Hippo pathway effector Yap promotes cardiac regeneration. *Proc. Natl. Acad. Sci. USA* **2013**, *110*, 13839–13844. [CrossRef]

78. Del Re, D.P.; Yang, Y.; Nakano, N.; Cho, J.; Zhai, P.; Yamamoto, T.; Zhang, N.; Yabuta, N.; Nojima, H.; Pan, D.; et al. Yes-associated protein isoform 1 (Yap1) promotes cardiomyocyte survival and growth to protect against myocardial ischemic injury. *J. Biol. Chem.* **2013**, *288*, 3977–3988. [CrossRef]

79. Wu, P.; Liu, Z.; Zhao, T.; Xia, F.; Gong, L.; Zheng, Z.; Chen, Z.; Yang, T.; Duan, Q. Lovastatin attenuates angiotensin II induced cardiovascular fibrosis through the suppression of YAP/TAZ signaling. *Biochem. Biophys. Res. Commun.* **2019**, *512*, 736–741. [CrossRef]

80. Schlegelmilch, K.; Mohseni, M.; Kirak, O.; Pruszak, J.; Rodriguez, J.R.; Zhou, D.; Kreger, B.T.; Vasioukhin, V.; Avruch, J.; Brummelkamp, T.R.; et al. Yap1 acts downstream of alpha-catenin to control epidermal proliferation. *Cell* **2011**, *144*, 782–795. [CrossRef]

81. Rognoni, E.; Walko, G. The Roles of YAP/TAZ and the Hippo Pathway in Healthy and Diseased Skin. *Cells* **2019**, *8*, 411. [CrossRef] [PubMed]

82. Zhang, H.; Pasolli, H.A.; Fuchs, E. Yes-associated protein (YAP) transcriptional coactivator functions in balancing growth and differentiation in skin. *Proc. Natl. Acad. Sci. USA* **2011**, *108*, 2270–2275. [CrossRef] [PubMed]

83. Lee, M.J.; Byun, M.R.; Furutani-Seiki, M.; Hong, J.H.; Jung, H.S. YAP and TAZ regulate skin wound healing. *J. Investig. Dermatol.* **2014**, *134*, 518–525. [CrossRef] [PubMed]

84. Piersma, B.; de Rond, S.; Werker, P.M.; Boo, S.; Hinz, B.; van Beuge, M.M.; Bank, R.A. YAP1 Is a Driver of Myofibroblast Differentiation in Normal and Diseased Fibroblasts. *Am. J. Pathol.* **2015**, *185*, 3326–3337. [CrossRef]

85. Toyama, T.; Looney, A.P.; Baker, B.M.; Stawski, L.; Haines, P.; Simms, R.; Szymaniak, A.D.; Varelas, X.; Trojanowska, M. Therapeutic Targeting of TAZ and YAP by Dimethyl Fumarate in Systemic Sclerosis Fibrosis. *J. Investig. Dermatol.* **2018**, *138*, 78–88. [CrossRef]

86. Darby, I.A.; Zakuan, N.; Billet, F.; Desmouliere, A. The myofibroblast, a key cell in normal and pathological tissue repair. *Cell. Mol. Life Sci.* **2016**, *73*, 1145–1157. [CrossRef] [PubMed]

87. Nieman, K.M.; Kenny, H.A.; Penicka, C.V.; Ladanyi, A.; Buell-Gutbrod, R.; Zillhardt, M.R.; Romero, I.L.; Carey, M.S.; Mills, G.B.; Hotamisligil, G.S.; et al. Adipocytes promote ovarian cancer metastasis and provide energy for rapid tumor growth. *Nat. Med.* **2011**, *17*, 1498–1503. [CrossRef]

88. Lu, P.; Weaver, V.M.; Werb, Z. The extracellular matrix: A dynamic niche in cancer progression. *J. Cell Biol.* **2012**, *196*, 395–406. [CrossRef]

89. Mueller, M.M.; Fusenig, N.E. Friends or foes—Bipolar effects of the tumour stroma in cancer. *Nat. Rev. Cancer* **2004**, *4*, 839–849. [CrossRef]

90. Quail, D.F.; Joyce, J.A. Microenvironmental regulation of tumor progression and metastasis. *Nat. Med.* **2013**, *19*, 1423–1437. [CrossRef]

91. Kalluri, R.; Zeisberg, M. Fibroblasts in cancer. *Nat. Rev. Cancer* **2006**, *6*, 392–401. [CrossRef]

92. Gaggioli, C.; Hooper, S.; Hidalgo-Carcedo, C.; Grosse, R.; Marshall, J.F.; Harrington, K.; Sahai, E. Fibroblast-led collective invasion of carcinoma cells with differing roles for RhoGTPases in leading and following cells. *Nat. Cell Biol.* **2007**, *9*, 1392–1400. [CrossRef]

93. Glentis, A.; Oertle, P.; Mariani, P.; Chikina, A.; El Marjou, F.; Attieh, Y.; Zaccarini, F.; Lae, M.; Loew, D.; Dingli, F.; et al. Cancer-associated fibroblasts induce metalloprotease-independent cancer cell invasion of the basement membrane. *Nat. Commun.* **2017**, *8*, 924. [CrossRef] [PubMed]

94. Calvo, F.; Ege, N.; Grande-Garcia, A.; Hooper, S.; Jenkins, R.P.; Chaudhry, S.I.; Harrington, K.; Williamson, P.; Moeendarbary, E.; Charras, G.; et al. Mechanotransduction and YAP-dependent matrix remodelling is required for the generation and maintenance of cancer-associated fibroblasts. *Nat. Cell Biol.* **2013**, *15*, 637–646. [CrossRef] [PubMed]

95. Chan, S.W.; Lim, C.J.; Guo, K.; Ng, C.P.; Lee, I.; Hunziker, W.; Zeng, Q.; Hong, W. A role for TAZ in migration, invasion, and tumorigenesis of breast cancer cells. *Cancer Res.* **2008**, *68*, 2592–2598. [CrossRef]

96. Foster, C.T.; Gualdrini, F.; Treisman, R. Mutual dependence of the MRTF-SRF and YAP-TEAD pathways in cancer-associated fibroblasts is indirect and mediated by cytoskeletal dynamics. *Genes Dev.* **2017**, *31*, 2361–2375. [CrossRef] [PubMed]

97. Yazdani, S.; Bansal, R.; Prakash, J. Drug targeting to myofibroblasts: Implications for fibrosis and cancer. *Adv. Drug Deliv. Rev.* **2017**, *121*, 101–116. [CrossRef] [PubMed]

98. Cernaro, V.; Lacquaniti, A.; Donato, V.; Fazio, M.R.; Buemi, A.; Buemi, M. Fibrosis, regeneration and cancer: What is the link? *Nephrol. Dial. Transplant. Off. Publ. Eur. Dial. Transpl. Assoc. Eur. Ren. Assoc.* **2012**, *27*, 21–27. [CrossRef] [PubMed]

99. Mehner, C.; Radisky, D.C. Triggering the landslide: The tumor-promotional effects of myofibroblasts. *Exp. Cell Res.* **2013**, *319*, 1657–1662. [CrossRef]

100. Otranto, M.; Sarrazy, V.; Bonte, F.; Hinz, B.; Gabbiani, G.; Desmouliere, A. The role of the myofibroblast in tumor stroma remodeling. *Cell Adhes. Migr.* **2012**, *6*, 203–219. [CrossRef] [PubMed]
101. Pan, D. The hippo signaling pathway in development and cancer. *Dev. Cell* **2010**, *19*, 491–505. [CrossRef] [PubMed]
102. Plouffe, S.W.; Hong, A.W.; Guan, K.L. Disease implications of the Hippo/YAP pathway. *Trends Mol. Med.* **2015**, *21*, 212–222. [CrossRef] [PubMed]
103. Bueno, R.; Stawiski, E.W.; Goldstein, L.D.; Durinck, S.; De Rienzo, A.; Modrusan, Z.; Gnad, F.; Nguyen, T.T.; Jaiswal, B.S.; Chirieac, L.R.; et al. Comprehensive genomic analysis of malignant pleural mesothelioma identifies recurrent mutations, gene fusions and splicing alterations. *Nat. Genet.* **2016**, *48*, 407–416. [CrossRef] [PubMed]
104. Oh, J.E.; Ohta, T.; Satomi, K.; Foll, M.; Durand, G.; McKay, J.; Le Calvez-Kelm, F.; Mittelbronn, M.; Brokinkel, B.; Paulus, W.; et al. Alterations in the NF2/LATS1/LATS2/YAP Pathway in Schwannomas. *J. Neuropathol. Exp. Neurol.* **2015**, *74*, 952–959. [CrossRef] [PubMed]
105. Evans, D.G. Neurofibromatosis 2 [Bilateral acoustic neurofibromatosis, central neurofibromatosis, NF2, neurofibromatosis type II]. *Genet. Med. Off. J. Am. Coll. Med Genet.* **2009**, *11*, 599–610. [CrossRef] [PubMed]
106. Endoh, H.; Yatabe, Y.; Shimizu, S.; Tajima, K.; Kuwano, H.; Takahashi, T.; Mitsudomi, T. RASSF1A gene inactivation in non-small cell lung cancer and its clinical implication. *Int. J. Cancer* **2003**, *106*, 45–51. [CrossRef]
107. Takahashi, Y.; Miyoshi, Y.; Takahata, C.; Irahara, N.; Taguchi, T.; Tamaki, Y.; Noguchi, S. Down-regulation of LATS1 and LATS2 mRNA expression by promoter hypermethylation and its association with biologically aggressive phenotype in human breast cancers. *Clin. Cancer Res. Off. J. Am. Assoc. Cancer Res.* **2005**, *11*, 1380–1385. [CrossRef] [PubMed]
108. Seidel, C.; Schagdarsurengin, U.; Blumke, K.; Wurl, P.; Pfeifer, G.P.; Hauptmann, S.; Taubert, H.; Dammann, R. Frequent hypermethylation of MST1 and MST2 in soft tissue sarcoma. *Mol. Carcinog.* **2007**, *46*, 865–871. [CrossRef] [PubMed]
109. Song, H.; Mak, K.K.; Topol, L.; Yun, K.; Hu, J.; Garrett, L.; Chen, Y.; Park, O.; Chang, J.; Simpson, R.M.; et al. Mammalian Mst1 and Mst2 kinases play essential roles in organ size control and tumor suppression. *Proc. Natl. Acad. Sci. USA* **2010**, *107*, 1431–1436. [CrossRef] [PubMed]
110. Zhou, D.; Zhang, Y.; Wu, H.; Barry, E.; Yin, Y.; Lawrence, E.; Dawson, D.; Willis, J.E.; Markowitz, S.D.; Camargo, F.D.; et al. Mst1 and Mst2 protein kinases restrain intestinal stem cell proliferation and colonic tumorigenesis by inhibition of Yes-associated protein (Yap) overabundance. *Proc. Natl. Acad. Sci. USA* **2011**, *108*, E1312–E1320. [CrossRef] [PubMed]
111. St John, M.A.; Tao, W.; Fei, X.; Fukumoto, R.; Carcangiu, M.L.; Brownstein, D.G.; Parlow, A.F.; McGrath, J.; Xu, T. Mice deficient of Lats1 develop soft-tissue sarcomas, ovarian tumours and pituitary dysfunction. *Nat. Genet.* **1999**, *21*, 182–186. [CrossRef] [PubMed]
112. Jiang, Z.; Li, X.; Hu, J.; Zhou, W.; Jiang, Y.; Li, G.; Lu, D. Promoter hypermethylation-mediated down-regulation of LATS1 and LATS2 in human astrocytoma. *Neurosci. Res.* **2006**, *56*, 450–458. [CrossRef] [PubMed]
113. Zender, L.; Spector, M.S.; Xue, W.; Flemming, P.; Cordon-Cardo, C.; Silke, J.; Fan, S.T.; Luk, J.M.; Wigler, M.; Hannon, G.J.; et al. Identification and validation of oncogenes in liver cancer using an integrative oncogenomic approach. *Cell* **2006**, *125*, 1253–1267. [CrossRef] [PubMed]
114. Overholtzer, M.; Zhang, J.; Smolen, G.A.; Muir, B.; Li, W.; Sgroi, D.C.; Deng, C.X.; Brugge, J.S.; Haber, D.A. Transforming properties of YAP, a candidate oncogene on the chromosome 11q22 amplicon. *Proc. Natl. Acad. Sci. USA* **2006**, *103*, 12405–12410. [CrossRef] [PubMed]
115. Tanas, M.R.; Sboner, A.; Oliveira, A.M.; Erickson-Johnson, M.R.; Hespelt, J.; Hanwright, P.J.; Flanagan, J.; Luo, Y.; Fenwick, K.; Natrajan, R.; et al. Identification of a disease-defining gene fusion in epithelioid hemangioendothelioma. *Sci. Transl. Med.* **2011**, *3*, 98ra82. [CrossRef]
116. Antonescu, C.R.; Le Loarer, F.; Mosquera, J.M.; Sboner, A.; Zhang, L.; Chen, C.L.; Chen, H.W.; Pathan, N.; Krausz, T.; Dickson, B.C.; et al. Novel YAP1-TFE3 fusion defines a distinct subset of epithelioid hemangioendothelioma. *Geneschromosomes Cancer* **2013**, *52*, 775–784. [CrossRef]
117. Ota, M.; Sasaki, H. Mammalian Tead proteins regulate cell proliferation and contact inhibition as transcriptional mediators of Hippo signaling. *Development* **2008**, *135*, 4059–4069. [CrossRef]

118. Nishioka, N.; Inoue, K.; Adachi, K.; Kiyonari, H.; Ota, M.; Ralston, A.; Yabuta, N.; Hirahara, S.; Stephenson, R.O.; Ogonuki, N.; et al. The Hippo signaling pathway components Lats and Yap pattern Tead4 activity to distinguish mouse trophectoderm from inner cell mass. *Dev. Cell* **2009**, *16*, 398–410. [CrossRef] [PubMed]

119. Gumbiner, B.M.; Kim, N.G. The Hippo-YAP signaling pathway and contact inhibition of growth. *J. Cell Sci.* **2014**, *127*, 709–717. [CrossRef]

120. Dupont, S.; Morsut, L.; Aragona, M.; Enzo, E.; Giulitti, S.; Cordenonsi, M.; Zanconato, F.; Le Digabel, J.; Forcato, M.; Bicciato, S.; et al. Role of YAP/TAZ in mechanotransduction. *Nature* **2011**, *474*, 179–183. [CrossRef] [PubMed]

121. Zhao, B.; Li, L.; Wang, L.; Wang, C.Y.; Yu, J.; Guan, K.L. Cell detachment activates the Hippo pathway via cytoskeleton reorganization to induce anoikis. *Genes Dev.* **2012**, *26*, 54–68. [CrossRef] [PubMed]

122. Driscoll, T.P.; Cosgrove, B.D.; Heo, S.J.; Shurden, Z.E.; Mauck, R.L. Cytoskeletal to Nuclear Strain Transfer Regulates YAP Signaling in Mesenchymal Stem Cells. *Biophys. J.* **2015**, *108*, 2783–2793. [CrossRef]

123. Aragona, M.; Panciera, T.; Manfrin, A.; Giulitti, S.; Michielin, F.; Elvassore, N.; Dupont, S.; Piccolo, S. A mechanical checkpoint controls multicellular growth through YAP/TAZ regulation by actin-processing factors. *Cell* **2013**, *154*, 1047–1059. [CrossRef] [PubMed]

124. Schwartz, M.A. Integrins and extracellular matrix in mechanotransduction. *Cold Spring Harb. Perspect. Biol.* **2010**, *2*, a005066. [CrossRef] [PubMed]

125. Serrano, I.; McDonald, P.C.; Lock, F.; Muller, W.J.; Dedhar, S. Inactivation of the Hippo tumour suppressor pathway by integrin-linked kinase. *Nat. Commun.* **2013**, *4*, 2976. [CrossRef]

126. Sabra, H.; Brunner, M.; Mandati, V.; Wehrle-Haller, B.; Lallemand, D.; Ribba, A.S.; Chevalier, G.; Guardiola, P.; Block, M.R.; Bouvard, D. beta1 integrin-dependent Rac/group I PAK signaling mediates YAP activation of Yes-associated protein 1 (YAP1) via NF2/merlin. *J. Biol. Chem.* **2017**, *292*, 19179–19197. [CrossRef]

127. Kim, N.G.; Gumbiner, B.M. Adhesion to fibronectin regulates Hippo signaling via the FAK-Src-PI3K pathway. *J. Cell Biol.* **2015**, *210*, 503–515. [CrossRef] [PubMed]

128. Nardone, G.; Oliver-De La Cruz, J.; Vrbsky, J.; Martini, C.; Pribyl, J.; Skladal, P.; Pesl, M.; Caluori, G.; Pagliari, S.; Martino, F.; et al. YAP regulates cell mechanics by controlling focal adhesion assembly. *Nat. Commun.* **2017**, *8*, 15321. [CrossRef]

129. Wada, K.; Itoga, K.; Okano, T.; Yonemura, S.; Sasaki, H. Hippo pathway regulation by cell morphology and stress fibers. *Development* **2011**, *138*, 3907–3914. [CrossRef]

130. Ikeda, S.; Cunningham, L.A.; Boggess, D.; Hawes, N.; Hobson, C.D.; Sundberg, J.P.; Naggert, J.K.; Smith, R.S.; Nishina, P.M. Aberrant actin cytoskeleton leads to accelerated proliferation of corneal epithelial cells in mice deficient for destrin (actin depolymerizing factor). *Hum. Mol. Genet.* **2003**, *12*, 1029–1037. [CrossRef]

131. Chakraborty, S.; Lakshmanan, M.; Swa, H.L.; Chen, J.; Zhang, X.; Ong, Y.S.; Loo, L.S.; Akincilar, S.C.; Gunaratne, J.; Tergaonkar, V.; et al. An oncogenic role of Agrin in regulating focal adhesion integrity in hepatocellular carcinoma. *Nat. Commun.* **2015**, *6*, 6184. [CrossRef]

132. Chakraborty, S.; Njah, K.; Pobbati, A.V.; Lim, Y.B.; Raju, A.; Lakshmanan, M.; Tergaonkar, V.; Lim, C.T.; Hong, W. Agrin as a Mechanotransduction Signal Regulating YAP through the Hippo Pathway. *Cell Rep.* **2017**, *18*, 2464–2479. [CrossRef]

133. Chakraborty, S.; Hong, W. Linking Extracellular Matrix Agrin to the Hippo Pathway in Liver Cancer and Beyond. *Cancers* **2018**, *10*, 45. [CrossRef]

134. Feng, M.; Gao, W.; Wang, R.; Chen, W.; Man, Y.G.; Figg, W.D.; Wang, X.W.; Dimitrov, D.S.; Ho, M. Therapeutically targeting glypican-3 via a conformation-specific single-domain antibody in hepatocellular carcinoma. *Proc. Natl. Acad. Sci. USA* **2013**, *110*, E1083–E1091. [CrossRef]

135. Kim, K.M.; Choi, Y.J.; Hwang, J.H.; Kim, A.R.; Cho, H.J.; Hwang, E.S.; Park, J.Y.; Lee, S.H.; Hong, J.H. Shear stress induced by an interstitial level of slow flow increases the osteogenic differentiation of mesenchymal stem cells through TAZ activation. *PLoS ONE* **2014**, *9*, e92427. [CrossRef]

136. Wang, K.C.; Yeh, Y.T.; Nguyen, P.; Limqueco, E.; Lopez, J.; Thorossian, S.; Guan, K.L.; Li, Y.J.; Chien, S. Flow-dependent YAP/TAZ activities regulate endothelial phenotypes and atherosclerosis. *Proc. Natl. Acad. Sci. USA* **2016**, *113*, 11525–11530. [CrossRef] [PubMed]

137. Sabine, A.; Bovay, E.; Demir, C.S.; Kimura, W.; Jaquet, M.; Agalarov, Y.; Zangger, N.; Scallan, J.P.; Graber, W.; Gulpinar, E.; et al. FOXC2 and fluid shear stress stabilize postnatal lymphatic vasculature. *J. Clin. Investig.* **2015**, *125*, 3861–3877. [CrossRef] [PubMed]

138. Elosegui-Artola, A.; Andreu, I.; Beedle, A.E.M.; Lezamiz, A.; Uroz, M.; Kosmalska, A.J.; Oria, R.; Kechagia, J.Z.; Rico-Lastres, P.; Le Roux, A.L.; et al. Force Triggers YAP Nuclear Entry by Regulating Transport across Nuclear Pores. *Cell* **2017**, *171*, 1397.e1314–1410.e1314. [CrossRef] [PubMed]

139. Sansores-Garcia, L.; Bossuyt, W.; Wada, K.; Yonemura, S.; Tao, C.; Sasaki, H.; Halder, G. Modulating F-actin organization induces organ growth by affecting the Hippo pathway. *Embo J.* **2011**, *30*, 2325–2335. [CrossRef]

140. Fan, R.; Kim, N.G.; Gumbiner, B.M. Regulation of Hippo pathway by mitogenic growth factors via phosphoinositide 3-kinase and phosphoinositide-dependent kinase-1. *Proc. Natl. Acad. Sci. USA* **2013**, *110*, 2569–2574. [CrossRef]

141. Wang, X.; Freire Valls, A.; Schermann, G.; Shen, Y.; Moya, I.M.; Castro, L.; Urban, S.; Solecki, G.M.; Winkler, F.; Riedemann, L.; et al. YAP/TAZ Orchestrate VEGF Signaling during Developmental Angiogenesis. *Dev. Cell* **2017**, *42*, 462–478. [CrossRef]

142. Rizvi, S.; Yamada, D.; Hirsova, P.; Bronk, S.F.; Werneburg, N.W.; Krishnan, A.; Salim, W.; Zhang, L.; Trushina, E.; Truty, M.J.; et al. A Hippo and Fibroblast Growth Factor Receptor Autocrine Pathway in Cholangiocarcinoma. *J. Biol. Chem.* **2016**, *291*, 8031–8047. [CrossRef]

143. Kapoor, A.; Yao, W.; Ying, H.; Hua, S.; Liewen, A.; Wang, Q.; Zhong, Y.; Wu, C.J.; Sadanandam, A.; Hu, B.; et al. Yap1 activation enables bypass of oncogenic Kras addiction in pancreatic cancer. *Cell* **2014**, *158*, 185–197. [CrossRef] [PubMed]

144. Jang, W.; Kim, T.; Koo, J.S.; Kim, S.K.; Lim, D.S. Mechanical cue-induced YAP instructs Skp2-dependent cell cycle exit and oncogenic signaling. *EMBO J.* **2017**, *36*, 2510–2528. [CrossRef] [PubMed]

145. Di Agostino, S.; Sorrentino, G.; Ingallina, E.; Valenti, F.; Ferraiuolo, M.; Bicciato, S.; Piazza, S.; Strano, S.; Del Sal, G.; Blandino, G. YAP enhances the pro-proliferative transcriptional activity of mutant p53 proteins. *EMBO Rep.* **2016**, *17*, 188–201. [CrossRef] [PubMed]

146. Mizuno, T.; Murakami, H.; Fujii, M.; Ishiguro, F.; Tanaka, I.; Kondo, Y.; Akatsuka, S.; Toyokuni, S.; Yokoi, K.; Osada, H.; et al. YAP induces malignant mesothelioma cell proliferation by upregulating transcription of cell cycle-promoting genes. *Oncogene* **2012**, *31*, 5117–5122. [CrossRef] [PubMed]

147. Zanconato, F.; Forcato, M.; Battilana, G.; Azzolin, L.; Quaranta, E.; Bodega, B.; Rosato, A.; Bicciato, S.; Cordenonsi, M.; Piccolo, S. Genome-wide association between YAP/TAZ/TEAD and AP-1 at enhancers drives oncogenic growth. *Nat. Cell Biol.* **2015**, *17*, 1218–1227. [CrossRef]

148. Miyazono, K.; Katsuno, Y.; Koinuma, D.; Ehata, S.; Morikawa, M. Intracellular and extracellular TGF-beta signaling in cancer: Some recent topics. *Front. Med.* **2018**, *12*, 387–411. [CrossRef]

149. Bora-Singhal, N.; Nguyen, J.; Schaal, C.; Perumal, D.; Singh, S.; Coppola, D.; Chellappan, S. YAP1 Regulates OCT4 Activity and SOX2 Expression to Facilitate Self-Renewal and Vascular Mimicry of Stem-Like Cells. *Stem Cells* **2015**, *33*, 1705–1718. [CrossRef] [PubMed]

150. Shao, D.D.; Xue, W.; Krall, E.B.; Bhutkar, A.; Piccioni, F.; Wang, X.; Schinzel, A.C.; Sood, S.; Rosenbluh, J.; Kim, J.W.; et al. KRAS and YAP1 converge to regulate EMT and tumor survival. *Cell* **2014**, *158*, 171–184. [CrossRef]

151. Cordenonsi, M.; Zanconato, F.; Azzolin, L.; Forcato, M.; Rosato, A.; Frasson, C.; Inui, M.; Montagner, M.; Parenti, A.R.; Poletti, A.; et al. The Hippo transducer TAZ confers cancer stem cell-related traits on breast cancer cells. *Cell* **2011**, *147*, 759–772. [CrossRef]

152. Lehmann, W.; Mossmann, D.; Kleemann, J.; Mock, K.; Meisinger, C.; Brummer, T.; Herr, R.; Brabletz, S.; Stemmler, M.P.; Brabletz, T. ZEB1 turns into a transcriptional activator by interacting with YAP1 in aggressive cancer types. *Nat. Commun.* **2016**, *7*, 10498. [CrossRef]

153. Tang, Y.; Feinberg, T.; Keller, E.T.; Li, X.Y.; Weiss, S.J. Snail/Slug binding interactions with YAP/TAZ control skeletal stem cell self-renewal and differentiation. *Nat. Cell Biol.* **2016**, *18*, 917–929. [CrossRef]

154. Bartucci, M.; Dattilo, R.; Moriconi, C.; Pagliuca, A.; Mottolese, M.; Federici, G.; Benedetto, A.D.; Todaro, M.; Stassi, G.; Sperati, F.; et al. TAZ is required for metastatic activity and chemoresistance of breast cancer stem cells. *Oncogene* **2015**, *34*, 681–690. [CrossRef] [PubMed]

155. Song, S.; Ajani, J.A.; Honjo, S.; Maru, D.M.; Chen, Q.; Scott, A.W.; Heallen, T.R.; Xiao, L.; Hofstetter, W.L.; Weston, B.; et al. Hippo coactivator YAP1 upregulates SOX9 and endows esophageal cancer cells with stem-like properties. *Cancer Res.* **2014**, *74*, 4170–4182. [CrossRef] [PubMed]

156. Ooki, A.; Del Carmen Rodriguez Pena, M.; Marchionni, L.; Dinalankara, W.; Begum, A.; Hahn, N.M.; VandenBussche, C.J.; Rasheed, Z.A.; Mao, S.; Netto, G.J.; et al. YAP1 and COX2 Coordinately Regulate Urothelial Cancer Stem-like Cells. *Cancer Res.* **2018**, *78*, 168–181. [CrossRef] [PubMed]

157. Basu-Roy, U.; Bayin, N.S.; Rattanakorn, K.; Han, E.; Placantonakis, D.G.; Mansukhani, A.; Basilico, C. Sox2 antagonizes the Hippo pathway to maintain stemness in cancer cells. *Nat. Commun.* **2015**, *6*, 6411. [CrossRef]

158. Zhao, X.; Sun, J.; Su, W.; Shan, H.; Zhang, B.; Wang, Y.; Shabanova, A.; Shan, H.; Liang, H. Melatonin Protects against Lung Fibrosis by Regulating the Hippo/YAP Pathway. *Int. J. Mol. Sci.* **2018**, *19*. [CrossRef]

159. Perumal, N.; Perumal, M.; Halagowder, D.; Sivasithamparam, N. Morin attenuates diethylnitrosamine-induced rat liver fibrosis and hepatic stellate cell activation by co-ordinated regulation of Hippo/Yap and TGF-beta1/Smad signaling. *Biochimie* **2017**, *140*, 10–19. [CrossRef]

160. Zhang, K.; Chang, Y.; Shi, Z.; Han, X.; Han, Y.; Yao, Q.; Hu, Z.; Cui, H.; Zheng, L.; Han, T.; et al. omega-3 PUFAs ameliorate liver fibrosis and inhibit hepatic stellate cells proliferation and activation by promoting YAP/TAZ degradation. *Sci. Rep.* **2016**, *6*, 30029. [CrossRef] [PubMed]

161. Zhou, Y.; Huang, X.; Hecker, L.; Kurundkar, D.; Kurundkar, A.; Liu, H.; Jin, T.H.; Desai, L.; Bernard, K.; Thannickal, V.J. Inhibition of mechanosensitive signaling in myofibroblasts ameliorates experimental pulmonary fibrosis. *J. Clin. Investig.* **2013**, *123*, 1096–1108. [CrossRef] [PubMed]

162. Bao, Y.; Nakagawa, K.; Yang, Z.; Ikeda, M.; Withanage, K.; Ishigami-Yuasa, M.; Okuno, Y.; Hata, S.; Nishina, H.; Hata, Y. A cell-based assay to screen stimulators of the Hippo pathway reveals the inhibitory effect of dobutamine on the YAP-dependent gene transcription. *J. Biochem.* **2011**, *150*, 199–208. [CrossRef] [PubMed]

163. Sorrentino, G.; Ruggeri, N.; Specchia, V.; Cordenonsi, M.; Mano, M.; Dupont, S.; Manfrin, A.; Ingallina, E.; Sommaggio, R.; Piazza, S.; et al. Metabolic control of YAP and TAZ by the mevalonate pathway. *Nat. Cell Biol.* **2014**, *16*, 357–366. [CrossRef] [PubMed]

164. Ding, N.; Hah, N.; Yu, R.T.; Sherman, M.H.; Benner, C.; Leblanc, M.; He, M.; Liddle, C.; Downes, M.; Evans, R.M. BRD4 is a novel therapeutic target for liver fibrosis. *Proc. Natl. Acad. Sci. USA* **2015**, *112*, 15713–15718. [CrossRef] [PubMed]

165. Yu, H.X.; Yao, Y.; Bu, F.T.; Chen, Y.; Wu, Y.T.; Yang, Y.; Chen, X.; Zhu, Y.; Wang, Q.; Pan, X.Y.; et al. Blockade of YAP alleviates hepatic fibrosis through accelerating apoptosis and reversion of activated hepatic stellate cells. *Mol. Immunol.* **2019**, *107*, 29–40. [CrossRef] [PubMed]

Review

The Hippo Pathway in Prostate Cancer

Omar Salem [1,2] and Carsten G. Hansen [1,2,*]

[1] Queen's Medical Research Institute, University of Edinburgh Centre for Inflammation Research, Edinburgh bioQuarter, 47 Little France Crescent, Edinburgh EH16 4TJ, UK; omar.m.salem@ed.ac.uk
[2] Institute for Regeneration and Repair, University of Edinburgh, Edinburgh bioQuarter, 5 Little France Drive, Edinburgh EH16 4UU, UK
* Correspondence: carsten.g.hansen@ed.ac.uk

Received: 19 March 2019; Accepted: 19 April 2019; Published: 23 April 2019

Abstract: Despite recent efforts, prostate cancer (PCa) remains one of the most common cancers in men. Currently, there is no effective treatment for castration-resistant prostate cancer (CRPC). There is, therefore, an urgent need to identify new therapeutic targets. The Hippo pathway and its downstream effectors—the transcriptional co-activators, Yes-associated protein (YAP) and its paralog, transcriptional co-activator with PDZ-binding motif (TAZ)—are foremost regulators of stem cells and cancer biology. Defective Hippo pathway signaling and YAP/TAZ hyperactivation are common across various cancers. Here, we draw on insights learned from other types of cancers and review the latest advances linking the Hippo pathway and YAP/TAZ to PCa onset and progression. We examine the regulatory interaction between Hippo-YAP/TAZ and the androgen receptor (AR), as main regulators of PCa development, and how uncontrolled expression of YAP/TAZ drives castration resistance by inducing cellular stemness. Finally, we survey the potential therapeutic targeting of the Hippo pathway and YAP/TAZ to overcome PCa.

Keywords: hippo pathway; YAP/TAZ; prostate cancer; castration resistance; signal cross-talk; feedback loops

---

1. Introduction

Prostate cancer (PCa) is worldwide one of the most prevalent cancers in men, with over one million new cases reported annually [1,2]. Initially, premalignant prostatic intraepithelial neoplasia (PIN) lesions form, which develop into advanced localized PCa followed by metastasis [3,4]. The prostate gland consists of luminal, basal, and neuroendocrine cells embedded in fibromuscular stroma (Figure 1) [4,5]. The most commonly reported PCa is acinar adenocarcinoma, which is androgen receptor (AR)-positive and arises from the prostate gland secretory luminal cell lineage [4,5]. A smaller subset of PCa develops from the neuroendocrine cell lineage [4,5]. Neuroendocrine tumors are classified as small-cell carcinoma and are more prevalent following recurrence [4,5] (Figure 1).

Early stages of prostate cancer are managed by surveillance, as well as classical approaches such as radiation therapy and surgery [6,7]. However, the first line of treatment of locally advanced or metastatic prostate cancer is androgen deprivation therapy (ADT) [8–10]. Although ADT is effective initially, patients develop castration-resistant prostate cancer (CRPC) within 1–3 years. CRPC is defined as PCa that progressed despite castrate serum testosterone levels (<50 ng/dL) [11,12].

Clinical management of CRPC is challenging, which is partly due to the molecular variation between patients [13]. Several mechanisms activate AR in CRPC patients [12,14]. These include AR mutations and amplification, which leads to AR hypersensitivity or promiscuity, causing the activation of AR in response to low androgen levels and non-androgenic steroids [15,16]. PCa expressing some AR splice variants also overcomes ADT. These alternative AR variants are constitutively active due to the loss of the C-terminal part of the AR ligand-binding domain [14,17]. Additionally, CRPC patients have relatively higher androgen levels compared to healthy males [18], which is due to intratumoral steroidogenesis, as well as altered adrenal steroid production [18,19]. Notably, ligand-independent activation of AR also plays prominent roles in CRPC [20]. Despite recent efforts to optimize current ADT strategies, CRPC remains a global burden. Advanced PCa is characterized by poor prognosis and high mortality rate, causing approximately 350,000 global deaths annually [1,2]. There is, therefore, an urgent need to unravel the complex mechanism underlying PCa development, progression, and ADT resistance in order to identify new druggable targets.

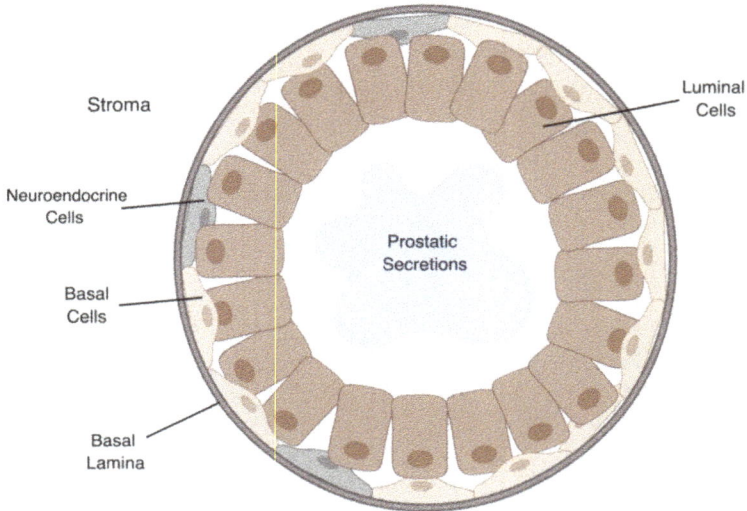

Figure 1. Representation of different cell types in the prostate gland.

The Hippo signaling pathway is a major player in stem cells and cancer biology [21,22]. The Hippo signaling cascade, identified through studies of tumor suppressors in the fruitfly, *Drosophila melanogaster* [23], is conserved across species, including humans [24]. It acts as a crucial regulator of cell growth and proliferation, organ development, cellular homeostasis, and regeneration [22,25]. The Hippo pathway is regulated by multiple signals such as, cell-density/polarity, mechanotransduction, nutrients, and via G-protein-coupled receptors [26–29]. Importantly, apparent kinase cascade independent regulation of Yes-associated protein (YAP)/ transcriptional coactivator with PDZ-binding motif (TAZ) also takes place [30–32] (Box 1). The upregulation of the Hippo pathway downstream effectors, YAP/TAZ, is central in a variety of solid tumors [21,25,29,33,34]. Prominently, the implications of elevated activity of YAP/TAZ in prostate cancer (PCa) are becoming apparent.

In this review article, we summarize the expanding evidence linking YAP and TAZ to PCa development, hormone inhibition resistance, and metastasis. Additionally, we highlight the role of the Hippo pathway in regulating prostate cancer stem cells and the importance of Hippo–YAP/TAZ as a potential therapeutic target for PCa, and we stress hitherto outstanding questions of how the dysregulated Hippo pathway drives PCa onset and development.

Box 1. Yes-associated protein (YAP)/PDZ-binding motif (TAZ) Regulation by the Canonical Hippo Pathway.

The Hippo pathway consists of an upstream serine-threonine kinase cascade. The chief kinases are MST1/2 (the mammalian Hippo homolog) and the MAP4K family of kinases, which phosphorylate and, in turn, activates large tumor suppressor (LATS1/2) [35–45]. When the Hippo kinases are "active", LATS1/2 phosphorylate and thereby inhibit the transcriptional co-activator YAP [46] and its paralog TAZ [47], causing their cytoplasmic retention by protein 14-3-3, AMOT, or degradation [30,48–51]. In contrast, when the kinase module is "inactive", dephosphorylation of YAP/TAZ occurs, which allows YAP/TAZ to translocate to the nucleus and regulate transcription. YAP/TAZ-mediated transcriptional regulation is predominantly via direct binding to the transcription factors TEAD1–TEAD4 [52–54]. As a consequence, the expression of multiple proliferative and antiapoptotic genes occurs, such as *connective tissue growth factor (CTGF)* and *cysteine-rich angiogenic factor (CYR61)* [52–54]. Additional kinases were also shown to directly phosphorylate and thereby regulate YAP/TAZ, such as SRC [55–58], Nuclear Dbf2-related 1/2 (NDR1/2) [59], c-Jun N-terminal kinase (JNK) [60,61], 5′ adenosine monophosphate-activated protein kinase (AMPK) [62–64], and Nemo-like kinase (NLK) [65,66]. Finally, kinase-independent regulation of YAP/TAZ is also taking place [30–32].

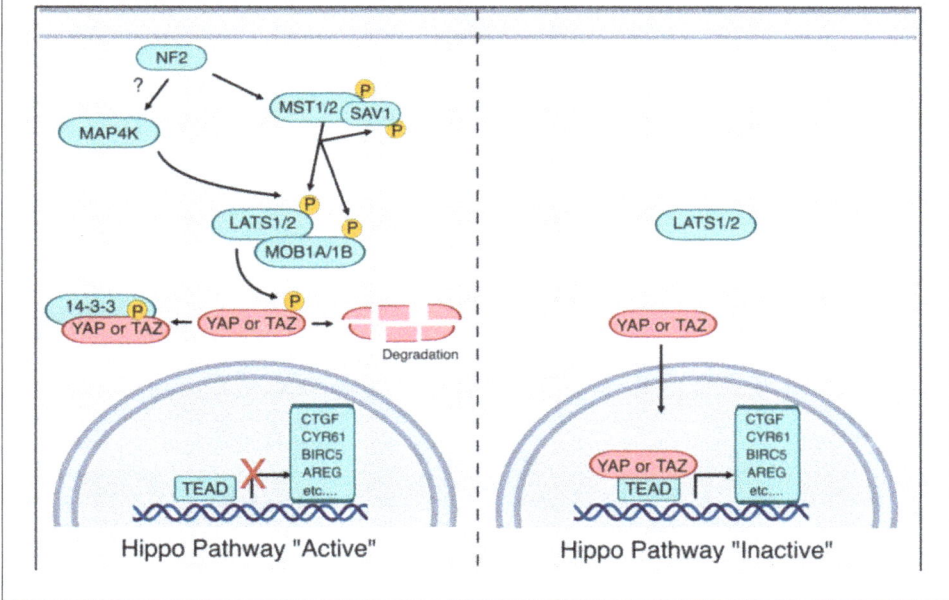

Hippo Pathway "Active"

Hippo Pathway "Inactive"

2. Hippo/YAP Key Players in Early Stages of Prostate Cancer

Elevated YAP activity is observed in most solid tumors [34], and hyperactive YAP induces the formation of several carcinomas including liver, lung, breast, sarcoma, and pancreas [21,22,33,67]. YAP is also identified as a clinical marker for PCa progression [68] and regulator of CRPC [69]. YAP levels correlate with patients' Gleason score, prostate-specific antigen (PSA) levels, and extraprostatic extensions [68,70] (Figure 2).

Additionally, exogenous overexpression of YAP in normal prostate epithelial cells induces colony formation and increased migration in three-dimensional (3D) cultures [71]. How YAP becomes hyperactivated and drives PCa initiation and development is currently not clear, but several mechanisms were recently implicated (Figure 3).

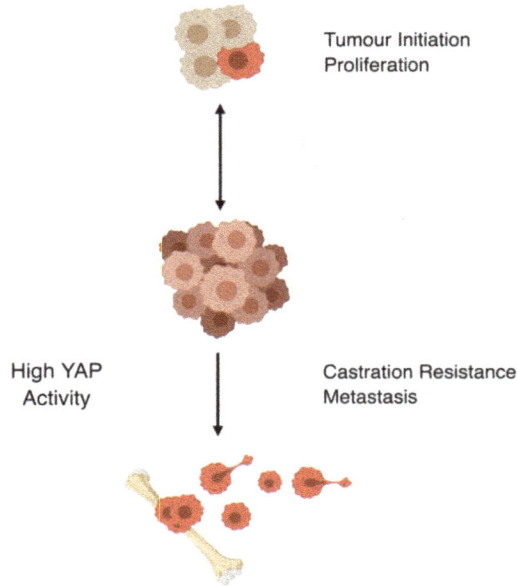

Figure 2. Schematic overview of YAP activity levels across different stages of prostate cancer (PCa). YAP regulates multiple stages of PCa [68,70,71].

2.1. E26 Transformation-Specific (ETS) Transcription Factors

ETS-regulated gene (ERG) is a transcription factor that belongs to the E26 transformation-specific (ETS) family and drives proliferation, apoptosis, and angiogenesis [72]. ERG overexpression in PCa results due to the fusion on chromosome 21q22 between the first exon of the androgen regulated gene *TMPRSS2* and the coding sequence of *ERG* [72,73]. This is a relatively frequent translocation, present in approximately 40–50% of PCa patients [74]. ERG overexpression results in the development of PCa tumors in aged mice [75]. Mechanistically, ERG induces *YAP* promoter activity in the hormone refractory PCa cell model (VCaP cells) [76], and ERG knockdown results in a decrease in YAP protein levels [75] (Figure 3). ERG both transactivates TEAD4 and directly binds to the *CTGF* promoter region, thereby inducing *CTGF* expression [75]. ETV1, an additional member of the ETS transcription factors, induces *YAP* expression in LNCaP cells by recruiting the lysine-specific demethylase (JMJD2A) to the *YAP* promoter [77] (Figure 3). ETV1-induced *YAP* expression in vivo causes PIN lesion formation, which, when combined with a single copy loss of phosphatase and tensin homolog (PTEN), progresses to malignant carcinoma [77]. PTEN is a negative regulator of the phosphoinositide 3-kinase/protein kinase B (PI3K/AKT) pathway, which controls proliferation and apoptosis [78]. PTEN deletions are identified in around 20% of primary PCa and 50% of advanced PCa [79].

2.2. Polarity Protein (Par3)

Epithelial cells are polarized cells with distinct functional apical and basolateral membrane domains [80]. Par3, among other polarity proteins, is a major regulator of epithelial cell structure and function [81,82]. Moreover, loss of Par3 is present in a variety of epithelial tumors [82]. Par3 loss leads to the formation of high-grade PIN lesions in vivo due to high YAP activity [82]. In this context, Par3 interrupts the NF2-derived recruitment of LATS1 to the plasma membrane [83]. Additionally, PIN lesions progress to PCa adenocarcinoma when combined with LATS1 loss in the Par3/LATS1 knockout (KO) murine model [83]. Contradictorily, Par3 sequesters the potent Hippo kinase cascade activator, kidney- and brain-expressed protein (KIBRA), and prevents it from forming a complex with NF2 [84].

As a result, KIBRA complexes with Par3 and atypical protein kinase C (aPKC). Knockdown of Par3 expression in PCa cells restores LATS1 and YAP phosphorylation levels, resulting in a lower migration rate in vitro and lower rate of metastasis in vivo [84]. The findings from both studies suggest that Par3 expression is lost during tumor initiation, but might be retained in advanced PCa triggering metastasis [83,84] (Figure 3). The regulation of Par3 in PCa is not fully understood and the interplay between the Hippo pathway and polarity proteins in PCa requires further investigation.

2.3. Heat Shock Proteins

Heat shock proteins (Hsps) are cellular stress modalities that regulate signaling and homeostasis [85]. Hsp expression is upregulated in response to chemotherapy and hormonal therapy [86]. The heat shock protein 27 (Hsp27) is elevated in a variety of tumors such as lung, breast, and cervical cancers [86]. Hsp27 is utilized by PCa tumor cells to resist apoptosis following androgen deprivation [87]. Hsp27 acts as a regulator of MST1 via promoting its ubiquitin-mediated degradation. As a result, LATS1 and MOB kinase activator 1 (MOB1) phosphorylation are reduced, causing YAP dephosphorylation and nuclear translocation [88] (Figure 3). However, it is worth noting that ablation of both MST1 and MST2 is needed both in vivo and in vitro to generally impair LATS1/2 activation [89–91], and, in some contexts, inhibition of the MAP4K family of kinases is also necessary to reduce overall LATS1/2 activity and thereby to increase YAP activity [37,42,45].

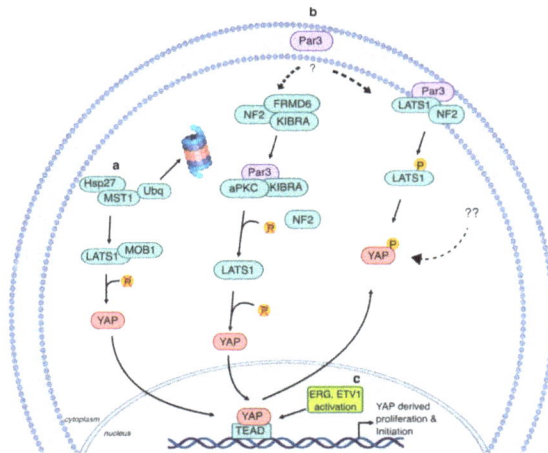

Figure 3. Mechanisms of YAP regulation in early stages of prostate cancer. **a.** Heat shock protein 27 (Hsp27) induces MST1 ubiquitin-mediated degradation, which in turn causes LATS1 and MOB1 dephosphorylation and thereby inactivation, consequently inducing YAP nuclear translocation [88]. **b.** Two different mechanisms were proposed by which polarity protein (Par3) regulates YAP; (1) Par3 inhibits YAP activity through inducing the recruitment of Neurofibromatosis type 2 (NF2/Merlin) and LATS1 to the membrane. As a result, LATS1 is activated, which induces YAP phosphorylation and cytoplasmic retention [83]. (2) Par3 induces YAP activation through the dissociation of kidney- and brain-expressed protein (KIBRA) from its canonical complex (KIBRA/NF2/ FERM domain-containing protein 6 (FRDM6)) and drives the recruitment of KIBRA to the Par3/aPKC/KIBRA complex. Thus, the interaction between KIBRA and LATS1 is disrupted, which induces LATS1 dephosphorylation and thereby YAP activation [84]. **c.** E26 transformation-specific (ETS) transcription factors trigger YAP induction. (1) ETS-regulated gene (ERG) activation drives YAP activation in old aged mice. ERG induces YAP and TEAD4 promoter activity and thereby triggers YAP target gene expression [75]. (2) ETS translocation variant 1 (ETV1) drives *YAP* activation by recruiting lysine specific demethylase (JMJD2A) to the *YAP* promoter [77].

3. The Hippo Pathway Promotes Castration Resistance and Metastasis in Prostate Cancer

3.1. Androgen Receptor—Regulator of CRPC Progression

The androgen receptor (AR) is a transcription factor that belongs to the superfamily of steroid receptor hormones [92]. AR signaling is essential for prostate development and homeostasis [93]. During absence of androgens, inactive AR resides in the cytoplasm bound to heat shock proteins [92]. Upon binding of dihydrotestosterone (DHT), AR dissociates and translocates to the nucleus to induce gene expression [92]. AR-mediated gene expression occurs via multiple AR coactivators and AR-mediated recognition of androgen response elements (AREs) on the target gene promoter. AREs consist of two common inverted hexameric half-sites (5′–AGAACA–3′) separated by three base pairs [94]. In healthy tissue, tight homeostatic androgen signaling between stromal and epithelial cells regulate the prostate gland function [93,95]. Disrupted AR signaling is a key event in PCa initiation, progression, and development of castration resistance [5,96,97]. However, to date, the exact molecular mechanism via which CRPC develops is yet to be fully explored [93].

3.2. AR and YAP Colocalization

The role of the Hippo pathway in CRPC development and AR regulation recently gained momentum (Figure 4). Coimmunoprecipitation and immunofluorescence microscopy revealed that AR and YAP colocalize to and interact in the nucleus [98]. This interaction is androgen-dependent in LNCaP cells, but androgen-independent in C4-2 cells [98]. C4-2 cells are a hormone-independent subline of LNCaP cells, representing a clinical CRPC in vitro cell model [99]. Downregulation of YAP signaling results in the suppression of AR target genes, suggesting that YAP is critical for AR activity [98].

Interestingly, C4-2 cells harbor low MST1 kinase signaling, and restoring MST1 expression in these cells results in impeding the YAP–AR nuclear interaction and AR activity [98,100]. One plausible mechanism for AR upregulation in CRPC patients is, therefore, MST epigenetic silencing [101]. Cellular myelocytomatosis (c-MYC), a transcription factor that regulates cellular growth and proliferation [102], is commonly overexpressed in PCa patients, which induces tumor initiation [103]. Mechanistically, c-MYC is a regulator of the enhancer of zeste homolog 2 (EHZ2), which is a subunit of polycomb repressive complex 2 (PRC2) [104]. EZH2, a histone methyltransferase, catalyzes the trimethylation of histone 3 at lysine 27 (H3K27me3) to regulate gene expression [105]. EZH2 functions both as a transcriptional activator and repressor for specific gene sets in a cell-context-dependent manner [106]. EZH2 acts as a coactivator of the androgen receptor in CRPC [107]. c-MYC induces EZH2 activity via suppressing microRNA (miR)-26a/b, which results in MST1 promoter silencing [101]. Treating C4-2 cells with JQ1 results in downregulating c-MYC, which in turn induces MST1 expression and decreases cell survival [101]. However, combining MST1 knockdown with either c-MYC inhibition by 10058-F4 or EZH2 inhibition by GSK126 restores cell survival [101].

LATS2 and AR were, using immunohistochemistry, reported to colocalize within healthy prostate epithelium patient samples [108]. When in the nucleus, LATS2 and AR form a protein complex, which binds to *prostate-specific antigen (PSA)* promoter and enhancer regions [108]. LATS2 suppresses AR activity through hindering the NH_2- and COOH-terminal interaction within the receptor [108]. The activation status of LATS2 was not examined, and whether AR is a direct substrate for LATS2 is unknown; it is, therefore, still an outstanding question if this LATS2-mediated regulation of AR transcription is phosphorylation-dependent [108]. Importantly, LATS2 levels negatively correlate with PCa tumor stage, a conserved phenomenon with several other types of carcinomas [108–110]. Paradoxically, *LATS2* is in a range of cell types and, in vivo, a YAP/TAZ–TEAD target gene [111–113], which forms an integral component of a feedback loop that keeps YAP/TAZ–TEAD activity levels in check [111–113]. Loss of LATS2 expression, but high YAP activity [75,98] and, therefore, impaired Hippo pathway feedback in high-grade PCa might, therefore, be a defining PCa hallmark. The relatively low LATS2 levels in PCa might be due to additional YAP/TAZ–TEAD-independent transcriptional regulation and/or post-transcriptional regulation of LATS2 protein. It will be critical to establish if negative feedback loops within the Hippo pathway are prevalent in healthy prostates and, if so, why these dynamic negative feedbacks might be defective in PCa. Therapeutically reinstating these negative feedback loops within the Hippo pathway might then be a viable option. Overall, these reports show that the Hippo kinase cascade and its effector YAP are regulators of AR nuclear localization and activity.

3.3. The Hippo Pathway, Tumor Microenvironment, and Immune Response Evasion

Cross-talk between the Hippo pathway and the tumor microenvironment is widespread across multiple solid tumors and regularly operates via a feed-forward loop that drives tumor progression [21,26,114,115]. YAP/TAZ is a signaling nexus and regulates cell–cell interaction and cell–stroma interaction through inducing the expression of a range of secretory proteins such as CYR61 and CTGF [31,54,115,116], as well as of components essential for mechanoresponsive plasma membrane organelles such as caveolae [117], and components and regulatory elements of focal adhesions such as integrins and cytoskeletal tension [54,118–122]. YAP/TAZ are well-established molecular sensors of the extracellular matrix (ECM), and both sense the stiffness and composition of the ECM [31,115,118]. In vitro experiments show that cells cultured on high ECM matrix stiffness result in increased YAP/TAZ nuclear localization and target gene expression [31,115,116,123,124]. In comparison, cells grown on low ECM stiffness have a higher cytoplasmic fraction of YAP/TAZ [31,115,116,123,124]. This is particularly important in PCa, as PCa is widely recognized for its rich tumor–stroma interaction [125,126].

Downregulation of α3 integrin causes PCa progression and promotes formation of metastatic lesions via altering YAP/TAZ activity. Mechanistically, loss of α3β1 in PCa results in the inhibition of the Abelson-related gene (Arg/abl2) tyrosine kinase cascade, which dephosphorylates the p190Rho-Guanosine triphosphate (GTP)ase activating protein-(GAP)/p120RAS-GAP (p190RhoGAP/ p120RAS-GAP complex [127]. Consequently, Rho signaling is activated, ultimately causing increased YAP/TAZ levels, which promotes cellular migration in vitro and metastasis in vivo [127]. However, it is not entirely clear whether RhoA in this instance acts via the Hippo–LATS kinase cascade or independently from it. Paradoxically, α3β1 loss inhibited skin tumorigenesis in vivo [128]. Importantly, ECM regulates multiple cellular cancer properties [129] and it remains unclear whether these effects are mediated via the Hippo pathway in PCa and importantly whether ECM stiffness is inducing YAP activity in PCa. Addressing these questions might partly explain the increased YAP expression levels in PCa patients.

The ability of tumor cells to evade immune response is widely recognized to be a hallmark of cancer progression [130]. Intriguingly, YAP is partly responsible for this in PCa [131]. In a PTEN/ SMAD4 knockout PCa mouse model, YAP levels are elevated [131]. In this model, YAP expression results in myeloid-derived suppressor cell (MDSC) recruitment via the CXCL5/CXCR2 axis. MDSCs mediate tumor immune response evasion through suppressing T-cell activation, proliferation, and viability [132]. YAP–TEAD directly binds to the *CXCL5* promoter, inducing *CXCL5* expression. Either MDSC depletion, or inhibition of YAP or CXCL5/CXCR2 activity halts tumor progression [131]. Similarly, YAP hyperactivation is observed in the Kras/p53 knockout pancreatic cancer mouse model, which stimulates chemokines expression and thereby recruitment of MDSCs to tumors [133]. Of note, YAP governs the recruitment of tumor-infiltrating type II macrophages (M2) in liver carcinoma, which promotes tumorigenesis by avoiding immune clearance [134] (Figure 4). However, how the Hippo pathway gets dysregulated and drives PCa tumor–stroma interactions is still not fully understood.

3.4. TAZ's Role in Metastasis

YAP and TAZ are modulators of cell motility and cytoskeletal dynamics in a feedback dependent manner [119]. TAZ in particular is a potent regulator of epithelial–mesenchymal transition (EMT) in most types of solid cancers, including ovarian cancer, glioma, and breast cancer [135–140]. The role of TAZ in PCa tumor progression and the regulatory nature between TAZ and AR is not well described.

TAZ overexpression induces malignant transformation of the non-cancerous prostate epithelial cells, RWPE-1 [141]. Knockdown of TAZ in PCa cells causes reduction in migratory rate in two-dimensional (2D) cultures, as well as lower metastatic rate when injected in vivo. Endogenous expression of TAZ is regulated by ETS transcription factors members ETV1/4/5 (Figure 4) [141]. ETV1/4/5 induce TAZ gene expression, which results in the expression of SH3 domain-binding protein 1 (SH3BP1) via TAZ–TEAD. SH3BP1 belongs to the RhoGAP protein family and regulates Rac signaling to modulate cytoskeletal dynamics and cell motility [142]. So far, the PCa stage-specific levels of TAZ in PCa tumor samples are yet to be investigated.

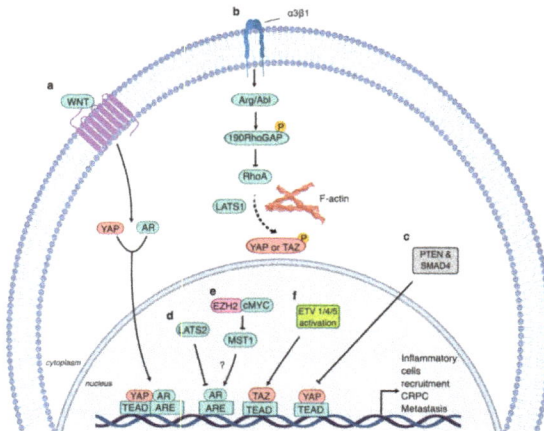

Figure 4. The Hippo pathway regulation of advanced prostate cancer. **a.** During androgen deprivation, Wingless (WNT) signaling drives the nuclear translocation of YAP and AR, resulting in YAP and AR target gene induction [143]. Additionally, YAP and AR colocalize in the nucleus and induce gene expression independently from WNT signaling or androgen availability [98]. **b.** α3β1 integrin stimulates the kinase activity of Arg/Abl, which phosphorylates 190RhoGAP, resulting in the inhibition of RhoA GTPases. Consequently, YAP/TAZ are phosphorylated via LATS1 activity and/or actin rearrangement and retained in the cytoplasm. α3β1 loss results in the disruption of this signaling cascade, inducing prostate cellular migration and metastasis [127]. **c.** PTEN and SMAD4 activity loss results in YAP hyperactivation. YAP signaling induces the recruitment of inflammatory cells [131]. **d.** LAST2 impedes AR receptor activity and restricts the binding of AR to the *prostate-specific antigen (PSA)* promoter [108]. **e.** EZH2 and c-MYC cooperate to induce methylation and silencing of *MST1*, which might induce AR activity [101]. **g.** ETV1/4/5 activate *TAZ*, triggering metastasis via the induction of *SH3BP1* [141].

4. The Hippo Pathway's Role in Prostate Cancer Stem Cells

The development of CRPC following androgen deprivation therapy (ADT) is often inevitable [144]. CRPC likely develops from the prostate cancer stem cells (PCSCs), a subset of cells within the tumor which regulate initiation, but importantly also recurrence [145]. PCSCs were successfully isolated from patient tissue samples on the basis of their $\alpha_2\beta_1{}^{hi}$CD133$^+$CD44$^+$ phenotype [146–149]. PCSCs have a high proliferation rate and increased ability of colony formation in 3D cultures, as well as an ability to form prostate-like structures when injected in immunocompromised mice compared to CD44$^-$ and CD133$^-$ cells [146–149]. The Hippo pathway regulates cancer stem cells (CSCs) within a variety of tumors [136,150,151]. Interestingly, PC3 and DU145 cells resistant to the chemotherapeutic agent docetaxel possess a CD44$^+$ phenotype. In this context, CD44 increases cellular migration rate in 2D cultures via inducing YAP, CYR61, and CTGF expression [152].

The stemness regulator microRNA, cluster miR-302–367, downregulates LATS2, which results in YAP dephosphorylation and nuclear translocation [153]. Additionally, miR-302–367 overexpression in LNCaP cells induces their capacity to form spheres in 2D cultures and xenograft tumors when injected into castrated mice [153]. Cyclic guanosine monophosphate (cGMP)-specific phosphodiesterase type 5 (PDE5) also induces stemness via the Hippo pathway [154]. Pharmacological PDE5 inhibition or inhibition via endogenous nitric oxide results in the activation of cGMP-dependent protein G (PKG); this activates MST1/LATS1 phosphorylation causing TAZ cytoplasmic retention and degradation [154]. AR further inhibits the transcriptional activity of YAP in LNCaP, as well as in the serially propagated castration-induced regression derived 22rv1 cells. Mechanistically, AR complexes with EZH2 and DNA methyltransferase 3 (DNMT3a) at the YAP promoter, causing its methylation and silencing [155]. In this sense, during androgen deprivation therapy, AR inhibition results in YAP transcriptional activation. YAP expression results in the transcription of stemness-stimulating genes in a TEAD-dependent manner, which induces sphere formation in vitro [155]. Additionally, inhibiting YAP activity in vivo prevents PCa recurrence in castrated TRAMP mice [155] (Figure 5).

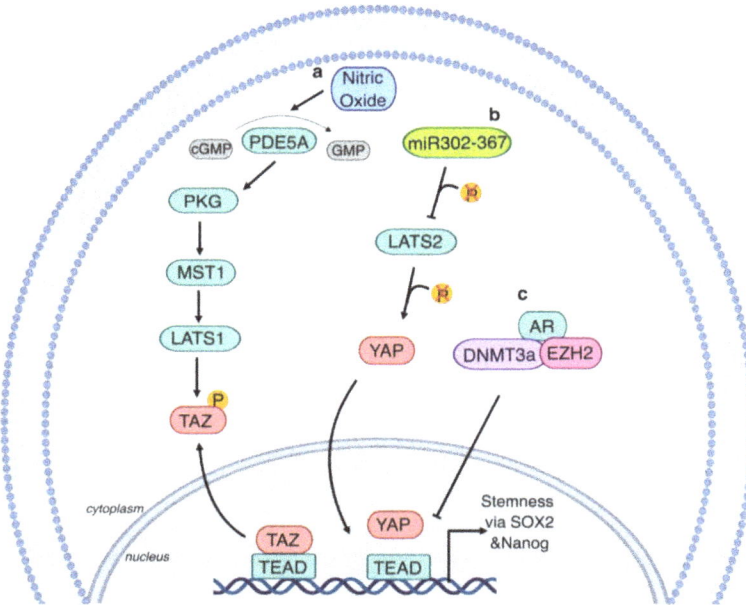

Figure 5. YAP/TAZ regulate prostate cancer stem cells (PCSCs). **a.** Inhibition of the stemness regulator cyclic GMP-specific phosphodiesterase type 5 (PDE5) by nitric oxide causes activation of cGMP-dependent protein G (PKG), which activates MST1/LATS1 and causes TAZ phosphorylation [154]. **b.** Stemness regulator microRNA (miR)-302–367 cluster induces LATS2 dephosphorylation which results in YAP nuclear translocation [153]. **c.** DNMT3a and EZH2 form a heterotrimeric complex with AR, which translocates to the *YAP* promoter inducing its silencing [155].

5. Targeting the Hippo Pathway for Prostate Cancer Therapy

5.1. Targeting YAP/TAZ–TEAD

The Hippo pathway is a critical regulator of several hallmarks of PCa. Targeting Hippo–YAP/TAZ clinically, therefore, has therapeutic potential. As YAP/TAZ are transcriptional coactivators that principally function via binding to the TEAD family of transcription factors [52,53,156–158], the most direct route to target the Hippo pathway is via this interaction [158]. Verteporfin is a small-molecule inhibitor of this YAP/TAZ–TEAD interaction [159]. Verteporfin suppresses CRPC tumor growth and PCSC proliferation, which ultimately also prevents recurrence [75,98,155]. Although Verteporfin is used for macular degeneration treatment, its future use for cancer therapeutics is hampered by Verteporfin's low solubility and low target affinity, which makes it generally toxic [160,161]. Vestigial-like 4 (VGLL4) is a tumor suppressor that competitively binds to TEAD via its tondu domain (TDU), thereby preventing YAP from mediating transcription [162,163]. VGLL4-mimicking peptide (super TDU) abrogates YAP binding to TEAD4, which has anti-tumor effects in gastric cancer patient-derived cells and in vivo in the gastric cancer mouse model driven by *Helicobacter pylori* infection [164]. A YAP-like peptide (17-mer) was designed aiming to impede YAP–TEAD binding. The 17-mer peptide has higher affinity for TEAD1 compared to YAP [165,166]. Although targeting the YAP–TEAD interaction appears to be the most straightforward route toward targeting the Hippo pathway, to date, none of the discovered agents are approved for cancer therapeutics.

5.2. Statins

Statins are a class of US food and drug administration (FDA)-approved drugs for hypercholesterolemia treatment [167]. Statins inhibit the enzyme, 3-hydroxy-3-methyl-glutaryl-coenzyme A (HMG-CoA) reductase, which prevents the conversion of HMG-CoA to mevalonic acid [168]. Subsequently, statins reduce the synthesis of geranylgeranyl pyrophosphate, which is required for Rho GTPase activity [169,170]. Statins induce YAP phosphorylation through Rho GTPase activity and actin rearrangement [169,170]. In vitro, statins induce gap 1 (G1) cell-cycle arrest and apoptosis in the PCa cell line, C4-2B [171]. Importantly, retrospective studies in a large Taiwanese cohort of statin-treated heart disease patients showed decreased incidence of PCa [172]. Furthermore, statins were recently identified to reduce PCa aggressiveness and metastasis incidence significantly in a retrospective study of a large cohort of Saskatchewan men [173]. Similarly, the occurrence of breast, ovarian, colorectal, and liver cancer is also reported to be lower in statin users [167]. Nonetheless, it remains unclear whether this statin-based clinical manifestation is mediated via the Hippo pathway.

5.3. Hippo Kinase Activators

The rapidly accelerated fibrosarcoma (RAF) family of serine/threonine kinases acts upstream of the MST kinases [174]. RAF-1 suppresses apoptosis by sequestering and preventing MST2 phosphorylation [174]. Inhibition of RAF-1, therefore, results in the activation of MST2. ISIS 1532 oligonucleotide was designed to target the 3' untranslated region of cRaf messenger RNA (mRNA) [175,176]. In preclinical trials, ISIS 1532 inhibited lung carcinoma in in vivo mouse models [175,176]. However, three phase II clinical trials in patients with advanced PCa, and ovarian and colon cancers showed no significant response, and the agent was withdrawn from further testing [176–179]. Targeting the Hippo pathway kinases proves challenging as it is regulated by a variety of external cues and interacts with multiple signaling pathways [161]. In essence, activators of the YAP/TAZ inhibitory kinases are needed, and designing kinase activators is in general more challenging than inhibitors [180,181].

6. Signaling Cross-Talk between the Hippo Pathway and Multiple Signaling Pathways

6.1. WNT Receptor Signaling

Upon androgen deprivation, WNT signaling stimulation triggers the nuclear translocation of AR and YAP to the nucleus, which induces AR-mediated gene expression independently from β-catenin translocation [143] (Figure 4). YAP and TAZ are downstream effectors of WNT/β-catenin [182,183]. Upon WNT stimulation, YAP and TAZ are released from the WNT destruction complex and translocate to the nucleus to induce transcription [182,183]. Additionally, WNT-mediated activation of YAP/TAZ can occur independently from β-catenin via the scaffold protein, adenomatous polyposis coli (APC), which facilitates SAV1 and LATS1 phosphorylation via glycogen synthase kinase 3 β (GSK-3β) activity [184]. Importantly, APC activation mutations are reported in 5% of PCa patients [185]. Contradictorily, knockout of APC in vivo in mouse models results in prostate tumor formation [186]. Impressive work using an array of CripsR knockout cell lines, as well as mouse models, showed that alternative WNT signlling (Wnt5a/b) activates YAP/TAZ via GPCR α12/13; these G-coupled proteins signal to activate RhoGTPases that inhibit LATS1/2 activity [139,187]. Whether APC or alternative WNT signaling activates YAP/TAZ in PCa and whether this mechanism is mediated via androgen receptor signaling remains unexplored.

6.2. Mechanistic Target of Rapamycin (mTOR) Signaling

The mTOR protein is a central cell growth regulator, which is regulated by growth factors, energy levels, and nutrients. When active, mTOR stimulates biosynthetic pathways including nucleotide, protein, and lipid synthesis, while inhibiting catabolic processes, such as autophagy [188,189]. Intriguingly, PCa tumor cells with high PI3K/AKT/mTOR activity are proposed as a mechanism for prostate tumors to surpass hormone inhibition therapy [190–192]. Additionally, speckle-type POZ (pox virus and zinc finger protein) protein (SPOP) mutations, the most common mutations in primary PCa (10%) [74], induce PCa tumorigenesis via PI3K/mTOR [193]. Interestingly, YAP activates mTORC signaling in breast epithelial MCF10A cells. Mechanistically, YAP suppresses PTEN activity via miR-29 induction [194]. Consequently, in a transgene YAP mouse model, mTOR was activated, causing skin hyperplasia [194]. One of the strongest regulators of mTOR activity is amino-acid sensing; when amino-acid availability of specific amino acids is low, mTOR is switched off [189]. YAP/TAZ–TEAD induce the expression of a range of cellular amino-acid transporters [195,196], including the high-affinity hetero dimeric leucine transporter, LAT1 (encoded by *SLC7A5* and *SLC3A3*). Expression of LAT1 results in increased uptake of leucine at nutrient-limiting conditions [195], as is prevalent in tumors. Consequently, the expression of amino-acid transporters activates mTOR [195–197]. These mechanisms thereby provide a metabolic advantage for tumor cells with hyperactive YAP/TAZ. Furthermore, integrin α3 controls YAP phosphorylation and nuclear localization via the focal adhesion kinase/cell division control protein 42/protein phosphatase 1A (FAK/Cdc42/PP1A) axis, which activates mTOR [198]. Although these studies were not carried out in the context of PCa, they did indicate that YAP/TAZ–TEAD activity might be triggering the activation of mTOR in PCa. Remarkably, PTEN is a negative regulator of YAP activity in the PCa PTEN/SMAD4 knockout mouse model [131]. It is, therefore, a relevant outstanding question as to whether there is a feedback loop between PTEN suppression and YAP activity in PCa. Additionally, it remains unclear whether resistance to hormone inhibition therapy occurs via positive selection of cells with high YAP activity, which in turn induce tumorigenesis synergistically via TEAD binding and PI3K/mTOR activation.

6.3. Activator Protein (AP-1)

The activator protein (AP-1) transcription factor consists of dimeric complexes, which include the DNA-binding protein families, cellular ju-nana (c-Jun), cellular FBJ osteosarcoma oncogene (c-Fos), activating transcription factor (ATF), and cellular musculoaponeurotic fibrosarcoma (c-MAF) proteins [199,200]. AP-1 activation is mediated via a range of paracrine signaling molecules, as well as by the mitogen-activated extracellular signaling responsive kinase kinases (MEKs) [199,201,202]. AP-1 regulates multiple cellular responses such as inflammation, proliferation, and apoptosis [200,203]. Importantly, genome-wide analysis revealed that YAP/TAZ–TEAD mediate tumorigenesis by co-occupying the same genomic region occupied by AP-1 [204,205]. YAP/TAZ–TEAD and the AP-1 interaction occurs with the aid of the p160 family of steroid receptor co-activators (SRC1-3) [206]. Interestingly, SRC-3 is overexpressed in PCa, which promotes cell proliferation via AR activation [207]. Treating LNCaP cells with DHT induces the activity of c-Jun and tumor necrosis factor alpha (TNF-α) promoter activity, which contains AP-1 binding sites. In this context, the activation of AP-1 due to AR induction might synergistically be prompting YAP/TAZ target gene transcription [208]. These findings provide evidence for cross-talk between YAP/TAZ–TEAD and AP-1. Nonetheless, further investigation is required to understand the role of this signaling cross-talk in PCa development and its role in regulating AR.

7. Conclusions and Perspectives

To date, there are no identified somatic mutations of the Hippo pathway components in PCa. Furthermore, although YAP was identified to be amplified in a subset of PCa [209], it is evident that YAP/TAZ is a much more widespread contributor to PCa development. YAP and TAZ play key roles in multiple stages of PCa initiation, development, and progression, as well as regulation of AR signaling. However, the mechanistic insights into how YAP/TAZ becomes hyperactivated, how YAP/TAZ interacts with the stroma and their precise role in PCa development are currently far from fully elucidated. Obtaining further fundamental understanding of the complexity of YAP/TAZ hyperactivation in PCa onset and development is, therefore, crucial for improving future clinical interventions and care for PCa patients [210,211].

- How does YAP drive CRPC development? Androgen receptor bypass is a contributing mechanism via which PCa cells develop castration resistance [212]. Androgen-deprived PCa cells activate a variety of hormone receptors such as glucocorticoid receptor (GR) and its targets in order to overcome androgen dependence [212]. Importantly, GR signaling activates YAP in MDA-MB-231 breast cancer cells [213]. Additionally, the perplexing ability of tumors to activate steroidogenesis pathways causing AR hypersensitivity is not completely understood. Of note, YAP regulates steroidogenesis in ovarian granulosa cells [214]. Whether YAP is involved in inducing CRPC via AR bypass and intratumoral steroidogenesis, and whether YAP is essential for CRPC PCa cell survival are, to a great extent, still unexplored questions.
- The estrogen receptor (ER) plays an important role in PCa [215,216]. ERα regulates proinflammatory and pro-proliferative targets and is associated with high Gleason score [215,216]. In comparison, ERβ receptor plays an anti-inflammatory, pro-apoptotic role [215,216]. Estradiol, the estrogen receptor agonist, activates the Hippo pathway in the breast SK-BR-3 cell line via G-protein-coupled estrogen receptor (GPER) [217]. Although anatomically distinct, the molecular and clinical similarities between breast and prostate cancer [217] highlight the importance of examining if a similar cross-talk mechanism is occurring in PCa.
- Activation of the Hippo kinase cascade module is a clear direction toward utilizing the Hippo pathway therapeutically [161,181]. However, an ongoing challenge of this route is the complexity of the Hippo pathway upstream regulators. Intriguingly, in PCa, it is unclear what causes the Hippo pathway dysregulation. Delineating the upstream regulators of the Hippo pathway in a PCa-specific context might, therefore, have direct clinical relevance. Importantly, YAP is upregulated in CRPC; therefore, developing YAP activity inhibitors is an equally important therapeutic direction. Successfully controlling YAP and/or TAZ activity state therapeutically would be an immense step toward developing a personalized therapeutic strategy in CRPC.

Author Contributions: Both authors wrote and commented on iterative drafts of the manuscript.

Funding: Work on-going in the Gram Hansen lab is supported by a University of Edinburgh Chancellor's Fellowship start-up fund, as well as by the Wellcome Trust University of Edinburgh Institutional Strategic Support Fund (ISSF3).

Acknowledgments: Lab members of the Gram Hansen lab, as well as Bin-Zhi Qian, are acknowledged for insightful and valuable discussions.

Conflicts of Interest: The authors declare no conflicts of interest. The funders had no role in the topics analyzed, in the writing of the manuscript, or in the decision to publish.

Abbreviations

ADT	Androgen deprivation therapy
AMOT	Angiomotin
AMPK	5′ adenosine monophosphate-activated protein kinase
AP-1	Activator protein 1
APC	Adenomatous polyposis coli
aPKC	Atypical protein kinase C
AR	Androgen receptor
Arg/abl2	Abelson-related gene
ATF	Activating transcription factor
CD133	Cluster of differentiation 133
CD44	Cluster of differentiation 44
Cdc42	Cell division control protein 42
c-Fos	FBJ osteosarcoma oncogene
cGMP	Cyclin guanosine monophosphate
c-Jun	Cellular ju-nana
c-MAF	Musculoaponeurotic fibrosarcoma
c-MYC	Cellular myelocytomatosis
CRPC	Castration-resistant prostate cancer
CTGF	Connective tissue growth factor
CXCL5	C–X–C motif chemokine 5
CXCR2	C–C chemokine receptor type 2
CYR61	Cysteine-rich angiogenic factor
DHT	Dihydrotestosterone
DNMT3a	DNA methyltransferase 3
ECM	Extracellular matrix
EMT	Epithelial–mesenchymal transition
ER	Estrogen receptor
ERG	ETS-regulated gene
ETS	E26 transformation-specific transcription factors
ETV1/4/5	E26 transformation-specific variant 1/4/5
EZH2	Enhancer of zeste homolog 2
FAK	Focal Adhesion Kinase
FDA	US food and drug administration
FRDM6	FERM domain-containing protein 6
GAP	Guanosine triphosphate (GTP)ase activating protein
GTP	Guanosine triphosphate
GR	Glucocorticoid receptor
HMG-CoA	3-hydroxy-3-methyl-glutaryl–coenzyme A
Hsp27	Heat shock protein 27
JMJD2A	Lysine-specific demethylase
JNK	c-Jun N-terminal kinase
KIBRA	Kidney- and brain-expressed protein
LATS1/2	Large tumor suppressor 1/2
M2	Tumor infiltration type II macrophages
MAP4K	MAP kinase kinase kinase kinases
MDSCs	Myeloid-derived suppressor cells
miR302-367	microRNA cluster 302-267
MOB1	MOB kinase activator 1
MST1/2	Mammalian Hippo homolog (Ste20-like kinases)
mTOR	Mammalian target of rapamycin
NF2/Merlin	Neurofibromatosis 2
NLK	Nemo-like kinase
Par3	Polarity protein 3

PCa	Prostate cancer
PCSCs	Prostate cancer stem cells
PDE5	Cyclic GMP-specific phosphodiesterase type 5
PI3K-AKT	Phosphoinositide 3-kinase/protein kinase B
PKG	cGMP-dependent protein G
POZ	Pox virus and zinc finger protein
PPA1	Protein phosphate 1
PRC2	Polycomb repressive complex 2
PSA	Prostate-specific antigen
PTEN	Phosphatase and tensin homolog
RAC	Ras-related C3 botulinum toxin substrate 1
RAF	Rapidly accelerated fibrosarcoma family of serine/threonine kinases
RhoGAP	Rho family of GTPases
SAV1	Protein salvador homolog 1
SH3BP1	SH3 domain-binding protein 1
Super TDU	VGLL4-mimicking peptide
TAZ	Transcriptional co-activator with PDZ-binding motif
TEAD1-4	TEA domain family member 1–4
VGLL4	Vestigial-like 4
WNT	Wingless
YAP	Yes-associated protein
17-mer	YAP-like peptide

References

1. Wong, M.C.S.; Goggins, W.B.; Wang, H.H.X.; Fung, F.D.H.; Leung, C.; Wong, S.Y.S.; Ng, C.F.; Sung, J.J.Y. Global Incidence and Mortality for Prostate Cancer: Analysis of Temporal Patterns and Trends in 36 Countries. *Eur. Urol.* **2016**, *70*, 862–874. [CrossRef] [PubMed]

2. Bray, F.; Ferlay, J.; Soerjomataram, I.; Siegel, R.L.; Torre, L.A.; Jemal, A. Global cancer statistics 2018: GLOBOCAN estimates of incidence and mortality worldwide for 36 cancers in 185 countries. *CA Cancer J. Clin.* **2018**, *68*, 394–424. [CrossRef] [PubMed]

3. Mcneal, J.; Kindrachuk, R.; Freiha, F.; Bostwick, D.; Redwine, E.; Stamey, T. Patterns of Progression in Prostate Cancer. *Lancet* **1986**, *327*, 60–63. [CrossRef]

4. Coleman, W.B. Molecular Pathogenesis of Prostate Cancer. In *Molecular Pathology*; Coleman, W., Tsongalis, G., Eds.; Elsevier: Berlin/Heidelberg, Germany, 2018; pp. 555–568, ISBN 9780128027615.

5. Shen, M.M.; Abate-Shen, C. Molecular genetics of prostate cancer: New prospects for old challenges. *Genes Dev.* **2010**, *24*, 1967–2000. [CrossRef]

6. De Reijke, T.M.; van Moorselaar, J.R. Ten-year Outcomes after Monitoring, Surgery, or Radiotherapy for Localized Prostate Cancer. *Eur. Urol.* **2017**, *71*, 491–492. [CrossRef] [PubMed]

7. Parker, C. Active surveillance: Towards a new paradigm in the management of early prostate cancer. *Lancet Oncol.* **2004**, *5*, 101–106. [CrossRef]

8. Huggins, C.; Hodges, C.V. Studies on Prostatic Cancer: I. The Effect of Castration, of Estrogen and of Androgen Injection on Serum Phosphatases in Metastatic Carcinoma of the Prostate. *J. Urol.* **1941**, *168*, 9–12.

9. Huggins, C. Endocrine-Induced Regression of Cancers. *Cancer Res.* **1967**, *27*, 1925–1930. [CrossRef]

10. Chen, Y.; Sawyers, C.L.; Scher, H.I. Targeting the androgen receptor pathway in prostate cancer. *Curr. Opin. Pharmacol.* **2008**, *8*, 440–448. [CrossRef] [PubMed]

11. Saad, F.; Chi, K.; Finelli, A.; Hotte, S.; Izawa, J.; Kapoor, A.; Kassouf, W.; Loblaw, A.; North, S.; Rendon, R.; et al. The 2015 CUA-CUOG Guidelines for the Management of Castration Resistant Prostate Cancer (CRPC). *Can. Urol. Assoc. J.* **2015**, *9*, 90. [CrossRef] [PubMed]

12. Mollica, V.; Di Nunno, V.; Cimadamore, A.; Lopez-Beltran, A.; Cheng, L.; Santoni, M.; Scarpelli, M.; Montironi, R.; Massari, F. Molecular Mechanisms Related to Hormone Inhibition Resistance in Prostate Cancer. *Cells* **2019**, *8*, 43. [CrossRef]

13. Vlachostergios, P.J.; Puca, L.; Beltran, H. Emerging Variants of Castration-Resistant Prostate Cancer. *Curr. Oncol. Rep.* **2017**, *19*, 1–17. [CrossRef]

14. Chandrasekar, T.; Yang, J.; Gao, A.; Evans, C.P. Mechanisms of resistance in castration-resistant prostate cancer (CRPC). *Transl. Androl. Urol.* **2015**, *4*, 365–380. [PubMed]

15. Visakorpi, T.; Hyytinen, E.; Koivisto, P.; Tanner, M.; Keinänen, R.; Palmberg, C.; Palotie, A.; Tammela, T.; Isola, J.; Kallioniemi, O. In vivo amplification of the androgen receptor gene and progression of human prostate cancer. *Nat. Genet.* **1995**, *9*, 401–406. [CrossRef]

16. Gregory, C.W.; Johnson, R.T.; Mohler, J.L.; French, F.S.; Wilson, E.M. Androgen receptor stabilization in recurrent prostate cancer is associated with hypersensitivity to low androgen. *Cancer Res.* **2001**, *61*, 2892–2898.

17. Kim, K.; Wongvipat, J.; Watson, P.A.; Balbas, M.D.; Chen, Y.F.; Sawyers, C.L.; Viale, A.; Socci, N.D. Constitutively active androgen receptor splice variants expressed in castration-resistant prostate cancer require full-length androgen receptor. *Proc. Natl. Acad. Sci. USA* **2010**, *107*, 16759–16765.

18. Montgomery, R.B.; Mostaghel, E.A.; Vessella, R.; Hess, D.L.; Kalhorn, T.F.; Higano, C.S.; True, L.D.; Nelson, P.S. Maintenance of Intratumoral Androgens in Metastatic Prostate Cancer: A Mechanism for Castration-Resistant Tumor Growth. *Cancer Res.* **2008**, *68*, 4447–4454. [CrossRef]

19. Stanbrough, M.; Bubley, G.J.; Ross, K.; Golub, T.R.; Rubin, M.A.; Penning, T.M.; Febbo, P.G.; Balk, S.P. Increased Expression of Genes Converting Adrenal Androgens to Testosterone in Androgen-Independent Prostate Cancer. *Cancer Res.* **2006**, *66*, 2815–2825. [CrossRef]

20. Wang, Q.; Li, W.; Zhang, Y.; Yuan, X.; Xu, K.; Yu, J.; Chen, Z.; Beroukhim, R.; Wang, H.; Lupien, M.; et al. Androgen Receptor Regulates a Distinct Transcription Program in Androgen-Independent Prostate Cancer. *Cell* **2009**, *138*, 245–256. [CrossRef]

21. Moroishi, T.; Hansen, C.G.; Guan, K.-L. The emerging roles of YAP and TAZ in cancer. *Nat. Rev. Cancer* **2015**, *15*, 73–79. [CrossRef] [PubMed]

22. Moya, I.M.; Halder, G. Hippo–YAP/TAZ signalling in organ regeneration and regenerative medicine. *Nat. Rev. Mol. Cell Biol.* **2019**, *20*, 211–226. [CrossRef]

23. Dong, J.; Feldmann, G.; Huang, J.; Wu, S.; Zhang, N.; Comerford, S.A.; Gayyed, M.F.; Anders, R.A.; Maitra, A.; Pan, D. Elucidation of a Universal Size-Control Mechanism in Drosophila and Mammals. *Cell* **2007**, *130*, 1120–1133. [CrossRef]

24. Pan, D. The Hippo Signaling Pathway in Development and Cancer. *Dev. Cell* **2010**, *19*, 491–505. [CrossRef] [PubMed]

25. Park, J.H.; Shin, J.E.; Park, H.W. The Role of Hippo Pathway in Cancer Stem Cell Biology. *Mol. Cells* **2018**, *41*, 83–92.

26. Hansen, C.G.; Moroishi, T.; Guan, K.L. YAP and TAZ: A nexus for Hippo signaling and beyond. *Trends Cell Biol.* **2015**, *25*, 499–513. [CrossRef] [PubMed]

27. Santinon, G.; Pocaterra, A.; Dupont, S. Control of YAP/TAZ Activity by Metabolic and Nutrient-Sensing Pathways. *Trends Cell Biol.* **2016**, *26*, 289–299. [CrossRef] [PubMed]

28. Totaro, A.; Panciera, T.; Piccolo, S. YAP/TAZ upstream signals and downstream responses. *Nat. Cell Biol.* **2018**, *20*, 888–899. [CrossRef]

29. Yu, F.X.; Zhao, B.; Guan, K.L. Hippo Pathway in Organ Size Control, Tissue Homeostasis, and Cancer. *Cell* **2015**, *163*, 811–828. [CrossRef] [PubMed]

30. Chan, S.W.; Lim, C.J.; Chong, Y.F.; Pobbati, A.V.; Huang, C.; Hong, W. Hippo Pathway-independent Restriction of TAZ and YAP by Angiomotin. *J. Biol. Chem.* **2011**, *286*, 7018–7026. [CrossRef] [PubMed]

31. Dupont, S.; Morsut, L.; Aragona, M.; Enzo, E.; Giulitti, S.; Cordenonsi, M.; Zanconato, F.; Le Digabel, J.; Forcato, M.; Bicciato, S.; et al. Role of YAP/TAZ in mechanotransduction. *Nature* **2011**, *474*, 179–184. [CrossRef] [PubMed]

32. Kofler, M.; Speight, P.; Little, D.; Di Ciano-Oliveira, C.; Szászi, K.; Kapus, A. Mediated nuclear import and export of TAZ and the underlying molecular requirements. *Nat. Commun.* **2018**, *9*, 4966. [CrossRef]

33. Zanconato, F.; Cordenonsi, M.; Piccolo, S. YAP/TAZ at the Roots of Cancer. *Cancer Cell* **2016**, *29*, 783–803. [CrossRef]

34. Steinhardt, A.A.; Gayyed, M.F.; Klein, A.P.; Dong, J.; Maitra, A.; Pan, D.; Montgomery, E.A.; Anders, R.A. Expression of Yes-associated protein in common solid tumors. *Hum. Pathol.* **2008**, *39*, 1582–1589. [CrossRef] [PubMed]

35. Schäfer, A.; Nousiainen, M.; Chan, E.H.Y.; Chalamalasetty, R.B.; Silljé, H.H.W.; Nigg, E.A. The Ste20-like kinase Mst2 activates the human large tumor suppressor kinase Lats1. *Oncogene* **2005**, *24*, 2076–2086.

36. Hao, Y.; Chun, A.; Cheung, K.; Rashidi, B.; Yang, X. Tumor Suppressor LATS1 Is a Negative Regulator of Oncogene YAP. *J. Biol. Chem.* **2008**, *283*, 5496–5509. [CrossRef]

37. Pan, D.; Liu, B.; Wang, W.; Zheng, Y.; Uster, E.; Deng, H. Identification of Happyhour/MAP4K as Alternative Hpo/Mst-like Kinases in the Hippo Kinase Cascade. *Dev. Cell* **2015**, *34*, 642–655.

38. Oh, H.; Irvine, K.D. In vivo regulation of Yorkie phosphorylation and localization. *Development* **2008**, *135*, 1081–1088. [CrossRef] [PubMed]

39. Harvey, K.F.; Pfleger, C.M.; Hariharan, I.K. The Drosophila Mst Ortholog, hippo, Restricts Growth and Cell Proliferation and Promotes Apoptosis. *Cell* **2003**, *114*, 457–467. [CrossRef]

40. Jianhang, J.; Zhang, W.; Wang, B.; Richard, T.; Jiang, J. The Drosophila Ste20 family kinase dMST functions as a tumor suppressor by restricting cell proliferation and promoting apoptosis. *Genes Dev.* **2003**, *17*, 2514–2519.

41. Justice, R.W.; Zilian, O.; Woods, D.F.; Noll, M.; Bryant, P.J. The Drosophila Tumor-Suppressor Gene Warts Encodes a Homolog of Human Myotonic-Dystrophy Kinase and Is Required for the Control of Cell-Shape and Proliferation. *Genes Dev.* **1995**, *9*, 534–546. [CrossRef]

42. Li, Q.; Li, S.; Mana-Capelli, S.; Roth Flach, R.J.; Danai, L.V.; Amcheslavsky, A.; Nie, Y.; Kaneko, S.; Yao, X.; Chen, X.; et al. The Conserved Misshapen-Warts-Yorkie Pathway Acts in Enteroblasts to Regulate Intestinal Stem Cells in Drosophila. *Dev. Cell* **2014**, *31*, 291–304. [CrossRef]

43. Pantalacci, S.; Tapon, N.; Léopold, P. The Salvador partner Hippo promotes apoptosis and cell-cycle exit in Drosophila. *Nat. Cell Biol.* **2003**, *5*, 921–927.

44. Udan, R.S.; Kango-Singh, M.; Nolo, R.; Tao, C.; Halder, G. Hippo promotes proliferation arrest and apoptosis in the Salvador/Warts pathway. *Nat. Cell Biol.* **2003**, *5*, 914–920. [CrossRef]

45. Meng, Z.; Moroishi, T.; Mottier-Pavie, V.; Plouffe, S.W.; Hansen, C.G.; Hong, A.W.; Park, H.W.; Mo, J.-S.; Lu, W.; Lu, S.; et al. MAP4K family kinases act in parallel to MST1/2 to activate LATS1/2 in the Hippo pathway. *Nat. Commun.* **2015**, *6*, 8357. [CrossRef] [PubMed]

46. Zhao, B.; Wei, X.; Li, W.; Udan, R.S.; Yang, Q.; Kim, J.; Xie, J.; Ikenoue, T.; Yu, J.; Li, L.; et al. Inactivation of YAP oncoprotein by the Hippo pathway is involved in cell contact inhibition and tissue growth control. *Genes Dev.* **2007**, *21*, 2747–2761. [CrossRef] [PubMed]

47. Lei, Q.-Y.; Zhang, H.; Zhao, B.; Zha, Z.-Y.; Bai, F.; Pei, X.-H.; Zhao, S.; Xiong, Y.; Guan, K.-L. TAZ Promotes Cell Proliferation and Epithelial-Mesenchymal Transition and Is Inhibited by the Hippo Pathway. *Mol. Cell. Biol.* **2008**, *28*, 2426–2436. [CrossRef] [PubMed]

48. Zhao, B.; Li, L.; Tumaneng, K.; Wang, C.Y.; Guan, K.L. A coordinated phosphorylation by Lats and CK1 regulates YAP stability through SCFβ-TRCP. *Genes Dev.* **2010**, *24*, 72–85. [CrossRef]

49. Liu, C.-Y.; Zha, Z.-Y.; Zhou, X.; Zhang, H.; Huang, W.; Zhao, D.; Li, T.; Chan, S.W.; Lim, C.J.; Hong, W.; et al. The Hippo Tumor Pathway Promotes TAZ Degradation by Phosphorylating a Phosphodegron and Recruiting the SCF β-TrCP E3 Ligase. *J. Biol. Chem.* **2010**, *285*, 37159–37169. [CrossRef] [PubMed]

50. Wang, S.; Xie, F.; Chu, F.; Zhang, Z.; Yang, B.; Dai, T.; Gao, L.; Wang, L.; Ling, L.; Jia, J.; et al. YAP antagonizes innate antiviral immunity and is targeted for lysosomal degradation through IKKI-mediated phosphorylation. *Nat. Immunol.* **2017**, *18*, 733–743. [CrossRef]

51. Schlegelmilch, K.; Mohseni, M.; Kirak, O.; Pruszak, J.; Rodriguez, J.R.; Zhou, D.; Kreger, B.T.; Vasioukhin, V.; Avruch, J.; Brummelkamp, T.R.; et al. Yap1 Acts Downstream of α-Catenin to Control Epidermal Proliferation. *Cell* **2011**, *144*, 782–795. [CrossRef]

52. Ota, M.; Sasaki, H. Mammalian Tead proteins regulate cell proliferation and contact inhibition as transcriptional mediators of Hippo signaling. *Development* **2008**, *135*, 4059–4069. [CrossRef] [PubMed]

53. Zhang, H.; Liu, C.-Y.; Zha, Z.-Y.; Zhao, B.; Yao, J.; Zhao, S.; Xiong, Y.; Lei, Q.-Y.; Guan, K.-L. TEAD Transcription Factors Mediate the Function of TAZ in Cell Growth and Epithelial-Mesenchymal Transition. *J. Biol. Chem.* **2009**, *284*, 13355–13362. [CrossRef]

54. Zhao, B.; Ye, X.; Yu, J.; Li, L.; Li, W.; Li, S.; Yu, J.; Lin, J.D.; Wang, C.-Y.; Chinnaiyan, A.M.; et al. TEAD mediates YAP-dependent gene induction and growth control. *Genes Dev.* **2008**, *22*, 1962–1971. [CrossRef] [PubMed]

55. Vassilev, A.; Kaneko, K.J.; Shu, H.; Zhao, Y.; DePamphilis, M.L. TEAD/TEF transcription factors utilize the activation domain of YAP65, a Src/Yes-associated protein localized in the cytoplasm. *Genes Dev.* **2001**, *15*, 1229–1241. [CrossRef]

56. Zaidi, S.K.; Sullivan, A.J.; Medina, R.; Ito, Y.; van Wijnen, A.J.; Stein, J.L.; Lian, J.B.; Stein, G.S. Tyrosine phosphorylation controls Runx2-mediated subnuclear targeting of YAP to repress transcription. *EMBO J.* **2004**, *23*, 790–799. [CrossRef]

57. Vlahov, N.; Scrace, S.; Soto, M.S.; Grawenda, A.M.; Bradley, L.; Pankova, D.; Papaspyropoulos, A.; Yee, K.S.; Buffa, F.; Goding, C.R.; et al. Alternate RASSF1 Transcripts Control SRC Activity, E-Cadherin Contacts, and YAP-Mediated Invasion. *Curr. Biol.* **2015**, *25*, 3019–3034. [CrossRef] [PubMed]

58. Byun, M.R.; Hwang, J.-H.; Kim, A.R.; Kim, K.M.; Park, J.I.; Oh, H.T.; Hwang, E.S.; Hong, J.-H. SRC activates TAZ for intestinal tumorigenesis and regeneration. *Cancer Lett.* **2017**, *410*, 32–40. [CrossRef] [PubMed]

59. Zhang, L.; Tang, F.; Terracciano, L.; Hynx, D.; Kohler, R.; Bichet, S.; Hess, D.; Cron, P.; Hemmings, B.A.; Hergovich, A.; et al. NDR Functions as a Physiological YAP1 Kinase in the Intestinal Epithelium. *Curr. Biol.* **2015**, *25*, 296–305. [CrossRef]

60. Tomlinson, V.; Gudmundsdottir, K.; Luong, P.; Leung, K.-Y.; Knebel, A.; Basu, S. JNK phosphorylates Yes-associated protein (YAP) to regulate apoptosis. *Cell Death Dis.* **2010**, *1*, e29. [CrossRef]

61. Codelia, V.A.; Sun, G.; Irvine, K.D. Regulation of YAP by Mechanical Strain through Jnk and Hippo Signaling. *Curr. Biol.* **2014**, *24*, 2012–2017. [CrossRef]

62. DeRan, M.; Yang, J.; Shen, C.-H.; Peters, E.C.; Fitamant, J.; Chan, P.; Hsieh, M.; Zhu, S.; Asara, J.M.; Zheng, B.; et al. Energy Stress Regulates Hippo-YAP Signaling Involving AMPK-Mediated Regulation of Angiomotin-like 1 Protein. *Cell Rep.* **2014**, *9*, 495–503. [CrossRef]

63. Wang, W.; Xiao, Z.-D.; Li, X.; Aziz, K.E.; Gan, B.; Johnson, R.L.; Chen, J. AMPK modulates Hippo pathway activity to regulate energy homeostasis. *Nat. Cell Biol.* **2015**, *17*, 490–499. [CrossRef]

64. Mo, J.; Meng, Z.; Kim, Y.C.; Park, H.W.; Hansen, C.G.; Kim, S. Cellular energy stress induces AMPK-mediated regulation of YAP and the Hippo pathway. *Nat. Cell Biol.* **2015**, *17*, 500–510. [CrossRef] [PubMed]

65. Moon, S.; Kim, W.; Kim, S.; Kim, Y.; Song, Y.; Bilousov, O.; Kim, J.; Lee, T.; Cha, B.; Kim, M.; et al. Phosphorylation by NLK inhibits YAP-14-3-3-interactions and induces its nuclear localization. *EMBO Rep.* **2017**, *18*, 61–71. [CrossRef]

66. Hong, A.W.; Meng, Z.; Yuan, H.; Plouffe, S.W.; Moon, S.; Kim, W.; Jho, E.; Guan, K. Osmotic stress-induced phosphorylation by NLK at Ser128 activates YAP. *EMBO Rep.* **2017**, *18*, 72–86. [CrossRef] [PubMed]

67. Mohamed, A.D.; Tremblay, A.M.; Murray, G.I.; Wackerhage, H. The Hippo signal transduction pathway in soft tissue sarcomas. *Biochim. Biophys. Acta Rev. Cancer* **2015**, *1856*, 121–129. [CrossRef]

68. Pinar, E.Z.; Filiz, K.C.; Seyma, O.; Ummuhan, D.; Esra, K. Increased expression of YAP1 in prostate cancer correlates with extraprostatic extension. *Cancer Biol. Med.* **2017**, *14*, 405.

69. Jiang, N.; Hjorth-Jensen, K.; Hekmat, O.; Iglesias-Gato, D.; Kruse, T.; Wang, C.; Wei, W.; Ke, B.; Yan, B.; Niu, Y.; et al. In vivo quantitative phosphoproteomic profiling identifies novel regulators of castration-resistant prostate cancer growth. *Oncogene* **2015**, *34*, 2764–2776. [CrossRef]

70. Sheng, X.; Li, W.B.; Wang, D.L.; Chen, K.H.; Cao, J.J.; Luo, Z.; He, J.; Li, M.C.; Liu, W.J.; Yu, C. YAP is closely correlated with castration-resistant prostate cancer, and downregulation of YAP reduces proliferation and induces apoptosis of PC-3 cells. *Mol. Med. Rep.* **2015**, *12*, 4867–4876. [CrossRef] [PubMed]

71. Zhang, L.; Yang, S.; Chen, X.; Stauffer, S.; Yu, F.; Lele, S.M.; Fu, K.; Datta, K.; Palermo, N.; Chen, Y.; et al. The Hippo Pathway Effector YAP Regulates Motility, Invasion, and Castration-Resistant Growth of Prostate Cancer Cells. *Mol. Cell. Biol.* **2015**, *35*, 1350–1362. [CrossRef] [PubMed]

72. Tomlins, S.A.; Laxman, B.; Varambally, S.; Cao, X.; Yu, J.; Helgeson, B.E.; Cao, Q.; Prensner, J.R.; Rubin, M.A.; Shah, R.B.; et al. Role of the TMPRSS2-ERG Gene Fusion in Prostate Cancer. *Neoplasia* **2008**, *10*, 177–188. [CrossRef] [PubMed]

73. Tomlins, S.A.; Rhodes, D.R.; Perner, S.; Dhanasekaran, S.M.; Mehra, R.; Sun, X.-W.; Varambally, S.; Cao, X.; Tchinda, J.; Kuefer, R.; et al. Recurrent Fusion of TMPRSS2 and ETS Transcription Factor Genes in Prostate Cancer. *Science* **2005**, *310*, 644–648. [CrossRef]

74. The Cancer Genome Atlas Research Network. The molecular taxonomy of primary prostate cancer. *Cell* **2015**, *163*, 1011–1025.

75. Nguyen, L.T.; Tretiakova, M.S.; Silvis, M.R.; Lucas, J.; Klezovitch, O.; Coleman, I.; Bolouri, H.; Kutyavin, V.I.; Morrissey, C.; True, L.D.; et al. ERG Activates the YAP1 Transcriptional Program and Induces the Development of Age-Related Prostate Tumors. *Cancer Cell* **2015**, *27*, 797–808. [CrossRef]

76. Korenchuk, S.; Lehr, J.E.; MClean, L.; Lee, Y.G.; Whitney, S.; Vessella, R.; Lin, D.L.; Pienta, K.J. VCaP, a cell-based model system of human prostate cancer. *In Vivo* **2001**, *15*, 163–168. [PubMed]

77. Kim, T.; Jin, F.; Shin, S.; Oh, S.; Lightfoot, S.A.; Grande, J.P.; Johnson, A.J.; van Deursen, J.M.; Wren, J.D.; Janknecht, R. Histone demethylase JMJD2A drives prostate tumorigenesis through transcription factor ETV1. *J. Clin. Invest.* **2016**, *126*, 706–720. [CrossRef]

78. Maehama, T.; Dixon, J.E. The Tumor Suppressor, PTEN/MMAC1, Dephosphorylates the Lipid Second Messenger, Phosphatidylinositol 3,4,5-Trisphosphate. *J. Biol. Chem.* **1998**, *273*, 13375–13378. [CrossRef]
79. Jamaspishvili, T.; Berman, D.M.; Ross, A.E.; Scher, H.I.; De Marzo, A.M.; Squire, J.A.; Lotan, T.L. Clinical implications of PTEN loss in prostate cancer. *Nat. Rev. Urol.* **2018**, *15*, 222–234. [CrossRef] [PubMed]
80. Dongre, A.; Weinberg, R.A. New insights into the mechanisms of epithelial–mesenchymal transition and implications for cancer. *Nat. Rev. Mol. Cell Biol.* **2019**, *20*, 69–84. [CrossRef] [PubMed]
81. Vasioukhin, V.; Bauer, C.; Yin, M.; Fuchs, E. Directed actin polymerization is the driving force for epithelial cell- cell adhesion. *Cell* **2000**, *100*, 209–219. [CrossRef]
82. Schäfer, R.; Iden, S.; Collard, J.G.; Hirose, T.; Ohno, S.; Song, J.-Y.; van Riel, W.E. Tumor Type-Dependent Function of the Par3 Polarity Protein in Skin Tumorigenesis. *Cancer Cell* **2012**, *22*, 389–403.
83. Zhou, P.-J.; Wang, X.; An, N.; Wei, L.; Zhang, L.; Huang, X.; Zhu, H.H.; Fang, Y.-X.; Gao, W.-Q. Loss of Par3 promotes prostatic tumorigenesis by enhancing cell growth and changing cell division modes. *Oncogene* **2019**, *38*, 2192–2205. [CrossRef]
84. Zhou, P.-J.; Xue, W.; Peng, J.; Wang, Y.; Wei, L.; Yang, Z.; Zhu, H.H.; Fang, Y.-X.; Gao, W.-Q. Elevated expression of Par3 promotes prostate cancer metastasis by forming a Par3/aPKC/KIBRA complex and inactivating the hippo pathway. *J. Exp. Clin. Cancer Res.* **2017**, *36*, 139. [CrossRef]
85. Ikwegbue, P.; Masamba, P.; Oyinloye, B.; Kappo, A. Roles of Heat Shock Proteins in Apoptosis, Oxidative Stress, Human Inflammatory Diseases, and Cancer. *Pharmaceuticals* **2017**, *11*, 2. [CrossRef]
86. Zoubeidi, A.; Gleave, M. Small heat shock proteins in cancer therapy and prognosis. *Int. J. Biochem. Cell Biol.* **2012**, *44*, 1646–1656. [CrossRef] [PubMed]
87. Rocchi, P.; So, A.; Kojima, S.; Signaevsky, M.; Beraldi, E.; Fazli, L.; Hurtado-coll, A.; Yamanaka, K.; Gleave, M. Heat Shock Protein 27 Increases after Androgen Ablation and Plays a Cytoprotective Role in Hormone-Refractory Prostate Cancer. *Cancer Res.* **2004**, *64*, 6595–6602. [CrossRef]
88. Vahid, S.; Thaper, D.; Gibson, K.F.; Bishop, J.L.; Zoubeidi, A. Molecular chaperone Hsp27 regulates the Hippo tumor suppressor pathway in cancer. *Sci. Rep.* **2016**, *6*, 31842. [CrossRef] [PubMed]
89. Zhou, D.; Payer, B.; Bardeesy, N.; Avruch, J.; Yin, Y.; Lauwers, G.Y.; Xia, F.; Park, J.-S.; Conrad, C.; Lee, J.T.; et al. Mst1 and Mst2 Maintain Hepatocyte Quiescence and Suppress Hepatocellular Carcinoma Development through Inactivation of the Yap1 Oncogene. *Cancer Cell* **2009**, *16*, 425–438. [CrossRef] [PubMed]
90. Song, H.; Mak, K.K.; Topol, L.; Yun, K.; Hu, J.; Garrett, L.; Chen, Y.; Park, O.; Chang, J.; Simpson, R.M.; et al. Mammalian Mst1 and Mst2 kinases play essential roles in organ size control and tumor suppression. *Proc. Natl. Acad. Sci. USA* **2010**, *107*, 1431–1436. [CrossRef]
91. Loforese, G.; Malinka, T.; Keogh, A.; Baier, F.; Simillion, C.; Montani, M.; Halazonetis, T.D.; Candinas, D.; Stroka, D. Impaired liver regeneration in aged mice can be rescued by silencing Hippo core kinases MST1 and MST2. *EMBO Mol. Med.* **2017**, *9*, 46–60. [CrossRef] [PubMed]
92. Modi, P.K.; Faiena, I.; Kim, I.Y. Androgen Receptor. In *Prostate Cancer*; Mydlo, J.H., Godec, C.J., Eds.; Elsevier: Berlin/Heidelberg, Germany, 2016; pp. 21–28, ISBN 9780128000779.
93. Zhou, Y.; Bolton, E.C.; Jones, J.O. Androgens and androgen receptor signaling in prostate tumorigenesis. *J. Mol. Endocrinol.* **2015**, *54*, R15–R29. [CrossRef] [PubMed]
94. Gewirth, D.T.; Dollins, D.E.; Shaffer, P.L.; Jivan, A.; Claessens, F. Structural basis of androgen receptor binding to selective androgen response elements. *Proc. Natl. Acad. Sci. USA* **2004**, *101*, 4758–4763.
95. Zhang, B.; Kwon, O.-J.; Henry, G.; Malewska, A.; Wei, X.; Zhang, L.; Brinkley, W.; Zhang, Y.; Castro, P.D.; Titus, M.; et al. Non-Cell-Autonomous Regulation of Prostate Epithelial Homeostasis by Androgen Receptor. *Mol. Cell* **2016**, *63*, 976–989. [CrossRef] [PubMed]
96. Tan, M.E.; Li, J.; Xu, H.E.; Melcher, K.; Yong, E.L. Androgen receptor: Structure, role in prostate cancer and drug discovery. *Acta Pharmacol. Sin.* **2015**, *36*, 3–23. [CrossRef] [PubMed]
97. Gao, W.; Bohl, C.E.; Dalton, J.T. Chemistry and structural biology of androgen receptor. *Chem. Rev.* **2005**, *105*, 3352–3370. [CrossRef]
98. Kuser-Abali, G.; Alptekin, A.; Lewis, M.; Garraway, I.P.; Cinar, B. YAP1 and AR interactions contribute to the switch from androgen-dependent to castration-resistant growth in prostate cancer. *Nat. Commun.* **2015**, *6*, 1–13. [CrossRef]
99. Thalmann, N.; Edmund, E.; Hopwood, V.L.; Pathak, S.; Von Eschenbach, A.; Chung, L.K. Androgen-independent Cancer Progression and Bone Metastasis in the LNCaP Model of Human Prostate Cancer. *Cancer Res.* **1994**, 2577–2582.

100. Cinar, B.; Collak, F.K.; Lopez, D.; Mukhopadhyay, N.K.; Akgul, S.; Freeman, M.R.; Kilicarslan, M.; Gioeli, D.G. MST1 Is a Multifunctional Caspase-Independent Inhibitor of Androgenic Signaling. *Cancer Res.* **2011**, *71*, 4303–4313. [CrossRef]

101. Kuser-Abali, G.; Alptekin, A.; Cinar, B. Overexpression of MYC and EZH2 cooperates to epigenetically silence MST1 expression. *Epigenetics* **2014**, *9*, 634–643. [CrossRef] [PubMed]

102. Dang, C.V.; O'Donnell, K.A.; Zeller, K.I.; Nguyen, T.; Osthus, R.C.; Li, F. The c-Myc target gene network. *Semin. Cancer Biol.* **2006**, *16*, 253–264. [CrossRef]

103. Gurel, B.; Iwata, T.; Koh, C.M.; Jenkins, R.B.; Lan, F.; Van Dang, C.; Hicks, J.L.; Morgan, J.; Cornish, T.C.; Sutcliffe, S.; et al. Nuclear MYC protein overexpression is an early alteration in human prostate carcinogenesis. *Mod. Pathol.* **2008**, *21*, 1156–1167. [CrossRef] [PubMed]

104. Koh, C.M.; Iwata, T.; Zheng, Q.; Bethel, C.; Yegnasubramanian, S.; De Marzo, A.M. Myc Enforces Overexpression of EZH2 in Early Prostatic Neoplasia via Transcriptional and Post-transcriptional Mechanisms. *Oncotarget* **2011**, *2*, 669–683. [CrossRef]

105. Cao, R.; Wang, L.; Wang, H.; Xia, L.; Erdjument-Bromage, H.; Tempst, P.; Jones, R.S.; Zhang, Y. Role of Histone H3 Lysine 27 Methylation in Polycomb-Group Silencing. *Science* **2002**, *298*, 1039–1043. [CrossRef]

106. Gan, L.; Yang, Y.; Li, Q.; Feng, Y.; Liu, T.; Guo, W. Epigenetic regulation of cancer progression by EZH2: From biological insights to therapeutic potential. *Biomark. Res.* **2018**, *6*, 1–10. [CrossRef]

107. Xu, K.; Wu, Z.J.; Groner, A.C.; He, H.H.; Cai, C.; Lis, R.T.; Wu, X.; Stack, E.C.; Loda, M.; Liu, T.; et al. EZH2 Oncogenic Activity in Castration-Resistant Prostate Cancer Cells Is Polycomb-Independent. *Science* **2012**, *338*, 1465–1469. [CrossRef]

108. Powzaniuk, M.; McElwee-Witmer, S.; Vogel, R.L.; Hayami, T.; Rutledge, S.J.; Chen, F.; Harada, S.; Schmidt, A.; Rodan, G.A.; Freedman, L.P.; et al. The LATS2/KPM Tumor Suppressor Is a Negative Regulator of the Androgen Receptor. *Mol. Endocrinol.* **2004**, *18*, 2011–2023. [CrossRef]

109. Stražišar, M.; Mlakar, V.; Glavač, D. LATS2 tumour specific mutations and down-regulation of the gene in non-small cell carcinoma. *Lung Cancer* **2009**, *64*, 257–262. [CrossRef]

110. Furth, N.; Pateras, I.S.; Rotkopf, R.; Vlachou, V.; Rivkin, I.; Schmitt, I.; Bakaev, D.; Gershoni, A.; Ainbinder, E.; Leshkowitz, D.; et al. LATS1 and LATS2 suppress breast cancer progression by maintaining cell identity and metabolic state. *Life Sci. Alliance* **2018**, *1*, e201800171. [CrossRef]

111. Moroishi, T.; Park, H.W.; Qin, B.; Chen, Q.; Meng, Z.; Plouffe, S.W.; Taniguchi, K.; Yu, F.-X.; Karin, M.; Pan, D.; et al. A YAP/TAZ-induced feedback mechanism regulates Hippo pathway homeostasis. *Genes Dev.* **2015**, *29*, 1271–1284. [CrossRef] [PubMed]

112. Feng, X.-H.; Li, L.; Yan, H.; Huang, J.; Ji, X.; Dai, X.; Liu, H.; Zhao, B.; Shen, S.; Guo, X. YAP activates the Hippo pathway in a negative feedback loop. *Cell Res.* **2017**, *27*, 1073.

113. He, C.; Lv, X.; Huang, C.; Hua, G.; Ma, B.; Chen, X.; Angeletti, P.C.; Dong, J.; Zhou, J.; Wang, Z.; et al. YAP1-LATS2 feedback loop dictates senescent or malignant cell fate to maintain tissue homeostasis. *EMBO Rep.* **2019**, *20*, e44948. [CrossRef]

114. Paszek, M.J.; Zahir, N.; Johnson, K.R.; Lakins, J.N.; Rozenberg, G.I.; Gefen, A.; Reinhart-King, C.A.; Margulies, S.S.; Dembo, M.; Boettiger, D.; et al. Tensional homeostasis and the malignant phenotype. *Cancer Cell* **2005**, *8*, 241–254. [CrossRef] [PubMed]

115. Calvo, F.; Ege, N.; Grande-Garcia, A.; Hooper, S.; Jenkins, R.P.; Chaudhry, S.I.; Harrington, K.; Williamson, P.; Moeendarbary, E.; Charras, G.; et al. Mechanotransduction and YAP-dependent matrix remodelling is required for the generation and maintenance of cancer-associated fibroblasts. *Nat. Cell Biol.* **2013**, *15*, 637–646. [CrossRef]

116. Wada, K.-I.; Itoga, K.; Okano, T.; Yonemura, S.; Sasaki, H. Hippo pathway regulation by cell morphology and stress fibers. *Development* **2011**, *138*, 3907–3914. [CrossRef]

117. Rausch, V.; Bostrom, J.R.; Park, J.; Bravo, I.R.; Feng, Y.; Hay, D.C.; Link, B.A.; Hansen, C.G. The Hippo Pathway Regulates Caveolae Expression and Mediates Flow Response via Caveolae. *Curr. Biol.* **2019**, *29*, 242–255e6. [CrossRef] [PubMed]

118. Nardone, G.; Oliver-De La Cruz, J.; Vrbsky, J.; Martini, C.; Pribyl, J.; Skládal, P.; Pešl, M.; Caluori, G.; Pagliari, S.; Martino, F.; et al. YAP regulates cell mechanics by controlling focal adhesion assembly. *Nat. Commun.* **2017**, *8*, 15321. [CrossRef]

119. Mason, D.E.; Collins, J.M.; Dawahare, J.H.; Nguyen, T.D.; Lin, Y.; Voytik-Harbin, S.L.; Zorlutuna, P.; Yoder, M.C.; Boerckel, J.D. YAP and TAZ limit cytoskeletal and focal adhesion maturation to enable persistent cell motility. *J. Cell Biol.* **2019**, *218*, 1369–1389. [CrossRef] [PubMed]

120. Fletcher, G.C.; Elbediwy, A.; Khanal, I.; Ribeiro, P.S.; Tapon, N.; Thompson, B.J. The Spectrin cytoskeleton regulates the Hippo signalling pathway. *EMBO J.* **2015**, *34*, 940–954. [CrossRef]

121. Zhao, B.; Li, L.; Wang, L.; Wang, C.-Y.; Yu, J.; Guan, K.-L. Cell detachment activates the Hippo pathway via cytoskeleton reorganization to induce anoikis. *Genes Dev.* **2012**, *26*, 54–68. [CrossRef]

122. Deng, H.; Wang, W.; Yu, J.; Zheng, Y.; Qing, Y.; Pan, D. Spectrin regulates Hippo signaling by modulating cortical actomyosin activity. *Elife* **2015**, *4*, 1–17. [CrossRef]

123. Aragona, M.; Panciera, T.; Manfrin, A.; Giulitti, S.; Michielin, F.; Elvassore, N.; Dupont, S.; Piccolo, S. A Mechanical Checkpoint Controls Multicellular Growth through YAP/TAZ Regulation by Actin-Processing Factors. *Cell* **2013**, *154*, 1047–1059. [CrossRef]

124. Tang, Y.; Rowe, R.G.; Botvinick, E.L.; Kurup, A.; Putnam, A.J.; Seiki, M.; Weaver, V.M.; Keller, E.T.; Goldstein, S.; Dai, J.; et al. MT1-MMP-Dependent Control of Skeletal Stem Cell Commitment via a β1-Integrin/YAP/TAZ Signaling Axis. *Dev. Cell* **2013**, *25*, 402–416. [CrossRef]

125. Shiao, S.L.; Chu, G.C.-Y.; Chung, L.W.K. Regulation of prostate cancer progression by the tumor microenvironment. *Cancer Lett.* **2016**, *380*, 340–348. [CrossRef] [PubMed]

126. Graham, N.; Qian, B.-Z. Mesenchymal Stromal Cells: Emerging Roles in Bone Metastasis. *Int. J. Mol. Sci.* **2018**, *19*, 1121. [CrossRef] [PubMed]

127. Varzavand, A.; Hacker, W.; Ma, D.; Gibson-Corley, K.; Hawayek, M.; Tayh, O.J.; Brown, J.A.; Henry, M.D.; Stipp, C.S. α3β1 Integrin Suppresses Prostate Cancer Metastasis via Regulation of the Hippo Pathway. *Cancer Res.* **2016**, *76*, 6577–6587. [CrossRef]

128. Secades, P.; van Hulst, L.; Sonnenberg, A.; Song, J.-Y.; Kreft, M.; Sachs, N. Loss of integrin α3 prevents skin tumor formation by promoting epidermal turnover and depletion of slow-cycling cells. *Proc. Natl. Acad. Sci. USA* **2012**, *109*, 21468–21473.

129. Pickup, M.W.; Mouw, J.K.; Weaver, V.M. The extracellular matrix modulates the hallmarks of cancer. *EMBO Rep.* **2014**, *15*, 1243–1253. [CrossRef]

130. Kerkar, S.P.; Restifo, N.P. Cellular Constituents of Immune Escape within the Tumor Microenvironment. *Cancer Res.* **2012**, *72*, 3125–3130. [CrossRef]

131. Wang, G.; Lu, X.; Dey, P.; Deng, P.; Wu, C.C.; Jiang, S.; Fang, Z.; Zhao, K.; Konaparthi, R.; Hua, S.; et al. Targeting YAP-Dependent MDSC Infiltration Impairs Tumor Progression. *Cancer Discov.* **2016**, *6*, 80–95. [CrossRef] [PubMed]

132. Talmadge, J.E.; Gabrilovich, D.I. History of myeloid-derived suppressor cells. *Nat. Rev. Cancer* **2013**, *13*, 739–752. [CrossRef] [PubMed]

133. Surana, R.; Yi, C.; Weiner, L.M.; Graham, G.T.; Shahbazian, D.; White, S.M.; Chen, H.; Zhang, W.; Murakami, S. Yes-associated protein mediates immune reprogramming in pancreatic ductal adenocarcinoma. *Oncogene* **2016**, *36*, 1232–1244.

134. Guo, X.; Zhao, Y.; Yan, H.; Yang, Y.; Shen, S.; Dai, X.; Ji, X.; Ji, F.; Gong, X.; Li, L.; et al. Single tumor-initiating cells evade immune clearance by recruiting type II macrophages. *Genes Dev.* **2017**, *31*, 247–259. [CrossRef]

135. Chan, S.W.; Lim, C.J.; Guo, K.; Ng, C.P.; Lee, I.; Hunziker, W.; Zeng, Q.; Hong, W. A Role for TAZ in Migration, Invasion, and Tumorigenesis of Breast Cancer Cells. *Cancer Res.* **2008**, *68*, 2592–2598. [CrossRef] [PubMed]

136. Cordenonsi, M.; Zanconato, F.; Azzolin, L.; Forcato, M.; Rosato, A.; Frasson, C.; Inui, M.; Montagner, M.; Parenti, A.R.; Poletti, A.; et al. The Hippo Transducer TAZ Confers Cancer Stem Cell-Related Traits on Breast Cancer Cells. *Cell* **2011**, *147*, 759–772. [CrossRef] [PubMed]

137. Bhat, K.P.L.; Salazar, K.L.; Balasubramaniyan, V.; Wani, K.; Heathcock, L.; Hollingsworth, F.; James, J.D.; Gumin, J.; Diefes, K.L.; Kim, S.H.; et al. The transcriptional coactivator TAZ regulates mesenchymal differentiation in malignant glioma. *Genes Dev.* **2011**, *25*, 2594–2609. [CrossRef]

138. Zanconato, F.; Battilana, G.; Forcato, M.; Filippi, L.; Azzolin, L.; Manfrin, A.; Quaranta, E.; Di Biagio, D.; Sigismondo, G.; Guzzardo, V.; et al. Transcriptional addiction in cancer cells is mediated by YAP/TAZ through BRD4. *Nat. Med.* **2018**, *24*, 1599–1610. [CrossRef] [PubMed]

139. Park, H.W.; Kim, Y.C.; Yu, B.; Moroishi, T.; Mo, J.S.; Plouffe, S.W.; Meng, Z.; Lin, K.C.; Yu, F.X.; Alexander, C.M.; et al. Alternative Wnt Signaling Activates YAP/TAZ. *Cell* **2015**, *162*, 780–794. [CrossRef]

140. Hong, J.-H.; Hwang, E.S.; Michael, T.M.; Amsterdam, A.; Tian, Y.; Kalmukova, R.; Mueller, E.; Benjamin, T.; Spiegelman, B.M.; Sharp, P.A.; et al. TAZ, a Transcriptional Modulator of Mesenchymal Stem Cell Differentiation. *Science* **2005**, *309*, 1074–1078. [CrossRef] [PubMed]

141. Liu, C.Y.; Yu, T.; Huang, Y.; Cui, L.; Hong, W. ETS (E26 transformation-specific) up-regulation of the transcriptional co-Activator TAZ promotes cell migration and metastasis in prostate cancer. *J. Biol. Chem.* **2017**, *292*, 9420–9430. [CrossRef]

142. Parrini, M.C.; Sadou-Dubourgnoux, A.; Aoki, K.; Kunida, K.; Biondini, M.; Hatzoglou, A.; Poullet, P.; Formstecher, E.; Yeaman, C.; Matsuda, M.; et al. SH3BP1, an Exocyst-Associated RhoGAP, Inactivates Rac1 at the Front to Drive Cell Motility. *Mol. Cell* **2011**, *42*, 650–661. [CrossRef]

143. Seo, W.I.; Park, S.; Gwak, J.; Ju, B.G.; Chung, J.I.; Kang, P.M.; Oh, S. Wnt signaling promotes androgen-independent prostate cancer cell proliferation through up-regulation of the hippo pathway effector YAP. *Biochem. Biophys. Res. Commun.* **2017**, *486*, 1034–1039. [CrossRef]

144. Yuan, X.; Cai, C.; Chen, S.; Chen, S.; Yu, Z.; Balk, S.P. Androgen receptor functions in castration-resistant prostate cancer and mechanisms of resistance to new agents targeting the androgen axis. *Oncogene* **2014**, *33*, 2815–2825. [CrossRef]

145. Wang, G.; Wang, Z.; Sarkar, F.H.; Wei, W. Targeting Prostate Cancer Stem Cells for Cancer Therapy. *Discov. Med.* **2012**, *13*, 135–142. [PubMed]

146. Collins, A.T.; Habib, F.K.; Maitland, N.J.; Neal, D.E. Identification and isolation of human prostate epithelial stem cells based on α2β1-integrin expression. *J. Cell Sci.* **2001**, *114*, 3865–3872. [PubMed]

147. Richardson, G.D.; Robson, C.N.; Lang, S.H.; Neal, D.E.; Maitland, N.J.; Collins, A.T. CD133, a novel marker for human prostatic epithelial stem cells. *J. Cell Sci.* **2004**, *117*, 3539–3545. [CrossRef]

148. Garraway, I.P.; Sun, W.; Tran, C.P.; Perner, S.; Zhang, B.; Goldstein, A.S.; Hahm, S.A.; Haider, M.; Head, C.S.; Reiter, R.E.; et al. Human prostate sphere-forming cells represent a subset of basal epithelial cells capable of glandular regeneration in vivo. *Prostate* **2009**, *70*, 491–501. [CrossRef] [PubMed]

149. Qin, W.; Zheng, Y.; Qian, B.-Z.; Zhao, M. Prostate Cancer Stem Cells and Nanotechnology: A Focus on Wnt Signaling. *Front. Pharmacol.* **2017**, *8*, 1–12. [CrossRef] [PubMed]

150. Mansukhani, A.; Coarfa, C.; Gunaratne, P.H.; Basilico, C.; Basu-Roy, U.; Lim, D.-S.; Seo, E. SOX2 Regulates YAP1 to Maintain Stemness and Determine Cell Fate in the Osteo-Adipo Lineage. *Cell Rep.* **2013**, *3*, 2075–2087.

151. Song, S.; Ajani, J.A.; Honjo, S.; Maru, D.M.; Chen, Q.; Scott, A.W.; Heallen, T.R.; Xiao, L.; Hofstetter, W.L.; Weston, B.; et al. Hippo coactivator YAP1 upregulates SOX9 and endows esophageal Cancer cells with stem-like properties. *Cancer Res.* **2014**, *74*, 4170–4182. [CrossRef]

152. Lai, C.-J.; Lin, C.-Y.; Liao, W.-Y.; Hour, T.-C.; Wang, H.-D.; Chuu, C.-P. CD44 Promotes Migration and Invasion of Docetaxel-Resistant Prostate Cancer Cells Likely via Induction of Hippo-Yap Signaling. *Cells* **2019**, *8*, 295. [CrossRef]

153. Guo, Y.; Cui, J.; Ji, Z.; Cheng, C.; Zhang, K.; Zhang, C.; Chu, M.; Zhao, Q.; Yu, Z.; Zhang, Y.; et al. miR-302/367/LATS2/YAP pathway is essential for prostate tumor-propagating cells and promotes the development of castration resistance. *Oncogene* **2017**, *36*, 6336–6347. [CrossRef] [PubMed]

154. Liu, N.; Mei, L.; Fan, X.; Tang, C.; Ji, X.; Hu, X.; Shi, W.; Qian, Y.; Hussain, M.; Wu, J.; et al. Phosphodiesterase 5/protein kinase G signal governs stemness of prostate cancer stem cells through Hippo pathway. *Cancer Lett.* **2016**, *378*, 38–50. [CrossRef] [PubMed]

155. Jiang, N.; Ke, B.; Hjort-Jensen, K.; Iglesias-Gato, D.; Wang, Z.; Chang, P.; Zhao, Y.; Niu, X.; Wu, T.; Peng, B.; et al. YAP1 regulates prostate cancer stem cell-like characteristics to promote castration resistant growth. *Oncotarget* **2017**, *8*, 115054–115067. [CrossRef] [PubMed]

156. Schmelzle, T.; Roma, G.; Ruchti, A.; Schübeler, D.; Clay, I.; Bardet, A.F.; Stein, C.; Bauer, A.; Bouwmeester, T.; Agarinis, C.; et al. YAP1 Exerts Its Transcriptional Control via TEAD-Mediated Activation of Enhancers. *PLOS Genet.* **2015**, *11*, e1005465.

157. Li, Z.; Zhao, B.; Wang, P.; Chen, F.; Dong, Z.; Yang, H. Structural insights into the YAP and TEAD complex service Structural insights into the YAP and TEAD complex. *Cell* **2010**, *3*, 8–187.

158. Holden, J.K.; Cunningham, C.N. Targeting the Hippo Pathway and Cancer through the TEAD Family of Transcription Factors. *Cancers* **2018**, *10*, 81. [CrossRef]

159. Liu-Chittenden, Y.; Huang, B.; Shim, J.S.; Chen, Q.; Lee, S.-J.; Anders, R.A.; Liu, J.O.; Pan, D. Genetic and pharmacological disruption of the TEAD-YAP complex suppresses the oncogenic activity of YAP. *Genes Dev.* **2012**, *26*, 1300–1305. [CrossRef]

160. Feng, W.; Dasari, V.R.; Gogoi, R.; Carey, D.J.; Nash, J.; Mazack, V. Verteporfin exhibits YAP-independent anti-proliferative and cytotoxic effects in endometrial cancer cells. *Oncotarget* **2017**, *8*, 28628–28640.

161. Lin, K.C.; Park, H.W.; Guan, K.-L. Deregulation and Therapeutic Potential of The Hippo Pathway in Cancer. *Annu. Rev. Cancer Biol* **2018**, *2*, 59–79. [CrossRef]

162. Zhang, L.; Li, P.; Ji, H.; Zhao, Y.; Yin, M.-X.; Lu, Y.; Wang, H.; Zhou, Z.; Zhang, W.; Lv, D.; et al. A novel partner of Scalloped regulates Hippo signaling via antagonizing Scalloped-Yorkie activity. *Cell Res.* **2013**, *23*, 1201–1214.

163. Zhang, W.; Gao, Y.; Li, P.; Shi, Z.; Guo, T.; Li, F.; Han, X.; Feng, Y.; Zheng, C.; Wang, Z.; et al. VGLL4 functions as a new tumor suppressor in lung cancer by negatively regulating the YAP-TEAD transcriptional complex. *Cell Res.* **2014**, *24*, 331–343. [CrossRef]

164. Jiao, S.; Wang, H.; Shi, Z.; Dong, A.; Zhang, W.; Song, X.; He, F.; Wang, Y.; Zhang, Z.; Wang, W.; et al. A Peptide Mimicking VGLL4 Function Acts as a YAP Antagonist Therapy against Gastric Cancer. *Cancer Cell* **2014**, *25*, 166–180. [CrossRef]

165. Zhang, Z.; Lin, Z.; Zhou, Z.; Shen, H.C.; Yan, S.F.; Mayweg, A.V.; Xu, Z.; Qin, N.; Wong, J.C.; Zhang, Z.; et al. Structure-Based Design and Synthesis of Potent Cyclic Peptides Inhibiting the YAP–TEAD Protein–Protein Interaction. *ACS Med. Chem. Lett.* **2014**, *5*, 993–998. [CrossRef]

166. Zhou, Z.; Hu, T.; Xu, Z.; Lin, Z.; Zhang, Z.; Feng, T.; Zhu, L.; Rong, Y.; Shen, H.; Luk, J.M.; et al. Targeting Hippo pathway by specific interruption of YAP-TEAD interaction using cyclic YAP-like peptides. *FASEB J.* **2015**, *29*, 724–732. [CrossRef] [PubMed]

167. Gronich, N.; Rennert, G. Beyond aspirin - Cancer prevention with statins, metformin and bisphosphonates. *Nat. Rev. Clin. Oncol.* **2013**, *10*, 625–642. [CrossRef] [PubMed]

168. Stancu, C.; Sima, A. Statins: Mechanism of action and effects. *J. Cell. Mol. Med.* **2001**, *5*, 378–387. [CrossRef] [PubMed]

169. Sorrentino, G.; Ruggeri, N.; Specchia, V.; Cordenonsi, M.; Mano, M.; Dupont, S.; Manfrin, A.; Ingallina, E.; Sommaggio, R.; Piazza, S.; et al. Metabolic control of YAP and TAZ by the mevalonate pathway. *Nat. Cell Biol.* **2014**, *16*, 357–366. [CrossRef]

170. Wang, Z.; Wu, Y.; Wang, H.; Zhang, Y.; Mei, L.; Fang, X.; Zhang, X.; Zhang, F.; Chen, H.; Liu, Y.; et al. Interplay of mevalonate and Hippo pathways regulates RHAMM transcription via YAP to modulate breast cancer cell motility. *Proc. Natl. Acad. Sci. USA* **2014**, *111*, E89–E98. [CrossRef]

171. Babcook, M.A.; Sramkoski, R.M.; Fujioka, H.; Daneshgari, F.; Almasan, A.; Shukla, S.; Nanavaty, R.R.; Gupta, S. Combination simvastatin and metformin induces G1-phase cell cycle arrest and Ripk1- and Ripk3-dependent necrosis in C4-2B osseous metastatic castration-resistant prostate cancer cells. *Cell Death Dis.* **2014**, *5*, e1536. [CrossRef]

172. Ho, W.; Choo, D.-W.; Wu, Y.-J.; Chan, T.-F.; Lin, Z.-F. Statins Use and the Risk of Prostate Cancer in Ischemic Heart Disease Patients in Taiwan. *Clin. Pharmacol. Ther.* **2019**. [CrossRef]

173. Van Rompay, M.I.; Solomon, K.R.; Nickel, J.C.; Ranganathan, G.; Kantoff, P.W.; McKinlay, J.B. Prostate cancer incidence and mortality among men using statins and non-statin lipid-lowering medications. *Eur. J. Cancer* **2019**, 1–9. [CrossRef]

174. O'Neill, E.; Rushworth, L.; Baccarini, M.; Kolch, W. Role of the Kinase MST2 in Suppression of Apoptosis by the Proto-Oncogene Product Raf-1. *Science* **2004**, *306*, 2267–2270. [CrossRef] [PubMed]

175. Monia, B.P.; Sasmor, H.; Johnston, J.F.; Freier, S.M.; Lesnik, E.A.; Muller, M.; Geiger, T.; Altmann, K.-H.; Moser, H.; Fabbro, D. Sequence-specific antitumor activity of a phosphorothioate oligodeoxyribonucleotide targeted to human C- raf kinase supports an antisense mechanism of action in vivo. *Proc. Natl. Acad. Sci. USA* **1996**, *93*, 15481–15484. [CrossRef]

176. Khazak, V.; Astsaturov, I.; Serebriiskii, I.G.; Golemis, E.A. Selective Raf inhibition in cancer therapy. *Expert Opin. Ther. Targets* **2007**, *11*, 1587–1609. [CrossRef] [PubMed]

177. Cripps, M.C.; Figueredo, A.T.; Amit, M.O.; Taylor, M.J.; Fields, A.L.; Holmlund, J.T.; McIntosh, L.W.; Geary, R.S.; Eisenhauer, E.A. Phase II randomized study of ISIS 3521 and ISIS 5132 in patients with locally advanced or metastatic colorectal cancer: A National Cancer Institute of Canada clinical trials group study. *Clin. Cancer Res.* **2002**, *8*, 2188–2192. [PubMed]

178. Oza, A.M.; Elit, L.; Swenerton, K.; Faught, W.; Ghatage, P.; Carey, M.; McIntosh, L.; Dorr, A.; Holmlund, J.T.; Eisenhauer, E. Phase II study of CGP 69846A (ISIS 5132) in recurrent epithelial ovarian cancer: An NCIC clinical trials group study (NCIC IND.116). *Gynecol. Oncol.* **2003**, *89*, 129–133. [CrossRef]

179. Tolcher, A.W.; Reyno, L.; Venner, P.M.; Ernst, S.D.; Moore, M.; Geary, R.S.; Chi, K.; Hall, S.; Walsh, W.; Eisenhauer, E. A randomized Phase II and pharmacokinetic study of the antisense oligonucleotides ISIS 3521 and ISIS 5132 in patients with hormone-refractory prostate cancer. *Clin. Cancer Res.* **2002**, *8*, 2530–2535.

180. Murray, B.; Hayward, M.M.; Kurumbail, R.G. CHAPTER 14. The Future of Kinase Therapeutics. In *Kinase Drug Discovery: Modern Approchase*; The Royal Society of Chemistry: London, UK, 2018; pp. 381–405, ISBN 9781788013093.

181. Ferguson, F.M.; Gray, N.S. Kinase inhibitors: The road ahead. *Nat. Rev. Drug Discov.* **2018**, *17*, 353–376. [CrossRef] [PubMed]

182. Azzolin, L.; Panciera, T.; Soligo, S.; Enzo, E.; Bicciato, S.; Dupont, S.; Bresolin, S.; Frasson, C.; Basso, G.; Guzzardo, V.; et al. YAP/TAZ Incorporation in the β-Catenin Destruction Complex Orchestrates the Wnt Response. *Cell* **2014**, *158*, 157–170. [CrossRef] [PubMed]

183. Azzolin, L.; Zanconato, F.; Bresolin, S.; Forcato, M.; Basso, G.; Bicciato, S.; Cordenonsi, M.; Piccolo, S. Role of TAZ as mediator of wnt signaling. *Cell* **2012**, *151*, 1443–1456. [CrossRef]

184. Cai, J.; Maitra, A.; Anders, R.A.; Taketo, M.M.; Pan, D. β-catenin destruction complex-independent regulation of Hippo-YAP signaling by APC in intestinal tumorigenesis. *Genes Dev.* **2015**, *29*, 1493–1506. [CrossRef]

185. Yardy, G.W.; Brewster, S.F. Wnt signalling and prostate cancer. *Prostate Cancer Prostatic Dis.* **2005**, *8*, 119–126. [CrossRef]

186. Bruxvoort, K.J.; Charbonneau, H.M.; Giambernardi, T.A.; Goolsby, J.C.; Qian, C.-N.; Zylstra, C.R.; Robinson, D.R.; Roy-Burman, P.; Shaw, A.K.; Buckner-Berghuis, B.D.; et al. Inactivation of Apc in the Mouse Prostate Causes Prostate Carcinoma. *Cancer Res.* **2007**, *67*, 2490–2496. [CrossRef]

187. Yu, F.; Zhao, B.; Panupinthu, N.; Jewell, J.L.; Lian, I.; Wang, L.H.; Zhao, J.; Tumaneng, K.; Li, H.; Fu, X.; et al. Regulation Of The Hippo-YAP Pathway By G-Protein- Coupled Receptor Signaling. *Cell* **2012**, *150*, 1–11. [CrossRef]

188. Rabanal-Ruiz, Y.; Korolchuk, V.I. mTORC1 and Nutrient Homeostasis: The Central Role of the Lysosome. *Int. J. Mol. Sci.* **2018**, *19*, 818. [CrossRef] [PubMed]

189. Kim, J.; Guan, K.-L. mTOR as a central hub of nutrient signalling and cell growth. *Nat. Cell Biol.* **2019**, *21*, 63–71. [CrossRef]

190. Carver, B.S.; Chapinski, C.; Wongvipat, J.; Hieronymus, H.; Chen, Y.; Chandarlapaty, S.; Arora, V.K.; Le, C.; Koutcher, J.; Scher, H.; et al. Reciprocal Feedback Regulation of PI3K and Androgen Receptor Signaling in PTEN-Deficient Prostate Cancer. *Cancer Cell* **2011**, *19*, 575–586. [CrossRef] [PubMed]

191. Mulholland, D.J.; Tran, L.M.; Li, Y.; Cai, H.; Morim, A.; Wang, S.; Plaisier, S.; Garraway, I.P.; Huang, J.; Graeber, T.G.; et al. Cell autonomous role of PTEN in regulating castration-resistant prostate cancer growth. *Cancer Cell* **2011**, *19*, 792–804. [CrossRef]

192. Crumbaker, M.; Khoja, L.; Joshua, A.M. AR signaling and the PI3K pathway in prostate cancer. *Cancers* **2017**, *9*, 34. [CrossRef]

193. Blattner, M.; Liu, D.; Robinson, B.D.; Huang, D.; Poliakov, A.; Gao, D.; Nataraj, S.; Deonarine, L.D.; Augello, M.A.; Sailer, V.; et al. SPOP Mutation Drives Prostate Tumorigenesis In Vivo through Coordinate Regulation of PI3K/mTOR and AR Signaling. *Cancer Cell* **2017**, *31*, 436–451. [CrossRef] [PubMed]

194. Tumaneng, K.; Schlegelmilch, K.; Russell, R.C.; Yimlamai, D.; Basnet, H.; Mahadevan, N.; Fitamant, J.; Bardeesy, N.; Camargo, F.D.; Guan, K.L. YAP mediates crosstalk between the Hippo and PI(3)K-TOR pathways by suppressing PTEN via miR-29. *Nat. Cell Biol.* **2012**, *14*, 1322–1329. [CrossRef] [PubMed]

195. Hansen, C.G.; Ng, Y.L.D.; Lam, W.-L.M.; Plouffe, S.W.; Guan, K.-L. The Hippo pathway effectors YAP and TAZ promote cell growth by modulating amino acid signaling to mTORC1. *Cell Res.* **2015**, *25*, 1299–1313. [CrossRef]

196. Park, Y.-Y.; Lee, J.-S.; Rupaimoole, R.; Jang, H.-J.; Mills, G.B.; Park, Y.N.; Jeong, W.; Rodriguez-Aguayo, C.; Sood, A.K.; Yoo, J.E.; et al. Yes-associated protein 1 and transcriptional coactivator with PDZ-binding motif activate the mammalian target of rapamycin complex 1 pathway by regulating amino acid transporters in hepatocellular carcinoma. *Hepatology* **2015**, *63*, 159–172. [CrossRef] [PubMed]

197. Nicklin, P.; Bergman, P.; Zhang, B.; Triantafellow, E.; Wang, H.; Nyfeler, B.; Yang, H.; Hild, M.; Kung, C.; Wilson, C.; et al. Bidirectional Transport of Amino Acids Regulates mTOR and Autophagy. *Cell* **2009**, *136*, 521–534. [CrossRef] [PubMed]

198. Hu, J.K.H.; Du, W.; Shelton, S.J.; Oldham, M.C.; DiPersio, C.M.; Klein, O.D. An FAK-YAP-mTOR Signaling Axis Regulates Stem Cell-Based Tissue Renewal in Mice. *Cell Stem Cell* **2017**, *21*, 91–106.e6. [CrossRef]

199. Whitmarsh, A.J.; Davis, R.J. Transcription factor AP-1 regulation by mitogen-activated protein kinase signal transduction pathways. *J. Mol. Med.* **1996**, *74*, 589–607. [CrossRef]

200. Eferl, R.; Wagner, E.F. AP-1: A double-edged sword in tumorigenesis. *Nat. Rev. Cancer* **2003**, *3*, 859–868. [CrossRef]

201. Rössler, O.G.; Henß, I.; Thiel, G. Transcriptional response to muscarinic acetylcholine receptor stimulation: Regulation of Egr-1 biosynthesis by ERK, Elk-1, MKP-1, and calcineurin in carbachol-stimulated human neuroblastoma cells. *Arch. Biochem. Biophys.* **2008**, *470*, 93–102. [CrossRef] [PubMed]

202. Thiel, G.; Rössler, O.G. Immediate-Early Transcriptional Response to Angiotensin II in Human Adrenocortical Cells. *Endocrinology* **2011**, *152*, 4211–4223. [CrossRef] [PubMed]

203. Kim, J.; Woolridge, S.; Biffi, R.; Borghi, E.; Lassak, A.; Ferrante, P.; Amini, S.; Khalili, K.; Safak, M. Members of the AP-1 Family, c-Jun and c-Fos, Functionally Interact with JC Virus Early Regulatory Protein Large T Antigen. *J. Virol.* **2003**, *77*, 5241–5252. [CrossRef]

204. Zanconato, F.; Forcato, M.; Battilana, G.; Azzolin, L.; Quaranta, E.; Bodega, B.; Rosato, A.; Bicciato, S.; Cordenonsi, M.; Piccolo, S. Genome-wide association between YAP/TAZ/TEAD and AP-1 at enhancers drives oncogenic growth. *Nat. Cell Biol.* **2015**, *17*, 1218–1227. [CrossRef] [PubMed]

205. Maglic, D.; Schlegelmilch, K.; Dost, A.F.; Panero, R.; Dill, M.T.; Calogero, R.A.; Camargo, F.D. YAP-TEAD signaling promotes basal cell carcinoma development via a c-JUN/AP1 axis. *EMBO J.* **2018**, *37*, e98642. [CrossRef]

206. Zhu, L.J.; Mercurio, A.M.; Li, H.; Cotton, J.L.; Ou, J.; Mao, J.; Liu, X.; Davis, R.J.; Li, Q.; Park, J.-S.; et al. Tead and AP1 Coordinate Transcription and Motility. *Cell Rep.* **2016**, *14*, 1169–1180.

207. Lin, S.-H.; Zhou, H.-J.; Tsai, M.-J.; Ittmann, M.; Erdem, H.; Ayala, G.; Yan, J.; Tsai, S.Y.; Luo, W. SRC-3 Is Required for Prostate Cancer Cell Proliferation and Survival. *Cancer Res.* **2017**, *65*, 7976–7983.

208. Thiel, G.; Welck, J.; Wissenbach, U.; Rössler, O.G. Dihydrotestosterone activates AP-1 in LNCaP prostate cancer cells. *Int. J. Biochem. Cell Biol.* **2019**, *110*, 9–20. [CrossRef]

209. Wanjala, J.; Taylor, B.S.; Chapinski, C.; Hieronymus, H.; Wongvipat, J.; Chen, Y.; Nanjangud, G.J.; Schultz, N.; Xie, Y.; Liu, S.; et al. Identifying Actionable Targets through Integrative Analyses of GEM Model and Human Prostate Cancer Genomic Profiling. *Mol. Cancer Ther.* **2015**, *14*, 278–288. [CrossRef] [PubMed]

210. Koo, K.M.; Mainwaring, P.N.; Tomlins, S.A.; Trau, M. Merging new-age biomarkers and nanodiagnostics for precision prostate cancer management. *Nat. Rev. Urol.* **2019**. [CrossRef] [PubMed]

211. Downing, A.; Wright, P.; Hounsome, L.; Selby, P.; Wilding, S.; Watson, E.; Wagland, R.; Kind, P.; Donnelly, D.W.; Butcher, H.; et al. Quality of life in men living with advanced and localised prostate cancer in the UK: A population-based study. *Lancet Oncol.* **2019**, *20*, 436–447. [CrossRef]

212. Arora, V.K.; Schenkein, E.; Murali, R.; Subudhi, S.K.; Wongvipat, J.; Balbas, M.D.; Shah, N.; Cai, L.; Efstathiou, E.; Logothetis, C.; et al. Glucocorticoid receptor confers resistance to antiandrogens by bypassing androgen receptor blockade. *Cell* **2013**, *155*, 1309–1322. [CrossRef]

213. Sorrentino, G.; Ruggeri, N.; Zannini, A.; Ingallina, E.; Bertolio, R.; Marotta, C.; Neri, C.; Cappuzzello, E.; Forcato, M.; Rosato, A.; et al. Glucocorticoid receptor signalling activates YAP in breast cancer. *Nat. Commun.* **2017**, *8*, 1–14. [CrossRef] [PubMed]

214. Fu, D.; Lv, X.; Hua, G.; He, C.; Dong, J.; Lele, S.M.; Li, D.W.-C.; Zhai, Q.; Davis, J.S.; Wang, C. YAP regulates cell proliferation, migration, and steroidogenesis in adult granulosa cell tumors. *Endocr. Relat. Cancer* **2014**, *21*, 297–310. [CrossRef] [PubMed]

215. Bonkhoff, H. Estrogen receptor signaling in prostate cancer: Implications for carcinogenesis and tumor progression. *Prostate* **2018**, *78*, 2–10. [CrossRef] [PubMed]

216. Di Zazzo, E.; Galasso, G.; Giovannelli, P.; Di Donato, M.; Castoria, G. Estrogens and Their Receptors in Prostate Cancer: Therapeutic Implications. *Front. Oncol.* **2018**, *8*, 1–7. [CrossRef] [PubMed]

217. Risbridger, G.P.; Davis, I.D.; Birrell, S.N.; Tilley, W.D. Breast and prostate cancer: More similar than different. *Nat. Rev. Cancer* **2010**, *10*, 205–212. [CrossRef] [PubMed]

Review

The Role of YAP and TAZ in Angiogenesis and Vascular Mimicry

Taha Azad [1], Mina Ghahremani [2] and Xiaolong Yang [1],*

[1] Department of Pathology and Molecular Medicine, Queen's University, Kingston, ON K7L 3N6, Canada; taha.biology@gmail.com
[2] Department of Biology, Queen's University, Kingston, ON K7L 3N6, Canada; m.gh.nbt@gmail.com
* Correspondence: yangx@queensu.ca; Tel.: +1-613-533-6000 (ext. 75998)

Received: 29 March 2019; Accepted: 30 April 2019; Published: 1 May 2019

Abstract: Angiogenesis, the formation of new blood vessels from pre-existing vasculature, is a physiological process that begins in utero and continues throughout life in both good health and disease. Understanding the underlying mechanism in angiogenesis could uncover a new therapeutic approach in pathological angiogenesis. Since its discovery, the Hippo signaling pathway has emerged as a key player in controlling organ size and tissue homeostasis. Recently, new studies have discovered that Hippo and two of its main effectors, Yes-associated protein (YAP) and its paralog transcription activator with PDZ binding motif (TAZ), play critical roles during angiogenesis. In this review, we summarize the mechanisms by which YAP/TAZ regulate endothelial cell shape, behavior, and function in angiogenesis. We further discuss how YAP/TAZ function as part of developmental and pathological angiogenesis. Finally, we review the role of YAP/TAZ in tumor vascular mimicry and propose directions for future work.

Keywords: Hippo pathway; cancer; angiogenesis; vascular mimicry; YAP; TAZ; MST1/2; LATS1/2

1. Introduction

1.1. Angiogenesis and Its Roles in Development and Pathogenesis

1.1.1. Angiogenesis

Angiogenesis is the formation of new blood vessels from pre-existing vasculature. There are two main types of angiogenesis: sprouting angiogenesis and intussusceptive angiogenesis [1,2]. In sprouting angiogenesis, enzymatic degradation of the capillary basement membrane weakens the endothelial cell–cell interaction, which allows endothelial cells to proliferate and sprout toward an angiogenic stimulator. During this stage, endothelial cells flatten to make a long, slender, tapering pseudopodium, called filopodia; these are structures with a highly expanded endoplasmic reticulum and Golgi apparatus, high in proteolytic enzyme production. Proteolytic enzymes secreted from filopodia cleave the extracellular matrix and open a pathway for the developing sprout [3]. Intussusceptive angiogenesis, on the other hand, does not rely on endothelial cell proliferation or migration. In this type of angiogenesis, which is considered a fast angiogenesis compared to sprouting angiogenesis, an existing vessel is split into two new vessels just by cellular reorganization. Based on recent studies using scanning electron micrographs of Mercox casts, it has been shown that both types of angiogenesis occur frequently [4]. However, as the intussusceptive angiogenesis process is faster, it normally occurs in the early stage of angiogenesis compared to sprouting angiogenesis.

A variety of growth signaling pathways regulate angiogenesis (Figure 1). So far, vascular endothelial growth-factor A (VEGF-A) is the best-understood pro-angiogenic signal. Loss of the single allele or minor inhibition of this ligand is lethal during the developmental angiogenesis. VEGF-A binds to two receptor tyrosine kinases (RTKs): vascular endothelial growth factor receptor 1 (VEGFR1/Flt1) and VEGFR2 (KDR). Usually, VEGF-A induces signaling pathways related to angiogenesis. For example, it increases the secretion of matrix-metalloproteinases (MMPs) as well as endothelial cell proliferation [5]. Interestingly, it seems that upstream signaling pathways of angiogenesis are highly redundant and several other RTKs such as the fibroblast growth factor receptor (FGFR), ephrin, and platelet-derived growth factor receptor (PDGFR) can show the same effects as VEGFRs. Two very similar outputs of these signaling pathways are: (1) hypoxia-inducible factor 1 (HIF-1) transcriptional activator hetrodimerization, which usually occurs under low oxygen conditions and is involved in the adaptation to both cellular and organismal hypoxia; and (2) MMP activation, which causes degradation of the extracellular matrix to release more growth factors and to make it easier for endothelial cells to migrate.

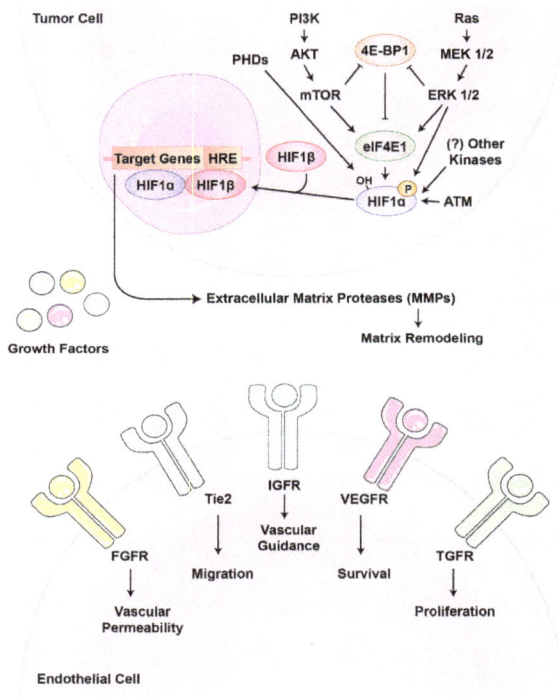

Figure 1. Tumor angiogenesis. In tumor cells, several signaling pathways such as phosphoinositide 3-kinase (PI3K) and mitogen-activated protein kinase (MAPK) can phosphorylate eIF4E1 to increase the HIF-1α translational level. Also, direct phosphorylation and hydroxylation are two important post-translational modifications that cause HIF-1 dimerization and promote its degradation, respectively. Activated HIF-1 then functions as a master switch to induce the expression of several growth factors such as VEGF, PDGF, and Ang2, as well as extracellular matrix proteases such as matrix metalloproteinase (MMP). MMPs allow the endothelial cells to escape into the interstitial matrix during sprouting angiogenesis. Also, secreted growth factors induce survival, proliferation, migration, and vascular permeability in endothelial cells. Abbreviations—PHD: prolyl hydroxylase domain protein; ATM: ataxia telangiectasia mutated; HRE: HIF-1 response elements; HIF-1: hypoxia-inducible factor 1; MMP: matrix metalloproteinase; VEGF: vascular endothelial growth-factor; VEGFR: VEGF receptor; PDGF: platelet-derived growth factor.

1.1.2. Angiogenesis in Development

Angiogenesis plays critical roles in embryogenesis and cardiovascular maturation [6,7]. The cardiovascular system is the first organ formed during vertebrate development. This complex vascular network provides different tissues with nutrition and immune system support. In primitive animals such as the fruit fly Drosophila melanogaster and the worm Caenorhabditis elegans, oxygen diffuses through their skin to all cells in their body. During evolution, animals became more complex and larger in size. Generally, modern metazoans have low surface area to volume ratio, which makes it impossible to distribute necessary oxygen and other nutrients using simple diffusion [8]. As oxygen diffusion is limited to about 100–200 μm thickness, all multicellular animals beyond this size need to recruit blood vessels by angiogenesis so that oxygen and nutrients can be distributed to their cells. Many cells with different properties and origins as well as numerous small angiogenic molecules are involved in this highly regulated process.

Shortly after gastrulation, angioblasts (endothelial progenitor cells) start differentiating from mesoderm layer in embryo. Angioblasts then unite together and form a very primitive network. This process is the first stage of the formation of the vascular network and is called vasculogenesis [9]. After uniting, angioblasts produce different growth factors to recruit other none-endothelial cells such as pericytes and smooth muscle cells to the primitive vasculature. The main role of non-endothelial cells is deposition of a specific basement membrane, for stabilization of the new made vasculature [9]. This primitive blood vessel network can be a foundation for more new vessel formation, at this point called angiogenesis. In summary, vasculogenesis is de novo vascular network formation in the embryo, whereas angiogenesis results in new blood vessels from pre-existing vessels.

VEGF-A and its receptor tyrosine kinase is the best-known angiogenic factors involved in angiogenesis. VEGF belongs to a gene family that includes several isoforms, namely VEGF-A, B, and C [10]. We will talk more about VEGF isoforms and their specific function in the following sections. Briefly, the VEGF-A gradient activates tip cells at the site of angiogenesis. These cells are very motile and guide the angiogenesis towards the angiogenic factors [11]. The Notch signaling pathway plays important role in regulation of VEGF-A response in endothelial cells to prevent excess angiogenesis. Adjacent stalk endothelial cells have high expression of Notch 1 and Notch 4. VEGF-A up-regulates the expression of delta-like ligand 4 (DLL-4) in tip cells, which binds to its receptor Notch 1 and Notch 4, and down-regulates VEGFR-2 expression in these cells [12]. At a short distance from tip cells, a critical step in the development of blood vessels and lumen formation and extension occurs. As it is difficult to mark the initial steps in tube formation, there is a lot of debate about its mechanism [13].

At the same time of angiogenesis the vessel start specifying into arteries, capillaries, and veins. It is assumed that the difference in blood flow is the main factor responsible for establishing arterial versus venous identity. It has been shown that the pattern of Eph receptor expression is different in arteries, capillaries, and veins. However, it is not clear whether or not the difference in Eph receptor expression is the result of arterial and venous formation or the cause [14]. Another important process happens is vascular anastomosis, which generates connection between angiogenic sprouts and blood vessels. Generally, there are two types of anastomosis, head-to-head and head-to-side. In head-to-head anastomosis, two tip cells and in head-to-side one tip cell and one functional blood vessel generate connections [13].

1.1.3. Angiogenesis in Pathogenesis

Angiogenesis needs to be accurately controlled during development and adulthood. Therefore, not surprisingly, an imbalance in angiogenesis contributes to many diseases, including cancer, age-related macular degeneration (AMD), diabetic retinopathy, immune disorders, inflammation, infection, and ischemia [15–17]. So far, the majority of the reports on angiogenesis in human diseases come from studies on the roles of angiogenesis in tumorigenesis. There is much evidence indicating a dependency of tumor growth and metastasis on neovascularization occurring within and surrounding a tumor; normally, tumors reach a "steady state" size of roughly 2 mm diameter when not fed by neovascularization [18].

In addition to supporting tumor growth, blood vessels offer a means for cancer cells to metastasize away from the primary tumor. Tumor cells survive in the circulation, target new organs, and induce angiogenesis where they seed, perpetuating a vicious cycle. After embryogenesis, endothelial cells in vasculature become largely quiescent and only activated in certain circumstances, such as wound healing. During tumor expansion, however, the angiogenic switch is turned on, activating quiescent endothelial cells to sprout new vessels [19]. Mounting evidence suggests that the angiogenic switch is activated when tumors cause an imbalance in ratio of pro- and anti-angiogenic molecules. Usually, tumors increase the pro-angiogenic factor in different ways. For example, increasing tumor size reduces the tumor oxygen exchange rate, which activates the hypoxia signaling pathway. As result of this activation, the expression of HIF increases, which subsequently results in upregulation of pro-angiogenic molecules VEGF and FGF, and MMPs to increase angiogenesis (Figure 1). In addition, it has been shown that up-regulation of some oncogene signaling pathways such as PI3K and MAPK can contribute to increasing or reducing angiogenesis regulators. In addition, increasing MMPs and other matrix degrading enzymes in aggressive tumors can also contribute to angiogenesis by releasing such pro-angiogenic molecules from extracellular matrix as FGF and VEGF ligands, which are tethered to the extracellular matrix through their heparin-binding motif. Moreover, thrombospondin 1 (TSP-1), which binds and suppresses transmembrane receptors in endothelial cells, is an important counterbalance in the angiogenic switch. However, the signaling pathways and mechanisms in tumors that down-regulate TSP-1 are not well known [20].

Blood vessels within tumors have distinct features different from those of normal blood vessels. Their blood flow is not as consistent as that in normal blood vessels, which causes micro-hemorrhage and sometimes bleeding in the tumors. Furthermore, these vessels are very active in branching and sprouting new vessels. Metabolism, cell proliferation, and even cell death in the NE vasculature endothelial cells are very high in tumor vessels [21]. Since the endothelial cells are almost always the genetically intact cells from somatic quiescent cells, such different vessel morphology and endothelial cell behavior and metabolism are likely the result of tumor cells that create an imbalance in angiogenic molecule levels [21]. Even though some signaling pathways related to VEGF, FGFR, and HIF-1 have been studied very well, many questions remain to be answered.

1.2. Yes-Associated Protein (YAP) and Paralog Transcription Activator with PDZ Binding Motif (TAZ)

YAP and its paralog are WW domain-containing transcriptional co-activators and major effectors of the Hippo signaling pathway, which regulates tissue homeostasis, organ size, cell death, stem cell renewal, and mechanotransduction [22–26]. The final output of the Hippo pathway is the inhibition of YAP/TAZ co-transactivational activity by large tumor suppressor (LATS)1/2 kinases, and thereby their cytoplasmic sequestration and ubiquitin ligase-dependent degradation (Figure 2) [27–29]. The four main components of the Hippo pathway, including Warts [30], Salvador [31], Hippo [32], and the adaptor protein Mats [33] were found during the screening of tumor suppressor genes in *Drosophila* [34]. Hippo, Sav, Wts, and Mats in *Drosophila* are conserved proteins and homologous to mammalian mammalian STE20-like protein kinas 1/2 (MST1/2), WW45, LATS1/2, and Mps one binder 1 (MOB1), respectively [35,36]. In mammalians MST1/2 serine/threonine (S/T) kinases play key role in the Hippo pathway, as it is able to phosphorylate and activate three other components, including LATS, MOB, and Salvador [37–39]. When LATS1/2 S/T kinases are activated, they bind to and phosphorylate YAP/TAZ at five different conserved HxH/R/KxxS/T (H, histidine; R, arginine; K, lysine; x, any amino acid) motifs, including YAP S127 and TAZ S89 [33,36,40,41]. LATS-dependent phosphorylation of YAP/TAZ produces an interaction site for phospho-protein-binding protein 14-3-3, which inhibits YAP/TAZ nuclear localization and its co-transactivation of downstream genes with transcription factors such as TEA domain family protein (TEAD) and AP1 (Figure 2).

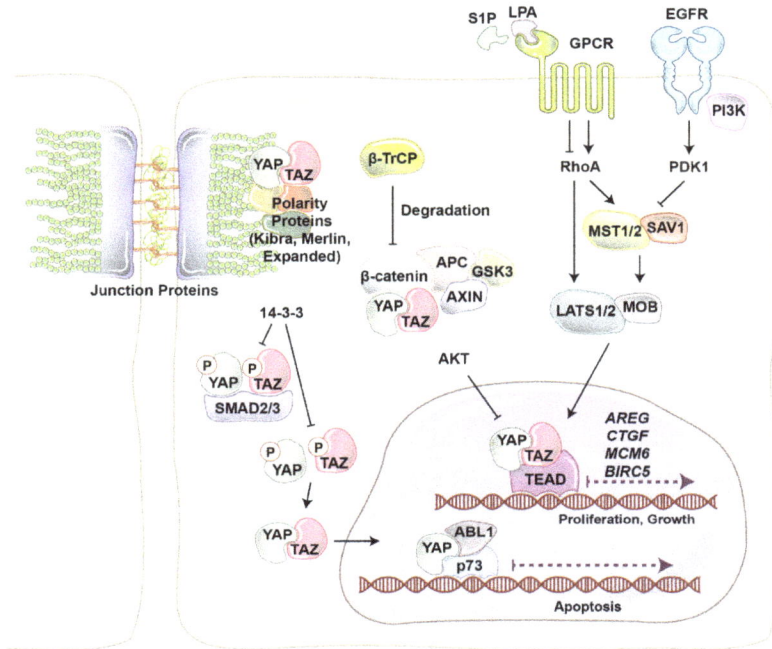

Figure 2. An overview of the regulation of YAP and TAZ transcriptional co-activators. YAP and TAZ are downstream mediators of numerous signaling pathways such as G-protein couple receptors (GPCRs) and epidermal growth factor (EGFR). YAP and TAZ localization is mainly regulated through phosphorylation by large tumor suppressor (LATS). The 14-3-3 phosphobinding protein interacts with and sequesters phosphorylated YAP and TAZ. YAP and TAZ localization is also regulated through physical interaction, for example with SMAD, β-catenin, and junction proteins. YAP: Yes-associated protein (YAP); TAZ: transcription activator with PDZ binding motif.

YAP/TAZ play a critical role in regulating many cellular behaviors in response to various internal and external stimuli [42]. For example, YAP/TAZ have been identified as conserved mechanotransducers for sensing diverse mechanical cues such as shear stress, cell shape, and extracellular matrix rigidity, and translating them into cell-specific transcriptional programs [43]. Cell extra-cellular matrix conformational change and mechanical stresses activate Rho GTPase mediated actin polymerization. Filamentous actin (F-actin) inhibits LATS activity and induces YAP/TAZ nuclear localization (Figure 2). Junction proteins can also regulate YAP/TAZ localization and activity [25]. Merlin (protein of the neurofibromatosis 2 (*NF2*) gene) directly interacts with angiomotin (AMOT) and α-catenin to recruit LATS kinase to adherent junction. Cross phosphorylation between AMOT and LATS at adherence junction results in YAP/TAZ phosphorylation and cytoplasmic retention. Scribble is a scaffold protein which recruits MST and LATS to basolateral junction and cause the same outcome. Junctions protein can also regulate YAP/TAZ activity just by sequestering them. It has been reported that AMOT and α-catenin can physically sequester YAP/TAZ in tight and adherent junctions [44,45].

YAP/TAZ also respond to extracellular cues such as hormones and growth factors. It has been shown that serum-borne lysophosphatidic acid (LPA) and sphingosine 1-phosphophate (S1P) act through a group of G-protein coupled receptors (GPCRs), G12/13-coupled receptors, to induce cell proliferation and migration. YAP/TAZ are necessary for G12/13-coupled receptors induced function. Rho GTPase is the main connector of GPCRs and YAP/TAZ. In addition, it has been discovered that epinephrine and glucagon can also regulate YAP/TAZ through a similar pathway [46]. In addition to

GPCRs, RTKs are other important cell membrane proteins that regulate YAP/TAZ function. Ligand binding induces RTK dimerization at the cell membrane [47]. Two kinase domains cross-phosphorylate each other, which causes increasing kinase activity. The activated kinase domains phosphorylate other sites and produce docking sites for intracellular signaling proteins. The activated RTK and signaling proteins form a signaling complex that broadcasts signals along other signaling pathways. It has been shown that PI3-kinase (PI3K), one of the main downstream signaling pathways of RTKs, induces YAP/TAZ nuclear localization through inhibition of LATS activity (Figure 2) [48,49]. Recently, we provided the first evidence that the Hippo pathway effectors TAZ and YAP are critical mediators of PI3K-induced mammary tumorigenesis and synergistically function together with PI3K in transformation of mammary cells [50].

2. Roles of YAP/TAZ in the Regulation of Endothelial Function during Angiogenesis

Angiogenesis is a complex process with a series of sequential events. Endothelial cells as building block of vasculatures play a critical role in this event. During early stage of angiogenesis, endothelial cells loosen their junctions with other cells, change their shape and increase their motility. Therefore, to gain a better understanding of angiogenesis, the regulation of endothelial cell shape and behaviors should be firstly studied. Endothelial cells can be in quiescent, proliferating, or differentiating state according to the stimuli they received from their environment. If endothelial cells are seeded into collagen-coated plates, they enter to a high proliferating state. However, soon after plating the same cells in Engelbreth–Holm–Swarm mouse sarcoma (matrigel), the cells stop proliferating and differentiate to tube-like structure within 8–12 h. The comparison of gene expression by endothelial cells in these two distinct states reveal several important regulators, such as CTGF, CYR61, and angiopoietin-2, all of which are known as transcriptional target of YAP/TAZ [51]. Significantly, many genes and signaling pathways have also been shown to regulate endothelial cell shape and behavior though YAP/TAZ.

Most in vitro cancer angiogenesis assays use endothelial cells. Bovine aortic endothelial cells (BAECs) and human umbilical vein endothelial cells (HUVECs) are two main cell types that are used in these assays. Although it is easy to harvest the aforementioned cells form blood vessels and then expand them in primary cell cultures, they do not represent high tumour blood vessel heterogeneity [52–55]. Furthermore, in vitro conditions rarely reflect the in vivo environment, as they cannot expose the cells to hemodynamic forces, which are necessary to activate several signaling pathways in endothelial cells. Due to these limitations, many scientists are focusing on developing new methods to study the function of key component of angiogenesis. Recently, Cuia et al. [56] developed a three-dimensional (3D) microfluidic angiogenesis model to study immune-vascular and cell-matrix interactions. By using this model they demonstrated that soluble immunosuppressive cytokines and endothelial–macrophage interactions are involved in regulating glioblastoma tumor angiogenesis. It was observed that endothelial–macrophage interactions regulate in vitro proangiogenic activity through integrin receptors and Src–PI3K–YAP signaling. In another study, a 3D culture model was developed in auxetic scaffolds to evaluate vascular differentiation [57]. By using this model it was confirmed that cytoplasmic YAP expression of cells from auxetic scaffolds contributes to the enhanced vascular differentiation and angiogenesis. Therefore, developing new models, which are closer to in vivo models, will facilitate uncovering unknown functions of the YAP/TAZ and the Hippo pathway in angiogenesis.

2.1. Angiomotin (AMOT) Family

The AMOT family contains three members: AMOT, AMOT like 1 (AMOTL1), and AMOT-like protein 2 (AMOTL2). AMOT is expressed in two different isoforms through alternative splicing, p130-AMOT, and p80-AMOT (lacks the N-terminal 400 amino acids present in the p130 isoform). It is a membrane associated protein that regulates endothelial cells motility and behavior during angiogenesis [58]. p130-AMOT contains PPXY motifs, which mediate its interaction with the WW domain of YAP/TAZ [44]. p130-AMOT can inhibit transactivating function of YAP/TAZ by either bringing MST, LATS, and YAP to tight junctions and increasing YAP phosphorylation or physically sequestering YAP into F-actin. AMOT like 1 (AMOTL1) has 62% homology with AMOT and compensates for the absence of AMOT in mice models and exhibits similar expression pattern, controls endothelial polarity and junction stability during sprouting angiogenesis [59]. Studying AMOTL1 uncovered a novel function for the first time for cytoplasmic YAP. The E3 ubiquitin ligase Nedd4 targets AMOTL1 for ubiquitin-dependent degradation at tight junctions. Interestingly, YAP recruits c-Abl to tight junctions to phosphorylate E3 ubiquitin ligase Nedd4 on tyrosine. This tyrosine phosphorylation reduces the affinity of Nedd4 to AMOTL1, resulting in increasing the AMOTL1 and tight junction stability [60]. Surprisingly, another component of the Hippo, LATS can also regulate AMOT. LATS regulates AMOT function through direct phosphorylation on the LATS substrate conserved sequence, H/X/R/XXS. This phosphorylation site is S175 of AMOT, and has critical role in interaction with F-actin. When LATS phosphorylates Ser-175, it disrupts AMOT interaction with F-actin and reduces F-actin stress fibers and focal adhesions, which subsequently inhibits endothelial cell migration in vitro and zebrafish embryonic angiogenesis in vivo [61]. Since LATS is the major regulator of YAP/TAZ and AMOT and its interaction with Hippo signaling pathway can also regulate cell shape and motility of non-endothelial cells, which play important roles in cancer biology [62], how the LATS–AMOT interaction can affect YAP/TAZ function in angiogenesis is a critical question for future studies (Figure 3).

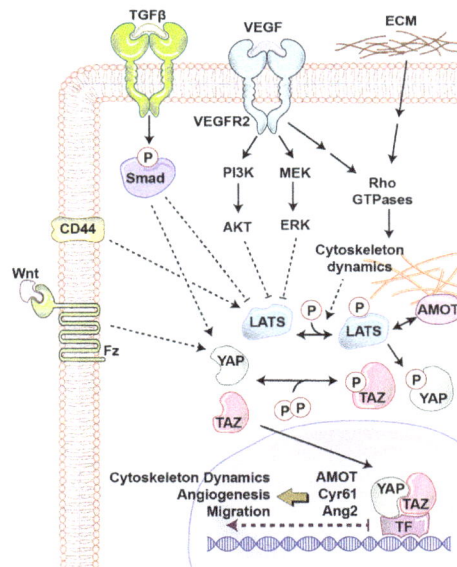

Figure 3. An overview of the regulation of YAP and TAZ transcriptional co-activators in angiogenesis. Several receptors regulate YAP and TAZ activity directly via LATS or other unknown pathways, which can affect angiogenesis. VEGFR regulates YAP and TAZ through three main pathways including the Rho GTPase, MAPK, and PI3K pathway. The TGF-β, Wnt, and CD44 pathways regulate YAP and TAZ as well as LATS activity through several not well-known mechanisms. Abbreviations—TF: transcription factor; ECM: extracellular matrix; Fz: frizzled receptor.

2.2. CD44

Vascular barrier integrity is another factor, which is mainly developed and regulated by endothelial cells during angiogenesis. This semi-selective barrier is disrupted in several diseases such as inflammation, atherosclerosis, and tumor angiogenesis. Endothelial cell-surface CD44 serves as a barrier regulatory receptor in endothelial cells [63]. Also, hyaluronan, a frequent glycosaminoglycan in the extracellular matrix, plays critical roles in angiogenesis, mainly through CD44 [64]. CD44 is also known for the regulation of matrix metalloproteinase level and activation, interactions with cortical membrane proteins, both of which plays a key role during angiogenesis. CD44 knock out mice show impaired barrier function, altered junctional organization, and dysfunctional endothelial cell junctions. It was observed that in these mice after angiogenesis, blood vessels express reduced levels of CD31 (also known as PECAM-1). Expression of CD31 restores barrier strength and endothelial cell morphology, which together with other evidence showing that CD44 regulates vascular endothelial barrier integrity and morphology via a CD31-dependent mechanism [65]. CD44 has been associated with one of upstream regulators of the Hippo pathway, Merlin [66]. Furthermore, CD44 deficiency causes reduction in CD31 and vascular endothelial (VE)-cadherin that has been connected with a reduction in tight junction, leading to Ajuba inhibition of LATS [67]. Another study showed independently that reduced CD31 expression induces Ajuba expression, resulting in LATS inactivation and YAP nuclear localization in cells originally derived from brain microvascular endothelial cells [68]. Additionally, it has been shown that CD44 regulates endothelial cell proliferation and survival by modulating cell adhesion molecules via the Hippo pathway [69,70]. Although these studies have shown CD44 as an upstream regulator of angiogenesis and the Hippo signaling pathway, how the CD44–Hippo interaction can affect YAP/TAZ function in angiogenesis is a critical question for future studies (Figure 3).

2.3. Extracellular Matrix (ECM)

One of the important factors that regulates endothelial cells' behavior in normal condition and during angiogenesis is the physical property of the microenvironment. This includes cell morphology, ECM stiffness or confined adhesiveness, and flow dependent endothelial cells regulation. In normal conditions, increasing cell density causes YAP/TAZ cytoplasmic localization. When the cells are confluent their morphology is different and covers less area. To answer the question whether YAP/TAZ localization is regulated by cell morphology, micropattern plates have been used. Each of these plates has many microdomains with different sizes and shapes. Culturing cells in these plates showed that in small domains, where cells were round, YAP was mostly cytoplasmic, whereas on larger domains, where cells were spread and flat, YAP was in nucleolus [71]. Endothelial cell spreading and ECM stiffness induce YAP/TAZ nuclear localization independent of the NF2/Hippo/LATS pathway, instead requiring Rho activity and the actomyosin cytoskeleton [72]. Nakajima et al. [73] studied the spatiotemporal localization and transcriptional activity of YAP in endothelial cells of in vivo and found blood flow regulates localization of YAP through mechanotransduction signaling during physiological condition and angiogenesis. They found LATS-independent YAP nuclear localization in HUVEC cells exposed to shear stress and endothelial cells in blood vessels with flow. They also found, during vessel remodeling (a type of intussusceptive angiogenesis), that YAP localized into the nucleus through a flow-regulated mechanism. Recently, by using micropatterns a new technique has been designed to mimic mechanosensitivity and study the influence of extracellular physical cues on endothelial mechanosensing during angiogenesis. Using, this novel method, they found potential divergent kinetics for two important transcriptional related proteins in endothelial cells, MRTF-A and YAP, during angiogenesis (Figure 3) [74].

2.4. Transforming Growth Factor-β (TGFβ)/Bone Morphogenic Protein (BMP)

Several signaling pathways play critical role in endothelial cells during angiogenesis. The most well-known one is VEGF/VEGFR pathway. We will discuss in detail about this pathway and its links to the Hippo pathway in the following sections. Transforming growth factor-β (TGFβ)/bone morphogenic protein (BMP) regulates proliferation, differentiation, migration, and survival depending on the cell type. In endothelial cells, TGFβ1, TGFβ2, and TGFβ3 are three most common ligands for serine/threonine kinase receptors, including TGFβ type II receptor, activin receptor-like kinase (ALK1) and ALK5. The latter two are known as TGFβ type I receptors. TGFβ receptor activation causes specific SMAD transcription factors phosphorylation and nuclear localization. In the nucleus SMAD interacts with other transcription factors, which leads to the transcription of a wide array of target genes. Several in vivo studies have shown the importance of the TGFβ signaling pathway in angiogenesis [75–77]. Interestingly, the Hippo pathway interacts with TGFβ pathway in several points. The RASS1FA scaffold protein regulates the TGFβ-induced YAP–SMAD interaction and SMAD cytoplasmic retention. However, when RASSF1A is degraded, TGF-β activation causes YAP/SMAD nuclear localization and gene transcription [78]. Also, in response to hypoxia and TGF-β stimuli, LATS is ubiquitinated and degraded, resulting in YAP dephosphorylation and subsequently activation [79]. Endoglin is a type III TGFβ receptor that is upregulated in endothelial cells during angiogenesis. It is shown that Endoglin crosstalk Hippo pathway that leads to the regulation of secreted matricellular proteins and chemokines, cell migration and morphology, and extracellular matrix remodeling [80]. Also, another transcriptomics study showed the importance of the Hippo signaling pathway in TGFβ-activated cells. Evaluation of long non-coding RNAs (lncRNAs) and mRNAs with the Arraystar Human lncRNA Expression Microarray after TGFβ in human endothelial cells showed many of the transcriptome changes can be through YAP/TAZ [81]. However, how the interaction between TGFβ and YAP/TAZ in human endothelial cells can regulate angiogenesis is unclear (Figure 3).

2.5. WNT Pathway

The WNT signaling pathway has been widely implicated in maintenance of the vascular system and angiogenesis. Wnt binds to the frizzled (Fz) and low-density lipoprotein receptor-related protein (LRP) complex at the cell surface. Cytoplasmic β-catenin levels are regulated by complex protein down-stream of Fz/LRP that include dishevelled (Dsh), glycogen synthase kinase-3β (GSK-3), axin, and adenomatous polyposis coli (APC). When Wnt binds Fz/LRP, β-catenin degradation is inhibited by GSK-3/APC/axin. Inversely, in the absence of Wnt protein, β-catenin is targeted for ubiquitination. β-catenin can regulate many features of endothelial cells such as gene transcription and cytoskeleton structure [82]. Fz can also interact with the heterotrimeric G proteins and led to induction of YAP/TAZ transcriptional activity during central nervous system angiogenesis (Figure 3) [83]. Activation of heterotrimeric G proteins, specifically Gα12/13, can activate Rho GTPases resulting in inhibition of LATS1/2 to promote YAP/TAZ activation and TEAD-mediated transcription [84]. The Wnt antagonist Dickkopf2, known as regulated by YAP, promotes angiogenesis in rodent and human endothelial cells through regulation of filopodial dynamics and angiogenic sprouting via LRP6-mediated APC/Asef2/Cdc42 signaling [85]. This evidence shows the potential collaboration between Wnt and Hippo pathway to regulate angiogenesis (Figure 3).

2.6. PROX1 Transcriptional Programing—Key Regulators of Blood Vessel and Lymphatic Endothelial Cell Trans-Differentiation

The lymphatic microvasculature collects the removal of interstitial fluid/proteins and brings them back to the circulatory system. Although its function is in close coordination with blood vasculature, lymphatic endothelial cells have their own specific function and molecular composition. Isolation of primary lymphatic and blood microvasculature endothelial cells from human has revealed some similarities as well as unique molecular properties that distinguish lymphatic and blood vascular endothelium from each other [86]. Several studies showed that both lymphatic and blood vessel endothelial cells have the ability to monitor and respond to shear stress through YAP/TAZ to control cell–cell junction, and prevent regression of valves and focal vascular lumen collapse [73,87–90]. Both blood vessels and lymphatic endothelial cells have the same origin. Differentiation of lymphatic endothelial cells and having different fate from blood vessel endothelial cells mainly depends on a transcription factor activity, called Prox1 [91]. Prox1 is involved in cell fate determination, gene transcriptional regulation, and progenitor cell regulation during lymphogenesis as well as trans-differentiation of blood vessel and lymphatic endothelial cells. It has been shown that Prox1 expression is negatively regulated by YAP/TAZ activity [92]. YAP/TAZ regulate lymphatic identity through Prox1 transcriptional programing, which is a critical checkpoint underlying lymphatic endothelial cells differentiation from blood vessel endothelial cells [92].

3. Roles of YAP/TAZ in Developmental Vasculogenesis and Angiogenesis

Vasculogenesis and angiogenesis are responsible for forming new blood network in the embryo. Vasculogenesis begins with the differentiation of precursor cells (angioblasts) into endothelial cells and the de novo formation of a primitive vascular network, which makes the heart and the first primitive vascular plexus inside the embryo. Then, angiogenesis remodels and expands this network [93]. At early stage of mouse embryo development, TAZ has no expression, and comes up at all stages after blastocyst. However, YAP expression is dynamic and widespread during mouse development from the beginning. Knock out mice reveal the importance of YAP for early development of both yolk sac vasculature and placenta [94]. YAP-knocked out embryos die in the first half of gestation because of severe defects in early yolk sac vascular plexus and placenta development. Interestingly, TAZ cannot compensate this YAP specific phenotype. YAP function in vascular development has been also explored in zebrafish. YAP null mutant zebrafish showed increased vessel collapse and regression [74]. During regression, blood circulation in the endothelial cells activates YAP/TAZ, which leads to induction of CTGF and actin polymerization [95–99].

Comparing sprouting front, remodeling plexus, arteries, and veins in postnatal mouse retinas showed YAP/TAZ have distinct expression and localization in endothelial cells according to their physiological conditions and developmental stage [100–107]. YAP is expressed evenly throughout the vasculature with mostly cytoplasmic localization. During angiogenesis, YAP localizes into nucleus at sprouting area. TAZ expression is variable throughout the vasculature with highest expression in sprouting fronts. Surprisingly, when YAP or TAZ were conditionally knocked out in endothelial cells, mild vascular defects were observed [100]. Deleting both YAP and TAZ in mice causes a dramatic defect in blood vessel development at the retinal vasculature, a 21% decrease in radial expansion, a 26% decrease in capillary density, and a 55% decrease in branching frequency. These results indicate that endothelial YAP and TAZ are critical for the vascular growth, branching, and regularity of blood vessel network. In another study, it was observed that overexpression of YAP-active form increased angiogenic sprouting, which is blocked by angiopoietin-2 (ANG-2) depletion or soluble Tie-2 treatment [101].

YAP and TAZ are known mainly for their co-transcriptional activity in the nucleus. However, cytoplasmic YAP/TAZ have also distinct functions in angiogenesis and vascular remodeling. Deletion of the LATS1/2 in mice endothelial cells results reduction in YAP/TAZ phosphorylation and cytoplasmic retention. The retinas of these mice show migration defect with reduced extension distance, which can be partially rescued by overexpressing YAP-S127D (cytoplasmic form of YAP). Intriguingly, deletion of YAP/TAZ, Cdc42, or LATS1/2 in endothelial cells has a very similar phenotype in the retina, implying the possibility of operation in a common pathway in angiogenic tip cell development. Cytoplasmic YAP/TAZ positively regulates the activity of the Rho family GTPase CDC42 activity, which causes proliferation, and migration of vascular endothelial cells during retinal angiogenesis [102].

VEGFRs and their ligands VEGFs regulate the functions of several types of cells in cardiovascular system such as hematopoietic precursors, as well as endothelial and lymphendothelial cells. In mammals, there are three RTKs, known as VEGFR1–3, which bind to five VEGF ligands, derived from alternative splicing. VEGFR-1 is one of main negative or positive regulators of VEGFR-2. VEGFR-2 and VEGFR-3 play crucial roles in angiogenesis and lymphogenesis, respectively. VEGF family members secrete 40-kDa dimeric glycoproteins including VEGFA, B, C, D and placenta growth factor (PLGF) [103]. PLGF and VEGFB just bind VEGFR-1. VEGFC, and D are known ligands for VEGFR-3 and regulate lymphogenesis. VEGFA, and B bind to both VEGFR-1 and -2 and are involved in regulating of endothelial cell mitosis, and survival, as well as angiogenesis and microvascular permeability. Alternative gene splicing of VEGFA makes several isoforms namely, $VEGF_{121}$, $VEGF_{165}$, $VEGF_{189}$, and $VEGF_{206}$ [104]. Ligand binding induces VEGFR dimerization and increasing kinase domain activity.

Cross phosphorylation between kinase domain produces docking site for intracellular signaling protein that broadcast signaling. PI3K and MAPK are two main signaling pathways downstream of VEGFRs. Recently, we made a novel bioluminescent-based biosensor to quantify the kinase activity of the large tumor suppressor (LATS), a central kinase in the Hippo signaling pathway [105,106]. Using this biosensor against a small kinase library, we found VEGFR activation by VEGF triggers PI3K/MAPK signaling, which subsequently activates the Hippo effectors YAP and TAZ (Figure 3). We demonstrated VEGF activation induces angiogenesis in vitro and in vivo, which was abolished by YAP/TAZ knockdown. In addition, we found treatment of YAP/TAZ knocked down cells with Cyr61 and ANG-2, known as YAP/TAZ pro-angiogenic targets, can partially rescue the phenotype. Wang et al. [107] identified that YAP/TAZ are essential co-transcriptional activators in endothelial cells and YAP/TAZ activity is controlled by VEGF during developmental angiogenesis. YAP/TAZ localization in endothelial cells is different during embryogenesis. In early embryos (E10.5 and E11.5) YAP/TAZ are predominantly in the nucleus. However, they are mostly localized in the cytoplasm in later developmental stages (E14.5). Interestingly, by using transgenic mouse cell line overexpressing VEGF, increased angiogenesis and nuclear YAP/TAZ were observed. Moreover, endothelia-specific deletion of YAP/TAZ caused severe vascular defects throughout the whole body as well as yolk sac vascularization. They also studied the roles of YAP/TAZ in retina and brain vascularization during postnatal angiogenesis. Deletion of YAP/TAZ in endothelial cells reduced cell proliferation but had no effect on blood vessel regression. However, YAP/TAZ knockout reduced vessel coverage, length, and branch points in the developing brain cortex. Significantly, they also showed that VEGF changes cytoskeleton dynamics by regulating Src family kinases (SFKs) and Rho GTPases, which results in YAP/TAZ nuclear localization. Interestingly, YAP/TAZ can also regulate VEGFR2 and its activity. YAP/TAZ knocked down in endothelial cell impairs proper VEGFR2 trafficking and its downstream signaling. ChIP-Seq analysis of endothelial cells after VEGF treatment and comparison with knocked down YAP/TAZ cells revealed that VEGF induces a YAP/TAZ-dependent transcriptome linked to cytoskeleton assembling genes.

Signal transducer and activator of transcription 3 (STAT3) is a transcription factor interacting with YAP/TAZ interacting proteins in endothelial cells [108]. Interleukin 6 (IL-6) treatment results in STAT3 phosphorylation by receptor-associated Janus kinase that induces its nuclear localization and regulates gene expression in the acute-phase inflammatory response. YAP binding extends IL-6-driven STAT3 accumulation in the nucleus and increases the expression of Ang-2, which promotes angiogenesis. Ang-2 blockage or selective STAT3 inhibitors attenuate retinal angiogenesis in Tie2Cre-mediated YAP transgenic mice.

Additionally, YAP/TAZ can also regulate endothelial cells and angiogenesis through regulating none-endothelial cells such as pericytes as well as vascular smooth muscle cells or mural cells. Pericytes are functionally important as if they are lost, blood vessels become hemorrhagic and hyper dilated, which cause edema, diabetic retinopathy, and embryonic lethality [109]. Recent studies have shown the importance of pericytes in tumor progression [110]. Kato et al. [111] showed recently that YAP/TAZ regulates several key functions of pericytes. Knocking down of YAP/TAZ in pericytes of mice reduces hepatocyte growth factor expression which leads to impaired activation of the c-Met receptor in alveolar cells [111]. They also showed YAP/TAZ regulates angiopoietin-1 expression in pericytes, which together with hepatocyte growth factor coordinates the behavior vascular cells during lung morphogenesis.

4. Roles of YAP/TAZ in Pathological Angiogenesis

In general angiogenesis related diseases can be categorized into two major groups depending on whether there is need for more angiogenesis or it should be inhibited to cure the disease. In the first category, therapeutic angiogenesis can help damaged tissue to be repaired, such as in limb ischemia, myocardial infarction, and arteriosclerosis. In second category, pathological angiogenesis should be inhibited to stop or slow down the progression of a disease such as in benign and malignant angiogenic tumors, as well as retinopathies [112].

One of the factors that can affect systematic angiogenesis is. Mesenchymal stromal cells (MSCs) are multipotent non-hematopoietic cells with multi-lineage potential to differentiate into various tissues, including osteoblasts (bone cells), chondrocytes (cartilage cells), myocytes (muscle cells), and adipocytes (fat cells which give rise to marrow adipose tissue). MSCs are also known to regulate angiogenesis in different organs especially in bone marrow. MSCs detect mechanosignals and regulate angiogenesis through secretion of several pro-angiogenic factors such as VEGF and placental growth factor (PGF). The amount of this pro-angiogenic factors secretion in MSCs cells is correlated with age. Interestingly, MSCs derived from children are more mechanosensitive, showing enhanced angiogenesis on stiff substrates. This difference is possibly due to activity in Hippo signaling as MSCs derived from children have lower Hippo signaling activity and higher YAP nuclear localization and transactivating activity [113]. Another factor affecting systematic angiogenesis is endothelial cell-derived exosomes (EDEs) that promote trans-endothelial migration and angiogenesis. Atherosclerotic cerebrovascular disease includes a variety of medical conditions such as elevated blood pressure [114]. Levels of YAP in EDE are higher and those of pS127-YAP are lower in patients with atherosclerotic cerebrovascular disease [114]. However, how this YAP protein level change in EDEs regulates angiogenesis is not completely understood.

Structural changes of the vascular wall and vascular remodeling is commonly correlated with hypertension. Vascular remodeling can be induced in mice by angiotensin II fusion in 2-weeks. Mice treated with angiotensin II show YAP up-regulation in their blood vessels. Interestingly, disrupting the YAP–TEAD interaction with verteporfin inhibits Ang II-induced vascular remodeling in mice [115].

Retinal neovascularization, such as age-related macular degeneration (AMD or ARMD), retinopathy of prematurity (ROP), and proliferative diabetic retinopathy (PDR), are the leading cause of blindness in developed countries. Although the mechanism of retinal neovascularization is not well understood, endothelial dysfunction as a result of hypoxic stimuli such as ischemia or inflammation is one of main players. Laser photocoagulation and VEGFR inhibitors are the two main existing treatments. Laser photocoagulation can destroy retinal neurons and affect normal visual function. Effects of VEGFR inhibitors are mostly short-term due to the development of resistance to anti-VEGF therapies [116]. Therefore, it is urgent to study the underlying cellular and molecular mechanisms of these diseases to find novel therapeutic approaches. AMD is the leading cause of sight loss, which can be classified as wet or dry. The wet AMD is characterized by choroidal neovascularization with abnormal blood vessels that grows behind the macula and leak fluid. The proliferation of choroidal endothelial cells is the key step in angiogenesis. The in vivo AMD mouse model is created by laser photocoagulation. In this model, YAP expression in choroidal endothelial cells is significantly higher and mostly localized in the nucleus. Choroidal endothelial cells with highly expressed YAP also express high protein level proliferating cell nuclear antigen (PCNA) with a higher mitosis rate. Intriguingly, knocking down YAP in these cells in combination therapy with ranibizumab, a VEGF monoclonal antibody, decreased the expression of PCNA, choroidal endothelial cell proliferation, and the incidence as well as leakage area of choroidal neovascularization [117]. YAP/TAZ conditionally knocked out mice show markedly attenuated choroidal neovascularization volume and suppressed vascular leakage compared with wild type mice [118]. Also, it has been shown that treatment of human retinal microvascular endothelial cells (HRMECs) with dimethyloxalylglycine (DMOG), to mimic hypoxia conditions, induces angiogenesis through YAP and STAT3 [119]. In endothelial cells and during hypoxia conditions total YAP, and p-STAT3 (Tyr705) increases, while pS127-YAP decreases. Un-phosphorylated YAP localizes p-STAT3 into the nucleus and promotes the transcription of VEGF. YAP also regulates proliferation, migration and tube formation of HRMECs in vitro. Additionally, YAP inhibition reduces retinal pathological neovascularization in mouse oxygen-induced retinopathy model.

High morbidity and mortality in diabetic patients are mainly due to impaired angiogenesis and wound healing. The most common complications of diabetes include having damaged blood vessels, increased permeability, and growth of new blood vessels in the retina. These symptoms together are known as diabetic retinopathy. Treatment for diabetic retinopathy is similar to AMD treatment and has low efficacy to arrest the progression of this disease. The Hippo signaling pathway is important for the progression of diabetic retinopathy and can be used to develop new therapeutic approaches. Diabetic rats show decreased levels of p-S127-YAP and p-MST, while TAZ and TEAD total proteins are increased in their retinas [120]. These changes in the Hippo signaling activity result in increasing pro-angiogenic factors secretion and retina new blood vessel formation. Another complication of diabetes is delayed wound healing due to impaired angiogenesis and poor dermal healing. Recently, it has been shown that substance P, a neurotransmitter composed of 11 amino acids, accelerates wound healing in diabetic mice through endothelial progenitor cell mobilization and YAP activation [121]. Additionally, a non-invasive method, low-intensity pulsed ultrasound can accelerate healing through activation of YAP and TAZ in endothelial cells. Low-intensity pulsed ultrasound treatment activates YAP and TAZ nuclear localization through LATS inhibition, which can promote angiogenesis and vascular remodeling [122].

Hyperglycemia and elevated free fatty acids in diabetic patients cause metabolic stress, which inhibits endothelial angiogenesis. The Hippo signaling pathway and YAP/TAZ play critical role in endothelial angiogenesis inhibition during metabolic stress. For example, treatment of endothelial cells with palmitic acid inhibits their cell proliferation, migration, and tube formation ability. After palmitic acid treatment, MST expression increases dramatically, which then activates LATS and results in YAP phosphorylation and exclusion from nucleolus. Interestingly, either knocking down MST or overexpression of YAP prevented palmitic-induced inhibition of angiogenesis. Palmitic acid treatment damages endothelial cells mitochondria and releases mitochondrial DNA. Mitochondrial DNA activates cytosolic DNA sensor cyclic GMP-AMP synthase (cGAS)-stimulator of interferon genes (STING)-interferon regulatory factor 3 (IRF-3). cGAS-STING-IRF3 signaling activates IRF3 transcription factor, which then directly interacts with MST promoter and increases its expression [123]. In addition to the mentioned functions, Hippo signaling also regulates endothelial metabolisms. Peroxisome proliferator-activated receptor gamma co-activator 1-alpha (PGC1α), a major player controlling glucose consumption in mitochondria, is regulated by YAP. Constitutively active YAP (YAP-S127A) up-regulates PGC1α expression, resulting in increased in vitro and vascular morphogenesis in the fibrin gel subcutaneously implanted on mice. Knocking down PGC1α inhibits YAP-S127A-induced angiogenesis [124].

5. Roles of YAP/TAZ in Tumor Angiogenesis and Vascular Mimicry

Like normal tissues, cancer cells need to have access to blood vessels to get oxygen and nutrition and evacuate their waste. Almost always, angiogenic switch activates quiescent vasculature near tumors to form new blood vessels. After angiogenic switch, tumors exhibit diverse patterns of neovascularization. Cancer cells in primary tumors use nearby vessels for entering to blood, known as intravasation. Those cells surviving in the blood extravasate and form micrometastases. Dormant micrometastases can form macroscopic tumors if they are able to activate tumor angiogenesis [125]. In the past decades many studies tried to find the mechanism by which cellular genes regulate angiogenesis during tumor progression. YAP and TAZ are known as two key regulators of tumor angiogenesis.

Astrocytoma, a type of cancer that can form in the brain or spinal cord, has very significant angiogenesis. Analysis of blood vessel density in diffuse astrocytomas, in particular glioblastoma, revealed that the levels of TAZ in endothelial cells are positively correlated with blood vessel density. TAZ expression is also correlated with vascular VEGFR2 level. Expression of both TAZ and VEGFR2 is up-regulated in endothelial cells of high grade glioblastoma, which indicates the importance of VEGFR2/TAZ signaling pathway in angiogenesis [126]. Comparing YAP and TAZ localization in normal brain and tumor-associated endothelial cells shows YAP and TAZ are highly accumulated in the nucleus [107]. Angiosarcoma is another type of cancers with high level of angiogenesis. Unexpectedly, YAP activation inhibits angiogenesis in angiosarcoma. Heterogonous angiosarcoma cells can be divided in to CD31 high or low expressing cells. CD31[low] cells are more tumorigenic and chemoresistant than

CD31^{high} cells due to more efficient reactive oxygen species (ROS) detoxification. CD31^{low} cells also have less angiogenesis ability with higher YAP nuclear localization [127].

YAP shows high activation in cholangiocarcinoma patient biopsies and human cholangiocarcinoma cell lines. Microarray expression profiling of cholangiocarcinoma cells with overexpressed or knocked-down YAP reveal that YAP regulates important genes in proliferation, apoptosis, and angiogenesis [128]. YAP/TEAD regulate pro-angiogenic microfibrillar-associated protein 5 (MFAP5) transcription in cholangiocarcinoma cells. Secreted MFAP5 induces angiogenesis of human micro-vascular endothelial cells. Also, YAP activation is correlated with high MFAP5 expression in both human cholangiocarcinoma and cholangiocarcinoma xenografts. Additionally, YAP activation and its correlation to angiogenesis have been reported in gastric cancer [129]. Cancer-associated MSCs are important component of tumor microenvironment. MSCs cells isolated from cancer patients demonstrate different features compared to bone marrow-derived MSCs. Cancer-associated MSCs have higher YAP expression. YAP knockout in MSCs inhibits their proliferation, migration, invasion, and pro-angiogenic ability by reducing the activation of β-catenin and its target genes.

Vasculogenic mimicry (VM) is a term that was first introduced by Maniotis et al. [130] in 1999 when they observed that tissue sections from aggressive human intraocular and metastatic melanomas do not show any significant necrosis and have patterned networks of interconnected loops of extra-cellular matrix. Vascular mimicry (VM) is the formation of blood supply system by tumor cells rather than endothelial cells, which is independent of typical modes of angiogenesis. High tumor VM reflects the plasticity of aggressive tumor cells to express vascular cell markers and tumor vasculature. High tumor VM is associated with a high tumor grade, shorter survival, invasion, metastasis and poor prognosis [131]. The first characterization of VM was in melanoma. In melanoma, VM is characterized as involving areas with high levels of laminin, collagen IV and VI, and heparin sulfate proteoglycans, and it should have plasma and red blood cells, confirming its function as a blood/nutrient supplier [132]. In addition to providing oxygen and nutrients, VM can provide a new route for escaping and metastasizing cancer cells. So far VM has been shown in variety of cancers including melanoma [133], breast and lung cancer [134,135], ovarian cancer [136], osteosarcoma [137], gastric cancer [138], bladder cancer [139], hepatocellular cancer [140], and colorectal cancer [141], etc.

Small population of cancer cells have stem cell characteristics, which are known as critical player in initiation, progression, metastasis, and recurrence of tumors. Cancer stem cells can be quiescent for several years to several decades. Upon activation, they can transdifferentiate to different tumor components including the vasculature. YAP is elevated in cancer stem cells derived from non-small cell lung cancer and regulates their ability to form angiogenic tubules [142]. YAP induces vasculogenic mimicry in cancer stem cells by regulating the embryonic stem cell transcription factor Sox2. YAP regulates Sox2 transcription independent of TEAD, through physical interaction with Oct4 by its WW domain. It has been shown that YAP also regulates vasculogenic mimicry of pancreatic ductal adenocarcinoma via interaction with TEAD [143]. YAP inhibition using verteporfin or short hairpin RNA (shRNA) suppresses vasculogenic mimicry by reducing Ang2, MMP2, VE-cadherin, and SMA expression in vitro and in vivo. Tube formation assay experiments showed when YAP is inhibited, pancreatic ductal adenocarcinoma cancer cells have less ability to induce HUVEC tube formation in vitro. Molecular analysis showed that in addition to Ang2 and MMP2, cyclin E1 decreased, which causes cell cycle arrest in the G1 phase. Also, poly-ADP ribose polymerase (PARP) western blot analysis showed more apoptotic cells in pancreatic cells. Evaluating of pancreatic adenocarcinoma xenograft mode with and without YAP inhibition showed reduction in tumour cells as well as VM content. In another study, we showed that YAP and TAZ regulate vasculogenic mimicry by upregulating pro-angiogenic factors such as Cyr61 and Ang2 in breast cancer cell lines [106]. In our study we inhibit YAP/TAZ by using verteporfin or short interference RNA (siRNA). Consistently with previous publication, YAP inhibition reduced VM in vitro. Also, we found that Cyr61 and Ang-2 may play critical roles in VM regulation, as an addback experiment using purified Ang-2 and Cyr61 can significantly rescue the phenotype.

6. Conclusion and Future Directions

In conclusion, YAP and TAZ play central roles in regulating angiogenesis during development and pathogenesis of many diseases (Table 1). YAP/TAZ can induce angiogenesis through activation of multiple downstream targets such as Cyr61, Ang2, and MMP2. YAP/TAZ may also mediate angiogenesis induced by CD44, ECM, TGFβ, AMOT, and RTKs such as VEGFR, FGFR, IGFR, TGFR, and Tie2 (Figure 3). On one hand, YAP/TAZ regulates key functions of endothelial cells and pro-angiogenic production of cancer cells during tumor angiogenesis. On another hand YAP/TAZ are involved in several important features of cancer cells such as cell proliferation, immune evasion, EMT, and the cancer stem cell phenotype [144]. Therefore, targeting YAP/TAZ offers a new and attractive strategy to treat aggressive tumors by targeting both angiogenesis and other cancer phenotypes at the same time. In addition, since targeting RTKs involved in angiogenesis (such as VEGFR, and FGFR) often results in short-term and limited effects on overall survival of cancer patients [145,146], combination treatments which target both RTKs and Hippo signaling can offer a new and attractive strategy to treat aggressive tumors. Moreover, since YAP/TAZ are also involved in the dysregulation of angiogenesis during the development of many other diseases such as diabetes and AMD, targeting YAP/TAZ will provide new therapeutic strategies to treat these diseases in the future.

Table 1. Summary of the role of YAP and TAZ function in physiological and pathological angiogenesis.

Main Finding	Ref.
Angiomotin is a negative regulator of YAP	[44]
Angiomotin-like 1 degradation by Nedd4 is regulated by YAP through c-ABL	[60]
YAP/TAZ are the main mediators of mechanotransduction in endothelial cells	[72]
BMP9 crosstalk with the Hippo pathway regulates endothelial cell matricellular response	[80]
mRNAs upregulated in ECs in response to TGFβ1 treatment are involved in hippo signaling	[81]
Flow-dependent endothelial YAP regulation contributes to vessel maintenance	[73]
YAP/TAZ negatively regulate prox1 during developmental and pathologic lymphangiogenesis	[92]
YAP disruption in mice causes defects in yolk sac vasculogenesis and chorioallantoic fusion	[94]
YAP/TAZ activity is essential for vascular regression via Ctgf and actin polymerization	[95]
Adherens junction and endothelial cell distribution in angiogenesis is regulated by YAP/TAZ	[100]
YAP regulates angiopoietin-2 expression in ECs	[101]
Vascular tip cell migration is regulated by YAP/TAZ-CDC42 signaling pathway	[102]
VEGFR is a regulator of YAP/TAZ in the Hippo pathway in angiogenesis through PI3K/MAPK pathways	[106]
VEGF activates YAP/TAZ via its effects on actin cytoskeleton	[107]
YAP promotes angiogenesis via Stat3	[108]
YAP mediates angiotensin II-induced vascular smooth muscle cell phenotypic modulation and hypertensive vascular remodelling	[115]
YAP inhibition ameliorates choroidal neovascularization	[117]
YAP/TAZ regulates vascular barrier maturation	[118]
YAP via interacting with STAT3 regulates VEGF-induced angiogenesis in retina	[119]
Substance P accelerates wound healing in type 2 diabetic mice through YAP activation	[121]
Ultrasound treatment accelerates angiogenesis by activating YAP/TAZ	[122]
Palmitic acid inhibits angiogenesis through YAP suppression	[123]
YAP1-TEAD1 controls angiogenesis and mitochondrial biogenesis through PGC1α.	[124]
Blood vascular density and VEGFR2 expression in astrocytomas is regulated by TAZ	[126]
Cell proliferation, chemoresistance, and angiogenesis in human cholangiocarcinoma is regulated by YAP	[128]
YAP regulates OCT4 activity and SOX2 expression to facilitate vascular mimicry	[142]
Verteporfin suppresses vasculogenic mimicry of pancreatic ductal adenocarcinoma	[143]

Funding: This work was supported by grants from the Canadian Institute of Health Research (CIHR#119325, 148629), and the Canadian Cancer Society (CRS)/Canadian Breast Cancer Foundation (CBCF) to XY.

Acknowledgments: We would like to thank the above funding agency for supporting our ongoing research on YAP/TAZ and Hippo pathway in cancer and for covering the costs to publish in open access.

Conflicts of Interest: The authors declare no conflict of interest.

References

1. Hillen, F.; Griffioen, A.W. Tumour vascularization: Sprouting angiogenesis and beyond. *Cancer Metastasis Rev.* **2007**, *26*, 489–502. [CrossRef] [PubMed]
2. Djonov, V.; Baum, O.; Burri, P.H. Vascular remodeling by intussusceptive angiogenesis. *Cell Tissue Res.* **2003**, *314*, 107–117. [CrossRef] [PubMed]
3. Lamalice, L.; Le Boeuf, F.; Huot, J. Endothelial cell migration during angiogenesis. *Circ. Res.* **2007**, *100*, 782–794. [CrossRef]
4. Caduff, J.H.; Fischer, L.C.; Burri, P.H. Scanning electron microscope study of the developing microvasculature in the postnatal rat lung. *Anat. Rec.* **1986**, *216*, 154–164. [CrossRef] [PubMed]
5. Otrock, Z.K.; Mahfouz, R.A.; Makarem, J.A.; Shamseddine, A.I. Understanding the biology of angiogenesis: Review of the most important molecular mechanisms. *Blood Cellsmoleculesand Dis.* **2007**, *39*, 212–220. [CrossRef] [PubMed]
6. Deveza, L.; Choi, J.; Yang, F.J. Therapeutic angiogenesis for treating cardiovascular diseases. *Theranostics* **2012**, *2*, 801. [CrossRef]
7. Couffinhal, T.; Dufourcq, P.; Daret, D.; Duplaa, C. The mechanisms of angiogenesis. Medical and therapeutic applications. *La Revue de Medecine Interne* **2001**, *22*, 1064–1082. [CrossRef]
8. Muñoz-Chápuli, R. Evolution of angiogenesis. *Int. J. Dev. Biol.* **2011**, *55*, 345–351. [CrossRef]
9. Breier, G. Angiogenesis in embryonic development—A review. *Placenta* **2000**, *21*, S11–S15. [CrossRef] [PubMed]
10. Klagsbrun, M.; D'Amore, P.A. Vascular endothelial growth factor and its receptors. *Cytokine Growth Factor Rev.* **1996**, *3*, 259–270. [CrossRef]
11. Chung, A.S.; Ferrara, N. Developmental and pathological angiogenesis. *Annu. Rev. Cell Dev. Biol.* **2011**, *27*, 563–584. [CrossRef]
12. Lobov, I.; Renard, R.; Papadopoulos, N.; Gale, N.; Thurston, G.; Yancopoulos, G.; Wiegand, S.J. Delta-like ligand 4 (Dll4) is induced by VEGF as a negative regulator of angiogenic sprouting. *Proc. Natl. Acad. Sci. USA* **2007**, *104*, 3219–3224. [CrossRef] [PubMed]
13. Betz, C.; Lenard, A.; Belting, H.-G.; Affolter, M. Cell behaviors and dynamics during angiogenesis. *Co. Biol.* **2016**, *143*, 2249–2260.
14. Flanagan, J.G.; Vanderhaeghen, P. The ephrins and Eph receptors in neural development. *Annu. Rev. Neurosci.* **1998**, *21*, 309–345. [CrossRef]
15. Creamer, D.; Sullivan, D.; Bicknell, R.; Barker, J. Angiogenesis in psoriasis. *Angiogenesis* **2002**, *5*, 231–236. [CrossRef] [PubMed]
16. Tuo, J.; Wang, Y.; Cheng, R.; Li, Y.; Chen, M.; Qiu, F.; Qian, H.; Shen, D.; Penalva, R.; Xu, H. Wnt signaling in age-related macular degeneration: Human macular tissue and mouse model. *J. Transl. Med.* **2015**, *13*, 330. [CrossRef] [PubMed]
17. Fallah, A.; Sadeghinia, A.; Kahroba, H.; Samadi, A.; Heidari, H.R.; Bradaran, B.; Zeinali, S.; Molavi, O. Therapeutic targeting of angiogenesis molecular pathways in angiogenesis-dependent diseases. *J. Cell. Physiol.* **2019**, *110*, 775–785. [CrossRef]
18. Zetter, P.B.R. Angiogenesis and tumor metastasis. *Annu. Rev. Med.* **1998**, *49*, 407–424. [CrossRef]
19. Bergers, G.; Benjamin, L.E. Tumorigenesis and the angiogenic switch. *Nat. Rev. Cancer* **2003**, *3*, 401–410. [CrossRef]
20. Lawler, J. Thrombospondin-1 as an endogenous inhibitor of angiogenesis and tumor growth. *J. Cell. Mol. Med.* **2002**, *6*, 1–12. [CrossRef]
21. Carmeliet, P.; Jain, R.K. Angiogenesis in cancer and other diseases. *Nature* **2000**, *407*, 249–257. [CrossRef]
22. Totaro, A.; Panciera, T.; Piccolo, S. YAP/TAZ upstream signals and downstream responses. *Nat. Cell Biol.* **2018**, *20*, 888–899. [CrossRef]
23. Moya, I.M.; Halder, G. Hippo–YAP/TAZ signalling in organ regeneration and regenerative medicine. *Nat. Rev. Mol. Cell Biol.* **2018**, *20*, 211–226. [CrossRef]

24. Maugeri-Saccà, M.; Barba, M.; Pizzuti, L.; Vici, P.; Di Lauro, L.; Dattilo, R.; Vitale, I.; Bartucci, M.; Mottolese, M.; De Maria, R. The Hippo transducers TAZ and YAP in breast cancer: Oncogenic activities and clinical implications. *J. Exp. Clin. Cancer Res.* **2015**, *17*, e14. [CrossRef]

25. Hansen, C.G.; Moroishi, T.; Guan, K.-L. YAP and TAZ: A nexus for Hippo signaling and beyond. *Trends Cell Biol.* **2015**, *25*, 499–513. [CrossRef]

26. Zhang, K.; Qi, H.-X.; Hu, Z.-M.; Chang, Y.-N.; Shi, Z.-M.; Han, X.-H.; Han, Y.-W.; Zhang, R.-X.; Zhang, Z.; Chen, T. YAP and TAZ take center stage in cancer. *Biochemistry* **2015**, *54*, 6555–6566. [CrossRef]

27. Csibi, A.; Blenis, J. Hippo-YAP and mTOR pathways collaborate to regulate organ size. *Nat. Cell Biol.* **2012**, *14*, 1244–1245. [CrossRef]

28. Tumaneng, K.; Russell, R.C.; Guan, K.-L. Organ size control by Hippo and TOR pathways. *Curr. Biol.* **2012**, *22*, R368–R379. [CrossRef]

29. Hong, W.; Guan, K.-L. The YAP and TAZ transcription co-activators: Key downstream effectors of the mammalian Hippo pathway. *Semin. Cell Dev. Biol.* **2012**, *23*, 785–793. [CrossRef]

30. Justice, R.W.; Zilian, O.; Woods, D.F.; Noll, M.; Bryant, P. The Drosophila tumor suppressor gene warts encodes a homolog of human myotonic dystrophy kinase and is required for the control of cell shape and proliferation. *Genes Dev.* **1995**, *9*, 534–546. [CrossRef]

31. Tapon, N.; Harvey, K.F.; Bell, D.W.; Wahrer, D.C.R.; Schiripo, T.A.; Haber, D.A.; Hariharan, I.K. salvador Promotes both cell cycle exit and apoptosis in Drosophila and is mutated in human cancer cell lines. *Cell* **2002**, *110*, 467–478. [CrossRef]

32. Wu, S.; Huang, J.; Dong, J.; Pan, D. hippo encodes a Ste-20 family protein kinase that restricts cell proliferation and promotes apoptosis in conjunction with salvador and warts. *Cell* **2003**, *114*, 445–456. [CrossRef]

33. Lai, Z.-C.; Wei, X.; Shimizu, T.; Ramos, E.; Rohrbaugh, M.; Nikolaidis, N.; Ho, L.-L.; Li, Y. Control of cell proliferation and apoptosis by mob as tumor suppressor, mats. *Cell* **2005**, *120*, 675–685. [CrossRef]

34. Pan, D. The hippo signaling pathway in development and cancer. *Dev. Cell* **2010**, *19*, 491–505. [CrossRef] [PubMed]

35. Saucedo, L.J.; Edgar, B.A. Filling out the Hippo pathway. *Nat. Rev. Mol. Cell Biol.* **2007**, *8*, 613–621. [CrossRef] [PubMed]

36. Zhao, B.; Wei, X.; Li, W.; Udan, R.S.; Yang, Q.; Kim, J.; Xie, J.; Ikenoue, T.; Yu, J.; Li, L. Inactivation of YAP oncoprotein by the Hippo pathway is involved in cell contact inhibition and tissue growth control. *Genes Dev.* **2007**, *21*, 2747–2761. [CrossRef] [PubMed]

37. Callus, B.A.; Verhagen, A.M.; Vaux, D.L. Association of mammalian sterile twenty kinases, Mst1 and Mst2, with hSalvador via C-terminal coiled-coil domains, leads to its stabilization and phosphorylation. *Febs J.* **2006**, *273*, 4264–4276. [CrossRef]

38. Praskova, M.; Xia, F.; Avruch, J. MOBKL1A/MOBKL1B phosphorylation by MST1 and MST2 inhibits cell proliferation. *Curr. Biol.* **2008**, *18*, 311–321. [CrossRef]

39. Chan, E.H.Y.; Nousiainen, M.; Chalamalasetty, R.B.; Schäfer, A.; Nigg, E.A.; Silljé, H.H.W. The Ste20-like kinase Mst2 activates the human large tumor suppressor kinase Lats1. *Oncogene* **2005**, *24*, 2076–2086. [CrossRef] [PubMed]

40. Yu, F.-X.; Guan, K.-L. The Hippo pathway: Regulators and regulations. *Genes Dev.* **2013**, *27*, 355–371. [CrossRef] [PubMed]

41. Hao, Y.; Chun, A.; Cheung, K.; Rashidi, B.; Yang, X. Tumor suppressor LATS1 is a negative regulator of oncogene YAP. *J. Biol. Chem.* **2008**, *283*, 5496–5509. [CrossRef] [PubMed]

42. Piccolo, S.; Dupont, S.; Cordenonsi, M. The biology of YAP/TAZ: Hippo signaling and beyond. *Physiol. Rev.* **2014**, *94*, 1287–1312. [CrossRef]

43. Panciera, T.; Azzolin, L.; Cordenonsi, M.; Piccolo, S. Mechanobiology of YAP and TAZ in physiology and disease. *Nat. Rev. Mol. Cell Biol.* **2017**, *18*, 758–770. [CrossRef] [PubMed]

44. Zhao, B.; Li, L.; Lu, Q.; Wang, L.H.; Liu, C.-Y.; Lei, Q.; Guan, K.-L. Angiomotin is a novel Hippo pathway component that inhibits YAP oncoprotein. *Genes Dev.* **2011**, *25*, 51–63. [CrossRef] [PubMed]

45. Yang, C.-C.; Graves, H.K.; Moya, I.M.; Tao, C.; Hamaratoglu, F.; Gladden, A.B.; Halder, G. Differential regulation of the Hippo pathway by adherens junctions and apical–basal cell polarity modules. *Proc. Natl. Acad. Sci. USA* **2015**, *112*, 1785–1790. [CrossRef] [PubMed]

46. Yu, F.-X.; Zhao, B.; Panupinthu, N.; Jewell, J.L.; Lian, I.; Wang, L.H.; Zhao, J.; Yuan, H.; Tumaneng, K.; Li, H. Regulation of the Hippo-YAP pathway by G-protein-coupled receptor signaling. *Cell* **2012**, *150*, 780–791. [CrossRef]

47. Gschwind, A.; Fischer, O.M.; Ullrich, A. The discovery of receptor tyrosine kinases: Targets for cancer therapy. *Nat. Rev. Cancer* **2004**, *4*, 361–370. [CrossRef]

48. van Rensburg, H.J.J.; Lai, D.; Azad, T.; Hao, Y.; Yang, X. TAZ enhances mammary cell proliferation in 3D culture through transcriptional regulation of IRS1. *Cell. Signal.* **2018**, *52*, 12–22. [CrossRef] [PubMed]

49. Fan, R.; Kim, N.-G.; Gumbiner, B.M. Regulation of Hippo pathway by mitogenic growth factors via phosphoinositide 3-kinase and phosphoinositide-dependent kinase-1. *Proc. Natl. Acad. Sci. USA* **2013**, *110*, 2569–2574. [CrossRef] [PubMed]

50. Zhao, Y.; Montminy, T.; Azad, T.; Lightbody, E.; Hao, Y.; SenGupta, S.; Asselin, E.; Nicol, C.; Yang, X. PI3K Positively Regulates YAP and TAZ in Mammary Tumorigenesis Through Multiple Signaling Pathways. *Mol. Cancer Res.* **2018**, *16*, 1046–1058. [CrossRef] [PubMed]

51. Glienke, J.; Schmitt, A.O.; Pilarsky, C.; Hinzmann, B.; Weiß, B.; Rosenthal, A.; Thierauch, K.H. Differential gene expression by endothelial cells in distinct angiogenic states. *Eur. J. Biochem.* **2000**, *267*, 2820–2830. [CrossRef]

52. Auerbach, R.; Lewis, R.; Shinners, B.; Kubai, L.; Akhtar, N. Angiogenesis assays: A critical overview. *Clin. Chem.* **2003**, *49*, 32–40. [CrossRef]

53. Auerbach, R.; Akhtar, N.; Lewis, R.L.; Shinners, B.L. Angiogenesis assays: Problems and pitfalls. *Cancer Metastasis Rev.* **2000**, *19*, 167–172. [CrossRef]

54. Staton, C.A.; Lewis, C.; Bicknell, R. *Angiogenesis Assays: A Critical Appraisal of Current Techniques*; John Wiley & Sons: Hoboken, NJ, USA, 2007.

55. Goodwin, A.M. In vitro assays of angiogenesis for assessment of angiogenic and anti-angiogenic agents. *Microvasc. Res.* **2007**, *74*, 172–183. [CrossRef]

56. Cui, X.; Morales, R.-T.T.; Qian, W.; Wang, H.; Gagner, J.-P.; Dolgalev, I.; Placantonakis, D.; Zagzag, D.; Cimmino, L.; Snuderl, M. Hacking macrophage-associated immunosuppression for regulating glioblastoma angiogenesis. *Biomaterials* **2018**, *161*, 164–178. [CrossRef]

57. Song, L.; Ahmed, M.F.; Li, Y.; Zeng, C.; Li, Y. Vascular Differentiation from Pluripotent Stem Cells in 3-D Auxetic Scaffolds. *J. Tissue Eng. Regen. Med.* **2018**, *12*, 1679–1689. [CrossRef]

58. Aase, K.; Ernkvist, M.; Ebarasi, L.; Jakobsson, L.; Majumdar, A.; Yi, C.; Birot, O.; Ming, Y.; Kvanta, A.; Edholm, D. Angiomotin regulates endothelial cell migration during embryonic angiogenesis. *Genes Dev.* **2007**, *21*, 2055–2068. [CrossRef]

59. Zheng, Y.; Vertuani, S.; Nystrom, S.; Audebert, S.; Meijer, I.; Tegnebratt, T.; Borg, J.-P.; Uhlén, P.; Majumdar, A.; Holmgren, L. Angiomotin-like protein 1 controls endothelial polarity and junction stability during sprouting angiogenesis. *Circ. Res.* **2009**, *105*, 260–270. [CrossRef]

60. Skouloudaki, K.; Walz, G. YAP1 recruits c-Abl to protect angiomotin-like 1 from Nedd4-mediated degradation. *PLoS ONE* **2012**, *7*, e35735. [CrossRef]

61. Dai, X.; She, P.; Chi, F.; Feng, Y.; Liu, H.; Jin, D.; Zhao, Y.; Guo, X.; Jiang, D.; Guan, K.-L.; Zhong, T.P.; Zhao, B. Phosphorylation of angiomotin by Lats1/2 kinases inhibits F-actin binding, cell migration and angiogenesis. *J. Biol. Chem.* **2013**, *288*, 34041–34051. [CrossRef]

62. Hong, S.A.; Son, M.W.; Cho, J.; Jang, S.H.; Lee, H.J.; Lee, J.H.; Cho, H.D.; Oh, M.H.; Lee, M.S. Low angiomotin-p130 with concomitant high Yes-associated protein 1 expression is associated with adverse prognosis of advanced gastric cancer. *Apmis* **2017**, *125*, 996–1006. [CrossRef]

63. Singleton, P.A.; Salgia, R.; Moreno-Vinasco, L.; Moitra, J.; Sammani, S.; Mirzapoiazova, T.; Garcia, J.G.N. CD44 Regulates Hepatocyte Growth Factor-mediated Vascular Integrity: ROLE OF c-Met, Tiam1/Rac1, DYNAMIN 2, AND CORTACTIN. *J. Biol. Chem.* **2007**, *282*, 30643–30657. [CrossRef] [PubMed]

64. Savani, R.C.; Cao, G.; Pooler, P.M.; Zaman, A.; Zhou, Z.; DeLisser, H.M. Differential involvement of the hyaluronan (HA) receptors CD44 and receptor for HA-mediated motility in endothelial cell function and angiogenesis. *J. Biol. Chem.* **2001**, *276*, 36770–36778. [CrossRef]

65. Flynn, K.M.; Michaud, M.; Canosa, S.; Madri, J.A. CD44 regulates vascular endothelial barrier integrity via a PECAM-1 dependent mechanism. *Angiogenesis* **2013**, *16*, 689–705. [CrossRef]

66. Stamenkovic, I.; Yu, Q. Merlin, a "magic" linker between the extracellular cues and intracellular signaling pathways that regulate cell motility, proliferation, and survival. *Curr. Protein Pept. Sci.* **2010**, *11*, 471–484. [CrossRef]

67. Badouel, C.; McNeill, H. SnapShot: The hippo signaling pathway. *Cell* **2011**, *145*, 484–484.e1. [CrossRef]
68. Tsuneki, M.; Madri, J.A. Adhesion molecule-mediated hippo pathway modulates hemangioendothelioma cell behavior. *Mol. Cell. Biol.* **2014**, *34*, 4485–4499. [CrossRef]
69. Tsuneki, M.; Madri, J.A. CD44 regulation of endothelial cell proliferation and apoptosis via modulation of CD31 and VE-cadherin expression. *J. Biol. Chem.* **2014**, *289*, 5357–5370. [CrossRef] [PubMed]
70. Xu, Y.; Stamenkovic, I.; Yu, Q. CD44 attenuates activation of the hippo signaling pathway and is a prime therapeutic target for glioblastoma. *Cancer Res.* **2010**, *70*, 2455–2464. [CrossRef]
71. Wada, K.-I.; Itoga, K.; Okano, T.; Yonemura, S.; Sasaki, H. Hippo pathway regulation by cell morphology and stress fibers. *Development* **2011**, *138*, 3907–3914. [CrossRef]
72. Dupont, S.; Morsut, L.; Aragona, M.; Enzo, E.; Giulitti, S.; Cordenonsi, M.; Zanconato, F.; Le Digabel, J.; Forcato, M.; Bicciato, S. Role of YAP/TAZ in mechanotransduction. *Nature* **2011**, *474*, 179–183. [CrossRef] [PubMed]
73. Nakajima, H.; Yamamoto, K.; Agarwala, S.; Terai, K.; Fukui, H.; Fukuhara, S.; Ando, K.; Miyazaki, T.; Yokota, Y.; Schmelzer, E. Flow-dependent endothelial YAP regulation contributes to vessel maintenance. *Dev. Cell* **2017**, *40*, 523–536.e6. [CrossRef] [PubMed]
74. Gegenfurtner, F.A.; Jahn, B.; Wagner, H.; Ziegenhain, C.; Enard, W.; Geistlinger, L.; Rädler, J.O.; Vollmar, A.M.; Zahler, S. Micropatterning as a tool to identify regulatory triggers and kinetics of actin-mediated endothelial mechanosensing. *J. Cell Sci.* **2018**, *131*, 212886. [CrossRef]
75. Dickson, M.C.; Martin, J.S.; Cousins, F.M.; Kulkarni, A.B.; Karlsson, S.; Akhurst, R.J. Defective haematopoiesis and vasculogenesis in transforming growth factor-beta 1 knock out mice. *Development* **1995**, *121*, 1845–1854.
76. Oshima, M.; Oshima, H.; Taketo, M.M. TGF-β receptor type II deficiency results in defects of yolk sac hematopoiesis and vasculogenesis. *Dev. Biol.* **1996**, *179*, 297–302. [CrossRef]
77. Larsson, J.; Goumans, M.J.; Sjöstrand, L.J.; van Rooijen, M.A.; Ward, D.; Levéen, P.; Xu, X.; ten Dijke, P.; Mummery, C.L.; Karlsson, S. Abnormal angiogenesis but intact hematopoietic potential in TGF-β type I receptor-deficient mice. *EMBO J.* **2001**, *20*, 1663–1673. [CrossRef] [PubMed]
78. Pefani, D.E.; Pankova, D.; Abraham, A.G.; Grawenda, A.M.; Vlahov, N.; Scrace, S. TGF-β Targets the Hippo Pathway Scaffold RASSF1A to Facilitate YAP/SMAD2 Nuclear Translocation. *Mol. Cell* **2016**, *63*, 156–166. [CrossRef] [PubMed]
79. Ma, B.; Cheng, H.; Gao, R.; Mu, C.; Chen, L.; Wu, S.; Chen, Q.; Zhu, Y. Zyxin-Siah2–Lats2 axis mediates cooperation between Hippo and TGF-β signalling pathways. *Nat. Commun.* **2016**, *7*, 11123. [CrossRef] [PubMed]
80. Young, K.; Tweedie, E.; Conley, B.; Ames, J.; FitzSimons, M.; Brooks, P.; Liaw, L.; Vary, C.P.H. BMP9 crosstalk with the hippo pathway regulates endothelial cell matricellular and chemokine responses. *PLoS ONE* **2015**, *10*, e0122892. [CrossRef] [PubMed]
81. Singh, K.K.; Matkar, P.N.; Quan, A.; Mantella, L.-E.; Teoh, H.; Al-Omran, M.; Verma, S. Investigation of TGFβ1-induced long noncoding RNAs in endothelial cells. *Int. J. Vasc. Med.* **2016**, *2016*, 2459687. [CrossRef] [PubMed]
82. Clevers, H. Wnt/β-Catenin Signaling in Development and Disease. *Cell* **2006**, *127*, 469–480. [CrossRef] [PubMed]
83. Hot, B.; Valnohova, J.; Arthofer, E.; Simon, K.; Shin, J.; Uhlén, M.; Kostenis, E.; Mulder, J.; Schulte, G. FZD10-Gα13 signalling axis points to a role of FZD10 in CNS angiogenesis. *Cell. Signal.* **2017**, *32*, 93–103. [CrossRef] [PubMed]
84. Park, H.W.; Kim, Y.C.; Yu, B.; Moroishi, T.; Mo, J.-S.; Plouffe, S.W.; Meng, Z.; Lin, K.C.; Yu, F.-X.; Alexander, C.M. Alternative Wnt signaling activates YAP/TAZ. *Cell* **2015**, *162*, 780–794. [CrossRef]
85. Min, J.-K.; Park, H.; Choi, H.-J.; Kim, Y.; Pyun, B.-J.; Agrawal, V.; Song, B.-W.; Jeon, J.; Maeng, Y.-S.; Rho, S.-S. The WNT antagonist Dickkopf2 promotes angiogenesis in rodent and human endothelial cells. *J. Clin. Investig.* **2011**, *121*, 1882–1893. [CrossRef] [PubMed]
86. Podgrabinska, S.; Braun, P.; Velasco, P.; Kloos, B.; Pepper, M.S.; Jackson, D.G.; Skobe, M. Molecular characterization of lymphatic endothelial cells. *Proc. Natl. Acad. Sci. USA* **2002**, *99*, 16069–16074. [CrossRef] [PubMed]
87. Reneman, R.S.; Arts, T.; Hoeks, A.P.G. Wall shear stress–an important determinant of endothelial cell function and structure–in the arterial system in vivo. *J. Vasc. Res.* **2006**, *3*, 251–269. [CrossRef] [PubMed]
88. Lee, H.J.; Diaz, M.F.; Price, K.M.; Ozuna, J.A.; Zhang, S.; Sevick-Muraca, E.M.; Hagan, J.P.; Wenzel, P.L. Fluid shear stress activates YAP1 to promote cancer cell motility. *Nat. Commun.* **2017**, *8*, 14122. [CrossRef]

89. Ivanov, K.I.; Agalarov, Y.; Valmu, L.; Samuilova, O.; Liebl, J.; Houhou, N.; Maby-El Hajjami, H.; Norrmén, C.; Jaquet, M.; Miura, N.; et al. Phosphorylation regulates FOXC2-mediated transcription in lymphatic endothelial cells. *Mol. Cell. Biol.* **2013**, *33*, 3749–3761. [CrossRef] [PubMed]

90. Sabine, A.; Bovay, E.; Demir, C.S.; Kimura, W.; Jaquet, M.; Agalarov, Y.; Zangger, N.; Scallan, J.P.; Graber, W.; Gulpinar, E. FOXC2 and fluid shear stress stabilize postnatal lymphatic vasculature. *J. Clin. Investig.* **2015**, *125*, 3861–3877. [CrossRef]

91. Johnson, N.C.; Dillard, M.E.; Baluk, P.; McDonald, D.M.; Harvey, N.L.; Frase, S.L.; Oliver, G. Lymphatic endothelial cell identity is reversible and its maintenance requires Prox1 activity. *Genes Dev.* **2008**, *22*, 3282–3291. [CrossRef]

92. Cho, H.; Kim, J.; Ahn, J.H.; Hong, Y.-K.; Mäkinen, T.; Lim, D.-S.; Koh, G.Y. YAP and TAZ Negatively Regulate Prox1 During Developmental and Pathologic Lymphangiogenesis. *Circ. Res.* **2019**, *124*, 225–242. [CrossRef]

93. Patan, S. Vasculogenesis and angiogenesis. *Cancer Treat. Res.* **2004**, *117*, 3–32.

94. Morin-Kensicki, E.M.; Boone, B.N.; Howell, M.; Stonebraker, J.R.; Teed, J.; Alb, J.G.; Magnuson, T.R.; O'Neal, W.; Milgram, S.L. Defects in yolk sac vasculogenesis, chorioallantoic fusion, and embryonic axis elongation in mice with targeted disruption of Yap65. *Mol. Cell. Biol.* **2006**, *26*, 77–87. [CrossRef]

95. Nagasawa-Masuda, A.; Terai, K. Yap/Taz transcriptional activity is essential for vascular regression via Ctgf expression and actin polymerization. *PLoS ONE* **2017**, *12*, e0174633. [CrossRef]

96. Brigstock, D.R. Regulation of angiogenesis and endothelial cell function by connective tissue growth factor (CTGF) and cysteine-rich 61 (CYR61). *Angiogenesis* **2002**, *5*, 153–165. [CrossRef]

97. Chaqour, B.; Goppelt-Struebe, M. Mechanical regulation of the Cyr61/CCN1 and CTGF/CCN2 proteins. *Fed. Eur. Biochem. Soc. J.* **2006**, *273*, 3639–3649. [CrossRef]

98. Jiang, W.G.; Watkins, G.; Fodstad, O.; Douglas-Jones, A.; Mokbel, K.; Mansel, R.E. Differential expression of the CCN family members Cyr61, CTGF and Nov in human breast cancer. *Endocr. Relat. Cancer* **2004**, *11*, 781–791. [CrossRef] [PubMed]

99. Lai, D.; Ho, K.C.; Hao, Y.; Yang, X. Taxol resistance in breast cancer cells is mediated by the hippo pathway component TAZ and its downstream transcriptional targets Cyr61 and CTGF. *Cancer Res.* **2011**, *71*, 2728–2738. [CrossRef]

100. Neto, F.; Klaus-Bergmann, A.; Ong, Y.T.; Alt, S.; Vion, A.-C.; Szymborska, A.; Carvalho, J.R.; Hollfinger, I.; Bartels-Klein, E.; Franco, C.A. YAP and TAZ regulate adherens junction dynamics and endothelial cell distribution during vascular development. *eLife* **2018**, *7*, e31037. [CrossRef] [PubMed]

101. Choi, H.-J.; Zhang, H.; Park, H.; Choi, K.-S.; Lee, H.-W.; Agrawal, V.; Kim, Y.-M.; Kwon, Y.-G. Yes-associated protein regulates endothelial cell contact-mediated expression of angiopoietin-2. *Nat. Commun.* **2015**, *6*, 6943. [CrossRef] [PubMed]

102. Sakabe, M.; Fan, J.; Odaka, Y.; Liu, N.; Hassan, A.; Duan, X.; Stump, P.; Byerly, L.; Donaldson, M.; Hao, J. YAP/TAZ-CDC42 signaling regulates vascular tip cell migration. *Proc. Natl. Acad. Sci. USA* **2017**, *114*, 10918–10923. [CrossRef]

103. Olsson, A.-K.; Dimberg, A.; Kreuger, J.; Claesson-Welsh, L. VEGF receptor signalling? In control of vascular function. *Nat. Rev. Mol. Cell Biol.* **2006**, *7*, 359. [CrossRef]

104. Ferrara, N.; Gerber, H.-P.; LeCouter, J. The biology of VEGF and its receptors. *Nat. Med.* **2003**, *9*, 669. [CrossRef]

105. Azad, T.; Nouri, K.; van Rensburg Janse, H.J.; Hao, Y.; Yang, X. Monitoring Hippo Signaling Pathway Activity Using a Luciferase-based Large Tumor Suppressor (LATS) Biosensor. *J. Vis. Exp.* **2018**, E58416. [CrossRef]

106. Azad, T.; van Rensburg, H.J.J.; Lightbody, E.D.; Neveu, B.; Champagne, A.; Ghaffari, A.; Kay, V.R.; Hao, Y.; Shen, H.; Yeung, B. A LATS biosensor screen identifies VEGFR as a regulator of the Hippo pathway in angiogenesis. *Nat. Commun.* **2018**, *9*, 1061. [CrossRef]

107. Wang, X.; Valls, A.F.; Schermann, G.; Shen, Y.; Moya, I.M.; Castro, L.; Urban, S.; Solecki, G.M.; Winkler, F.; Riedemann, L. YAP/TAZ orchestrate VEGF signaling during developmental angiogenesis. *Dev. Cell* **2017**, *42*, 462–478.e7. [CrossRef]

108. He, J.; Bao, Q.; Zhang, Y.; Liu, M.; Lv, H.; Liu, Y.; Yao, L.; Li, B.; Zhang, C.; He, S. Yes-associated protein promotes angiogenesis via signal transducer and activator of transcription 3 in endothelial cells. *Circ. Res.* **2018**, *122*, 591–605. [CrossRef]

109. Bergers, G.; Song, S. The role of pericytes in blood-vessel formation and maintenance. *Neuro Oncol.* **2005**, *7*, 452–464. [CrossRef]

110. Paiva, A.E.; Lousado, L.; Guerra, D.A.; Azevedo, P.O.; Sena, I.F.; Andreotti, J.P.; Santos, G.S.; Gonçalves, R.; Mintz, A.; Birbrair, A. Pericytes in the premetastatic niche. *Cancer Res.* **2018**, *78*, 2779–2786. [CrossRef]

111. Kato, K.; Diéguez-Hurtado, R.; Hong, S.P.; Kato-Azuma, S.; Adams, S.; Stehling, M.; Trappmann, B.; Wrana, J.L.; Koh, G.Y.; Adams, R.H. Pulmonary pericytes regulate lung morphogenesis. *Nat. Commun.* **2018**, *9*, 2448. [CrossRef]

112. Tímár, J.; Döme, B.; Fazekas, K.; Janovics, Á.; Paku, S. Angiogenesis-dependent diseases and angiogenesis therapy. *Pathol. Oncol. Res.* **2001**, *7*, 85–94. [CrossRef] [PubMed]

113. Barreto, S.; Gonzalez-Vazquez, A.; Cameron, A.R.; Cavanagh, B.; Murray, D.J.; O'Brien, F.J. Identification of the mechanisms by which age alters the mechanosensitivity of mesenchymal stromal cells on substrates of differing stiffness: Implications for osteogenesis and angiogenesis. *Acta Biomater.* **2017**, *53*, 59–69. [CrossRef] [PubMed]

114. Goetzl, E.J.; Schwartz, J.B.; Mustapic, M.; Lobach, I.V.; Daneman, R.; Abner, E.L.; Jicha, G.A. Altered cargo proteins of human plasma endothelial cell–derived exosomes in atherosclerotic cerebrovascular disease. *FASEB J.* **2017**, *31*, 3689–3694. [CrossRef] [PubMed]

115. Lin, M.; Yuan, W.; Su, Z.; Lin, C.; Huang, T.; Chen, Y.; Wang, J. Yes-associated protein mediates angiotensin II-induced vascular smooth muscle cell phenotypic modulation and hypertensive vascular remodelling. *Cell Prolif.* **2018**, *51*, e12517. [CrossRef]

116. Bharadwaj, A.S.; Appukuttan, B.; Wilmarth, P.A.; Pan, Y.; Stempel, A.J.; Chipps, T.J.; Benedetti, E.E.; Zamora, D.O.; Choi, D.; David, L.L. Role of the retinal vascular endothelial cell in ocular disease. *Prog. Retin. Eye Res.* **2013**, *32*, 102–180. [CrossRef] [PubMed]

117. Yan, Z.; Shi, H.; Zhu, R.; Li, L.; Qin, B.; Kang, L.; Chen, H.; Guan, H. Inhibition of YAP ameliorates choroidal neovascularization via inhibiting endothelial cell proliferation. *Mol. Vis.* **2018**, *24*, 83–93.

118. Kim, J.; Kim, Y.H.; Kim, J.; Bae, H.; Lee, D.-H.; Kim, K.H.; Hong, S.P.; Jang, S.P.; Kubota, Y.; Kwon, Y.-G. YAP/TAZ regulates sprouting angiogenesis and vascular barrier maturation. *J. Clin. Investig.* **2017**, *127*, 3441–3461. [CrossRef]

119. Zhu, M.; Liu, X.; Wang, Y.; Chen, L.; Wang, L.; Qin, X.; Xu, J.; Li, L.; Tu, Y.; Zhou, T. YAP via interacting with STAT3 regulates VEGF-induced angiogenesis in human retinal microvascular endothelial cells. *Exp. Cell Res.* **2018**, *373*, 155–163. [CrossRef]

120. Hao, G.-M.; Lv, T.-T.; Wu, Y.; Wang, H.-L.; Xing, W.; Wang, Y.; Li, C.; Zhang, Z.-J.; Wang, Z.-L.; Wang, W. The Hippo signaling pathway: A potential therapeutic target is reversed by a Chinese patent drug in rats with diabetic retinopathy. *BMC Complementary Altern. Med.* **2017**, *17*, 187. [CrossRef]

121. Um, J.; Yu, J.; Park, K.S. Substance P accelerates wound healing in type 2 diabetic mice through endothelial progenitor cell mobilization and Yes-associated protein activation. *Mol. Med. Rep.* **2017**, *15*, 3035–3040. [CrossRef]

122. Xu, X.-M.; Xu, T.-M.; Wei, Y.-B.; Gao, X.-X.; Sun, J.-C.; Wang, Y.; Kong, Q.-J.; Shi, J.-G. Low-Intensity Pulsed Ultrasound Treatment Accelerates Angiogenesis by Activating YAP/TAZ in Human Umbilical Vein Endothelial Cells. *Ultrasound Med. Biol.* **2018**, *44*, 2655–2661. [CrossRef] [PubMed]

123. Yuan, L.; Mao, Y.; Luo, W.; Wu, W.; Xu, H.; Wang, X.L.; Shen, Y.H. Palmitic acid dysregulates the Hippo-YAP pathway and inhibits angiogenesis by inducing mitochondrial damage and activating the cytosolic DNA sensor cGAS-STING-IRF3 signaling. *J. Biol. Chem.* **2017**, *292*, 15002–15015. [CrossRef]

124. Mammoto, A.; Muyleart, M.; Kadlec, A.; Gutterman, D.; Mammoto, T. YAP1-TEAD1 signaling controls angiogenesis and mitochondrial biogenesis through PGC1α. *Microvasc. Res.* **2018**, *119*, 73–83. [CrossRef]

125. Hanahan, D.; Weinberg, R.A. Hallmarks of cancer: The next generation. *Cell* **2011**, *144*, 646–674. [CrossRef] [PubMed]

126. Xu, C.; Mao, L.; Xiong, J.; Wen, J.; Wang, Y.; Geng, D.; Liu, Y. TAZ Expression on Endothelial Cells Is Closely Related to Blood Vascular Density and VEGFR2 Expression in Astrocytomas. *J. Neuropathol. Exp. Neurol.* **2019**, *78*, 172–180. [CrossRef] [PubMed]

127. Venkataramani, V.; Küffer, S.; Cheung, K.C.P.; Jiang, X.; Trümper, L.; Wulf, G.G.; Ströbel, P. CD31 expression determines redox status and chemoresistance in human angiosarcomas. *Clin. Cancer Res.* **2018**, *24*, 460–473. [CrossRef]

128. Marti, P.; Stein, C.; Blumer, T.; Abraham, Y.; Dill, M.T.; Pikiolek, M.; Orsini, V.; Jurisic, G.; Megel, P.; Makowska, Z. YAP promotes proliferation, chemoresistance, and angiogenesis in human cholangiocarcinoma through TEAD transcription factors. *Hepatology* **2015**, *62*, 1497–1510. [CrossRef] [PubMed]

129. Pan, Z.; Tian, Y.; Zhang, B.; Zhang, X.; Shi, H.; Liang, Z.; Wu, P.; Li, R.; You, B.; Yang, L. YAP signaling in gastric cancer-derived mesenchymal stem cells is critical for its promoting role in cancer progression. *Int. J. Oncol.* **2017**, *51*, 1055–1066. [CrossRef]

130. Maniotis, A.J.; Folberg, R.; Hess, A.; Seftor, E.A.; Gardner, L.M.; Pe'er, J.; Trent, J.M.; Meltzer, P.S.; Hendrix, M.J. Vascular channel formation by human melanoma cells in vivo and in vitro: Vasculogenic mimicry. *Am. J. Pathol.* **1999**, *155*, 739–752. [CrossRef]

131. Delgado-Bellido, D.; Serrano-Saenz, S.; Fernández-Cortés, M.; Oliver, F.J. Vasculogenic mimicry signaling revisited: Focus on non-vascular VE-cadherin. *Mol. Cancer* **2017**, *16*, 65. [CrossRef]

132. Hendrix, M.J.; Seftor, E.A.; Seftor, R.E.; Chao, J.T.; Chien, D.S.; Chu, Y.W. Tumor cell vascular mimicry: Novel targeting opportunity in melanoma. *Pharmacol. Ther.* **2016**, *159*, 83–92. [CrossRef] [PubMed]

133. Hendrix, M.J.; Seftor, E.A.; Hess, A.R.; Seftor, R.E. Vasculogenic mimicry and tumour-cell plasticity: Lessons from melanoma. *Nat. Rev. Cancer* **2003**, *3*, 411. [CrossRef]

134. Shirakawa, K.; Kobayashi, H.; Heike, Y.; Kawamoto, S.; Brechbiel, M.W.; Kasumi, F.; Iwanaga, T.; Konishi, F.; Terada, M.; Wakasugi, H. Hemodynamics in vasculogenic mimicry and angiogenesis of inflammatory breast cancer xenograft. *Cancer Res.* **2002**, *62*, 560–566.

135. Williamson, S.C.; Metcalf, R.L.; Trapani, F.; Mohan, S.; Antonello, J.; Abbott, B.; Leong, H.S.; Chester, C.P.; Simms, N.; Polanski, R. Vasculogenic mimicry in small cell lung cancer. *Nat. Commun.* **2016**, *7*, 13322. [CrossRef]

136. Sood, A.K.; Seftor, E.A.; Fletcher, M.S.; Gardner, L.M.; Heidger, P.M.; Buller, R.E.; Seftor, R.E.; Hendrix, M.J. Molecular determinants of ovarian cancer plasticity. *Am. J. Pathol.* **2001**, *158*, 1279–1288. [CrossRef]

137. Cai, X.S.; Jia, Y.W.; Mei, J.; Tang, R.Y. Tumor blood vessels formation in osteosarcoma: Vasculogenesis mimicry. *Chin. Med. J.* **2004**, *117*, 94–98.

138. Li, M.; Gu, Y.; Zhang, Z.; Zhang, S.; Zhang, D.; Saleem, A.F.; Zhao, X.; Sun, B. Vasculogenic mimicry: A new prognostic sign of gastric adenocarcinoma. *Pathol. Oncol. Res.* **2010**, *16*, 259–266. [CrossRef] [PubMed]

139. Streeter, E.H.; Harris, A.L. Angiogenesis in bladder cancer—Prognostic marker and target for future therapy. *Surg. Oncol.* **2002**, *11*, 85–100. [CrossRef]

140. Sun, T.; Sun, B.C.; Zhao, X.L.; Zhao, N.; Dong, X.Y.; Che, N.; Yao, Z.; Ma, Y.M.; Gu, Q.; Zong, W.K. Promotion of tumor cell metastasis and vasculogenic mimicry by way of transcription coactivation by Bcl-2 and Twist1: A study of hepatocellular carcinoma. *Hepatology* **2011**, *54*, 1690–1706. [CrossRef] [PubMed]

141. Baeten, C.I.; Hillen, F.; Pauwels, P.; de Bruine, A.P.; Baeten, C.G. Prognostic role of vasculogenic mimicry in colorectal cancer. *Dis. Colon Rectum* **2009**, *52*, 2028–2035. [CrossRef] [PubMed]

142. Bora-Singhal, N.; Nguyen, J.; Schaal, C.; Perumal, D.; Singh, S.; Coppola, D.; Chellappan, S. YAP1 regulates OCT4 activity and SOX2 expression to facilitate self-renewal and vascular mimicry of stem-like cells. *Stem Cells* **2015**, *33*, 1705–1718. [CrossRef]

143. Wei, H.; Wang, F.; Wang, Y.; Li, T.; Xiu, P.; Zhong, J.; Sun, X.; Li, J. Verteporfin suppresses cell survival, angiogenesis and vasculogenic mimicry of pancreatic ductal adenocarcinoma via disrupting the YAP-TEAD complex. *Cancer Sci.* **2017**, *108*, 478–487. [CrossRef]

144. van Rensburg, H.J.J.; Azad, T.; Ling, M.; Hao, Y.; Snetsinger, B.; Khanal, P.; Minassian, L.M.; Graham, C.H.; Rauh, M.J.; Yang, X. The Hippo pathway component TAZ promotes immune evasion in human cancer through PD-L1. *Cancer Res.* **2018**, *78*, 1457–1470. [CrossRef]

145. Cross, M.J.; Claesson-Welsh, L. FGF and VEGF function in angiogenesis: Signalling pathways, biological responses and therapeutic inhibition. *Trends Pharmacol. Sci.* **2001**, *22*, 201–207. [CrossRef]

146. Zhao, Y.; Adjei, A.A. Targeting angiogenesis in cancer therapy: Moving beyond vascular endothelial growth factor. *Oncologist* **2015**, *20*, 660–673. [CrossRef]

![cells logo]

![MDPI logo]

Brief Report

Somatic Mutations of *lats2* Cause Peripheral Nerve Sheath Tumors in Zebrafish

Zachary J. Brandt [1], Paula N. North [2] and Brian A. Link [1,*]

[1] Department of Cell Biology, Neurobiology and Anatomy, Medical College of Wisconsin, Milwaukee, WI 53226, USA

[2] Department of Pediatrics, Medical College of Wisconsin, Milwaukee, WI 53226, USA

* Correspondence: blink@mcw.edu

Received: 25 June 2019; Accepted: 22 August 2019; Published: 25 August 2019

Abstract: The cellular signaling pathways underlying peripheral nerve sheath tumor (PNST) formation are poorly understood. Hippo signaling has been recently implicated in the biology of various cancers, and is thought to function downstream of mutations in the known PNST driver, *NF2*. Utilizing CRISPR-Cas9 gene editing, we targeted the canonical Hippo signaling kinase Lats2. We show that, while germline deletion leads to early lethality, targeted somatic mutations of zebrafish *lats2* leads to peripheral nerve sheath tumor formation. These peripheral nerve sheath tumors exhibit high levels of Hippo effectors Yap and Taz, suggesting that dysregulation of these transcriptional co-factors drives PNST formation in this model. These data indicate that somatic *lats2* deletion in zebrafish can serve as a powerful experimental platform to probe the mechanisms of PNST formation and progression.

Keywords: Lats2; Hippo; Yap; Taz; peripheral nerve sheath tumor; schwannoma; zebrafish

1. Introduction

Peripheral nerve sheath tumors (PNSTs) are relatively uncommon, making up just 8.6% of all primary brain or central nervous system tumors reported in 2011–2015 [1]. Based on these numbers, it was estimated that an additional 7500 new cases would be reported in 2018, representing one in every 50,000 people [1]. PNSTs are comprised of multiple tumor classifications including schwannoma, neurofibroma, perineurioma, and malignant peripheral nerve sheath tumors (MPNSTs) [2,3]. Nearly all of these tumors are benign (99.2%); however, they can result in loss of nerve function and/or neuropathic pain [4]. Although rare, the malignant subset of these tumors, termed MPNSTs, have poor prognosis for patient survival [5,6].

Very few genetic alterations are well recognized as drivers of PNST formation, and our understanding of these alterations and their underlying molecular mechanisms is far from complete. It is generally accepted that schwannomas, neurofibromas, and MPNTSs are derived from Schwann cells [7]. The subset of PNSTs with known genetic causes is termed neurofibromatosis. Neurofibromatosis commonly refers to three distinct and genetically determined dominant disorders, all leading to PNST formation with distinct characteristics [8]. These three disorders are referred to as neurofibromatosis 1 (NF1), neurofibromatosis 2 (NF2), and schwannomatosis. The aberrant genes that drive these disorders are *NF1*, *NF2*, and *INI1/SMARCB1*, respectively [8]. While the specific genetic variants causing these disorders are well defined, the molecular mechanisms by which they drive tumorigenesis remain incompletely understood and represent a key area of research. While approximately 90% of NF1 cases are sporadic, both syndromic and sporadic tumors can result from loss-of-function mutations in *NF1* [9]. *NF1* encodes neurofibromin 1, and acts as a Ras GTPase-activating protein, suggesting that improper regulation of Ras-related growth and proliferation may be an underlying cause in NF1 cases [10]. The *NF2* gene encodes for the tumor suppressor Merlin [11,12]. While the exact function of Merlin is not well understood, it has been shown to function as an upstream regulator of the Hippo signaling pathway, with direct implications in Schwann cells [13,14]. Finally, in schwannomatosis, the mutated gene of interest, *INI1/SMARCB1*, encodes a subunit of the SWI/SNF protein complexes that is known to function in chromatin remodeling [8]. In all three of these cases, the phenotypic variability is common and a high mutational load is noted [8]. This suggests that genetic modifiers as well as cooperating mutations exist and can alter the phenotypic outcomes of PNSTs. In support of this, while approximately half of MPNST cases are associated with *NF1* mutations, only 10% of NF1 cases eventually develop MPNST [15]. The other half of MPNST cases are sporadic, highlighting the importance of identifying other potential drivers of tumorigenesis.

As a signaling pathway downstream of Merlin/NF2, Hippo-Yap/Taz signaling is an obvious candidate for further study in PNST biology, and a logical pathway for targeted therapeutics. The Hippo pathway is primarily defined as a phosphorylation cascade composed of kinases Serine/Threonine Kinase 3 and 4 (Stk3 in zebrafish, Mst1/2 in mammals), and Large Tumor Suppressor Kinases 1 and 2 (Lats1/2), along with their respective adaptor/scaffolding proteins Salvador Family WW Domain Containing Protein 1 (Sav1 in zebrafish, WW45 in mammals), and MOB kinase activators 1 and 2 (Mob1a/b). After phosphorylation by Stk3/4, Lats1/2 regulates the main effectors of Hippo signaling transcriptional co-activators Yes-associated protein 1 (Yap) and its paralog, WW domain containing transcription regulator 1 (WWTR1/Taz), via phosphorylation. This phosphorylation leads to cytoplasmic sequestration of Yap/Taz through 14-3-3 protein binding, as well as subsequent ubiquitination and proteasomal degradation. Non-phosphorylated Yap/Taz translocate to the nucleus and, as they cannot bind DNA directly, associate with several families of transcription factors, most commonly TEA-Domain (TEAD) transcription factors, and both activate and repress expression of various genes.

Several studies have demonstrated roles for Hippo signaling in normal Schwann cell biology. Conditional knockout of *Nf2* in Schwann cells of mice led to hypomyelination and an increased Schwann cell number. This conditional knockout also affected injury response, as impaired axonal regeneration and remyelination were observed following sciatic nerve crush. Extending these findings to Yap signaling, the authors found that either mono- or bi-allelic deletion of *Yap* was able to rescue the impaired axonal regeneration in *Nf2* conditional knockouts [14]. Yap and Taz have since been identified as crucial to proper Schwann cell development. Several groups have shown that Yap and Taz can function redundantly in Schwann cells and are necessary for proper proliferation and myelination [16–18].

In recent years, Hippo signaling has also been shown to play a key role in aberrant Schwann cell biology, including PNST formation. Some of the first evidence of Hippo pathway involvement in PNSTs came through study of NF2. While working to uncover the molecular mechanisms of *NF2* mutant tumorigenesis, Li et. al. found that E3 ubiquitin ligase, CRL4^{DCAF1} promotes tumorigenesis of *NF2*-mutant cells by inactivating Lats1/2 and, hence, activating Yap [19]. Further study of Hippo signaling in the context of *NF2* mutation revealed that genetic and pharmacological inhibition of Yap led to decreased tumor cell proliferation and survival in *NF2* mutant cells and was able to reduce schwannoma tumor growth in mouse transplant models [20]. At the same time, several discovery-based studies of nerve sheath tumors identified alterations in Hippo pathway signaling. A proteomic screen for factors affected within sporadic human schwannomas revealed the activation of several receptors, including PDGFRβ, Her3, and Her2. Immunostaining and in vitro work in human schwannoma cell lines went on to suggest that expression of these receptors was under the control of Yap, and that proliferation in these tumors was linked to Yap signaling [21]. Furthermore, whole-exome sequencing of inherited and sporadic schwannoma reported *LATS1* mutations, suggesting that Hippo signaling could play a role in both types of schwannoma [22]. A more comprehensive and targeted study of sporadic schwannomas provided compelling evidence for the role of Hippo-Yap/Taz signaling in these tumors. Targeted sequencing of *LATS1* and *LATS2* in sporadic schwannoma revealed mutations in 1% and 2% of cases, respectively, suggesting that their mutation may be rare. However, promoter methylation of *LATS1* and *LATS2* was seen in 17% and 30% of cases, respectively. Overall, 76% of cases had at least one alteration in the *NF2*, *LATS1*, or *LATS2* gene. Of those cases, 43% of the tumors contained nuclear Yap expression by immunohistochemistry [23]. Transcriptomic data of vestibular schwannoma tumors from those patients also revealed deregulation of Hippo signaling [24]. These alterations may not be confined to schwannomas, as a case report of whole-exome sequencing in a single NF1 patient suggested that mutations in Hippo pathway-associated genes were overrepresented [25]. Strong evidence of Yap activation has also been seen in low-grade meningiomas and embryonal rhabdomyosarcoma [26,27]. *LATS2* mutations were also found in 11% of malignant pleural mesotheliomas and, of particular interest, co-occurring mutations of *NF2* and *LATS2* correlated with poor prognosis [28]. Consistently, these studies have found evidence for Lats1/2 activity in PNST formation, suggesting an important function in nerve sheath tumorigenesis. Cementing this role, combined conditional deletion of *Lats1* and *Lats2* within the Schwann cell lineage led to nerve-associated tumors in 100% of mice, and the development of a mouse model for MPNST [29]. The development of malignant tumors in this model also suggests that Lats1/2 and Yap/Taz signaling may be involved in malignant transformation of PNSTs. Consistent with this thought, a Yap/Taz conserved gene signature was more highly elevated in human MPNST samples than in NF1 or normal nerve samples [29]. Overall, these studies highlight a clear role for Hippo-Yap/Taz signaling in PNSTs.

Zebrafish have emerged as a reliable and advantageous model organism for the study of various cancers [30,31]. This includes several models of MPNST [32]. An MPNST model in *nf1*- and *p53*-deficient zebrafish successfully demonstrates several advantages of zebrafish in this context. The lack of a fully developed immune system in early zebrafish larvae facilitated the transplantation of tumor cells without host rejection, while the large reproductive capacity of the fish allowed the investigators to test the effect of various drugs on tumor growth [33]. As many tumor suppressor genes play essential roles in developmental differentiation and proliferation, the germline mutation of these genes can lead to embryonic or larval lethality and difficulty in studying their role in cancer biology. Somatic inactivation via transcription activator-like effector nucleases (TALEN) and clustered regularly interspaced short palindromic repeats and caspase9 (CRISPR-Cas9) has proved valuable for the development of retinoblastoma and medulloblastoma models in zebrafish [34–36]. These advantages suggest that the development of a zebrafish model of Hippo pathway-associated PNST would be beneficial.

In this study, we provide evidence supporting the role for Hippo-Yap/Taz signaling in PNST formation by presenting a novel model of PNSTs in zebrafish. Somatic inactivation of the Hippo kinase Lats2 by mosaic CRISPR-Cas9 gene editing resulted in obvious tumorigenesis. Histological examination of the resulting tumors identified them as resembling PNST in morphology. This diagnosis was confirmed by immunostaining of known markers Sox10 and S100. Immunohistochemical analysis also revealed the tumors to be highly proliferative and to contain high levels of Hippo pathway effectors Yap and Taz.

2. Materials and Methods

2.1. CRISPR Design

Clustered regularly interspaced short palindromic repeats (CRISPR) guides were designed against regions of zebrafish *lats2* exon5 and exon6 and *lats1* exon4 and exon7 using ZiFiT Targeter Version 4.2 (http://zifit.partners.org/ZiFiT, in the public domain). The targeted sequences were, 5′-GGAGCTAGTTATGGGGCTGA-3′ for *lats2* exon5, 5′-GGCATTGGGGCCTTTGGTG-3′ for *lats2* exon6, 5′-CCTCCCTATTCCATGCACC-3′ for *lats1* exon4, and 5′-GGGACTCTCGGGCGACGCAC-3′ for *lats1* exon7. CRISPR gRNA templates were generated by cloning annealed oligonucleotides with appropriate overhangs into *Bsa*I-digested pDR274 plasmid. CRISPR gRNAs were synthesized using a MEGAshortscript T7 Transcription Kit and purified using a mirVana miRNA Isolation Kit (Ambion, Austin, TX, USA). Zebrafish codon-optimized *cas9* was synthesized using a mMESSAGE mMACHINE Kit (Ambion) and polyadenylated using a Poly(A) Tailing Kit (Ambion). CRISPR gRNAs and *cas9* mRNA were co-injected into 1- to 4-cell zebrafish embryos from wild type ZDR fish maintained internally in the Link lab, at 12.5 ng/μL and 300 ng/μL, respectively, and surviving embryos were raised to adulthood before outcrossing to identify the founder fish carrying germline edits in *lats2*. Offspring from these fish were raised to adulthood, then fin-clipped for genotyping (see below for details). The resulting 1710-bp deletion mutant described here was identified via sequencing (Retrogen, San Diego, California, USA). This mutant allele is designated *lats2*^{mw87} (c.1048-2605del), where 1048–2605 denotes the deleted nucleotides of transcript: lats2-204 ENSDART00000139620.3 from GRCz11 genomic build.

2.2. Genotyping

Genomic DNA was extracted from zebrafish tissue using a Puregene Core Kit (Qiagen, Hilden, Germany). The genomic region containing the *lats2*^{mw87} mutation was amplified by PCR. The PCR protocol utilized primers flanking the expected deletion. The thermocycle conditions for detecting presence of a large deletion allele were designed using extension times allowing amplification of an amplicon ~750 bp in size, the expected size of an allele containing our targeted large deletion, but not for the WT allele amplicon of 2455 bp (Figure 1A). When detecting the presence of a WT allele, we either increased the extension time and screened for the presence of a 2455-bp amplicon, as well as the 750-bp amplicon, or included a third internal primer located within the sequence deleted in the *lats2*^{mw87} allele. A list of primers used for genotyping is provided below.

lats2 Exon4 Forward Primer: 5'-CCTGAAACAGACTGGTAGC-3'
lats2 Exon6 Reverse Primer: 5'-TTGAGTTGTGAGTCCATCGG-3'
lats2 Exon5 Large Deletion Internal Primer: 5'-CATGTTTGTGGAGTAAGCAC-3'
lats1 Exon4 Forward Primer: 5'-CAAGCGCTATTCTGGGAACT-3'
lats1 Exon7 Reverse Primer: 5'-AAACTGAGCCAAGTCCTCCT-3'

2.3. Paraffin Histology

Adult fish used for paraffin histology were fixed in 10% neutral buffer formalin overnight at 4 °C. Large adult fish were cut open along their belly on the anterior to posterior axis to allow for better penetration of the fixative. Samples were then processed in paraffin on a Sakura VIP5 automated tissue processor (Sakura Finetek Europe, Flemingweg, The Netherlands) for histology and immunohistochemistry. After paraffin embedding, samples were sectioned at 4 μm (Microm HM355S, ThermoFisher Scientific, Waltham, Massachutesetts, USA) onto poly-l-lysine coated slides and air-dried at 45 °C overnight for any subsequent immunohistochemistry or routine H&E staining. Brightfield light microscopy images were taken using a NanoZoomer 2.0-HT (Hamamatsu Photonics K.K., Hamamatsu City, Shizuoka, Japan).

2.4. Immunohistochemistry

An optimal immunostaining protocol was developed with the use of a Leica-Bond Max Immunostaining platform. All slides were dewaxed prior to their optimal antigen retrieval protocol. All antibodies used a citrate buffer epitope retrieval (Leica Epitope Retrieval Solution 1, AR 9661). PCNA (sc56, 1:6000, Santa Cruz Biotechnology, Dallas, TX, USA), Yap1 (ab81183, 1:100, Abcam, Cambridge, UK), Sox10 (GTX128374, 1:100, Genetex, Irvine, CA, USA), Yap/Taz (D24E4, 8418, 1:200, Cell Signaling Technology, Danvers, MA, USA), and S100 (PA0900, ready to use, Leica, Wetzlar, Germany) were detected and visualized using Bond Polymer Refine Detection System (DS9800, Leica) with the addition of a DAB Enhancer (AR9432, Leica), with a Modified F protocol (primary antibody incubation: 15 min at room temperature). All slides were counter-stained with hematoxylin and cover-slipped using a synthetic mounting media. Omission of the primary antibody served as the negative control.

2.5. Nomenclature

HUGO gene nomenclature for gene names and symbols was used within the text.

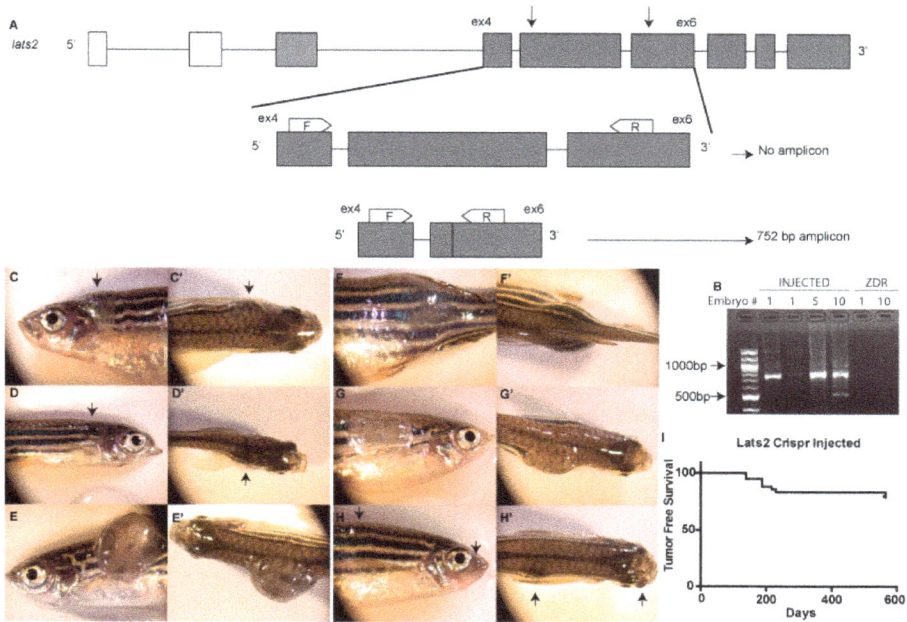

Figure 1. Injection of two *lats2* gRNAs leads to deletions and tumorigenesis. (**A**) Schematic depicting the zebrafish *lats2* gene and the PCR-based genotyping assay used to identify and differentiate between successful large deletion alleles and WT alleles. Arrows indicate the target sites of the gRNAs used. F refers to a forward primer; R refers to a reverse primer. (**B**) Agarose gel separation of PCR amplicons generated based on the genotyping assay depicted in (**A**). Amplicons of the correct size, indicative of a large genomic deletion, are present in *lats2* CRISPR-injected embryos, but absent in uninjected negative controls. Numbers above lanes represent the number of embryos used as the template DNA for each PCR sample. First lane contains a 100-bp ladder (NEB). (**C–H′**) Six examples of *lats2* CRISPR-injected fish that developed tumors by 6 mpf. Arrows denote tumors when not obviously visible. (**I**) Kaplan-Meier plot of tumor-free survival in *lats2* CRISPR-injected fish.

3. Results and Discussion

We targeted zebrafish *lats2* for deletion with CRISPR-Cas9 genetic editing techniques. Utilizing simultaneous injection of two CRISPR guide RNAs (gRNAs) targeting exons 5 and 6 of zebrafish *lats2*, we generated large deletions early in the coding sequence. These deletions remove the coding sequence that includes the serine threonine kinase domain, which is critical for the Lats2 protein function in canonical Hippo pathway phosphorylation. Providing evidence for large deletions of human *LATS2* in a clinical setting, the engineered mutations in zebrafish are similar to the large deletions reported in patients with malignant pleural mesotheliomas [28]. To determine whether we are able to create large deletions in zebrafish with this strategy, we designed a genotyping protocol utilizing primers flanking the expected deletion. The thermocycle conditions for our PCR amplification were designed using extension times allowing of an amplicon ~750 bp in size, the expected size of an allele containing our targeted large deletion, but not for the WT allele amplicon of 2455 bp (Figure 1A). By increasing the extension time, or including a primer within the predicted deletion site, WT alleles were amplified, ruling out false negatives. Genotyping both single and pooled F0-injected embryos demonstrated that large deletions were efficiently generated with simultaneous injection of two CRISPR gRNAs, along with *cas9* mRNA (Figure 1B).

Using this strategy, we were able to identify F0 adults with germline transmission of *lats2* large deletions. This led to the generation of F1 fish with a *lats2* mutant allele containing an in -frame 1710-bp deletion. This deletion corresponds to a 570-amino-acid deletion in the Lats2 protein. We refer to this mutant allele as *lats2*^{mw87}. This was the only germline-transmitting mutant allele recovered. A previously reported *lats2* nonsense mutant allele containing a 16-bp deletion in exon 3 (termed *lats2*^{ncv108}) has been described. Fish homozygous for this allele were reported as viable, with no obvious defects [37]. Interestingly, we find that homozygous *lats2*^{mw87} fish generated from heterozygous incrosses appear viable, with no overt phenotypes at five days post-fertilization (dpf); however, by 60 dpf we fail to recover Mendelian ratios of *lats2*^{mw87} homozygous mutants, with a significant discrepancy between the observed and expected genotype proportions ($p < 0.0025$; Chi-square analysis; Table 1). The few *lats2*^{mw87} homozygous mutants that did survive to adulthood appeared to be largely normal; however, all died prior to reaching 12 mpf, with no identifiable cause of death, and we were unable to breed these animals. One explanation for the potential differences in viability between these *lats2* alleles could be the recently described genetic compensation observed in several cases of mutant alleles containing indels and a premature termination codon [38,39]. In these cases, it appears that the nonsense-mediated decay caused by premature termination codons is able to initiate compensatory increases in gene expression of known orthologues, and thus mask potential phenotypes. As the *lats2*^{ncv108} is a nonsense frameshift mutation, it may trigger this compensatory mechanism, with *lats1* being a prime candidate. Our in-frame deletion would not be predicted to trigger this effect. Alternatively, our large deletion could result in a truncated Lats2 protein that functions in a dominant negative fashion and thus caused the more severe decrease in viability we observed. Although they are not the focus of this study, the discrepancies between these two mutant alleles could provide interesting insight into Lats2 function and warrant further investigation.

Table 1. *lats2*$^{WT/mw87}$ in-cross survival and Chi-square analysis.

	lats2$^{WT/WT}$	*lats2*$^{WT/mw87}$	*lats2*$^{mw87/mw87}$
Expected	14.25 (25%)	28.50 (50%)	14.25 (25%)
Observed	19 (33.33%)	35 (61.40%)	3 (5.26%)
Chi-square		11.95	
Degrees Freedom		2	
p value (two tailed)		0.0025	

As F0 fish, mosaic for deletion of *lats2*, grew to adulthood, we found that as early as three months post-fertilization (mpf), fish developed large tumors (Figure 1C–H'). These tumors varied in location across both the head and the torso of the fish. The tumors observed also varied in the degree of pigmentation they contained; some appeared primarily opaque with scattered pigmentation, and others were more darkly pigmented. By 6 mpf, ~14% (13/96) of surviving *lats2* CRISPR-injected fish developed visible tumors. At 12 mpf the percentage of tumor incidence increased to ~24% (23/96) (Figure 1I). Notably, we identified tumors only by gross observation of free swimming fish in tanks. Therefore, the tumor incidence could likely be higher than the numbers reported here. Similarly, the initial onset of these tumors was likely to occur earlier than our first detection of obvious tumors. Indeed, tumors showed variability in size, making it likely that we were unable to identify smaller, internal tumors. Most often we observed one prominent tumor on an affected fish. However, we did identify some individuals where two separate tumors were visible. Of note, we did not observe any tumor formation in the few *lats2*^{mw87} homozygous mutants that survived to adulthood.

We isolated six fish containing tumors of varying sizes and locations for detailed histology. We utilized hematoxylin- and eosin-stained paraffin sections for this analysis. Histology revealed that every tumor exhibited very similar spindle cell morphology, and extensively infiltrated both muscle and bone. Spindled cells were observed wrapping around muscle cell fibers in each fish analyzed (Figure 2A–F'). Histological analysis suggested that these were PNSTs. More specifically, the whorl pattern seen in the tumors, termed a "Schwannian whorl," is a distinct characteristic of cellular schwannomas [40]. Tumors also contained varied numbers of pigment-bearing cells, scattered throughout the tumor area. While rare, cases of PNSTs with pigmented cells have been reported [41]. Given that both melanocytes and Schwann cells arise from Sox10-positive neural crest lineage, the tumors could originate from an early neural crest lineage. As Hippo signaling has been implicated in cancers of multiple type and origin, we wondered whether other tissues might also be affected. To assess this, we dissected fish containing visible tumors and assessed all tissues for tumor nodules. We also performed histological analysis and were unable to identify any tumors other than the described PNSTs. As we did not perform this detailed analysis on every *lats2* CRISPR-injected fish, it is possible that we missed rare tumors from other origins or tissues. However, the fact that all identified tumors were PNSTs suggests that Hippo pathway signaling is of particular importance in PNST biology.

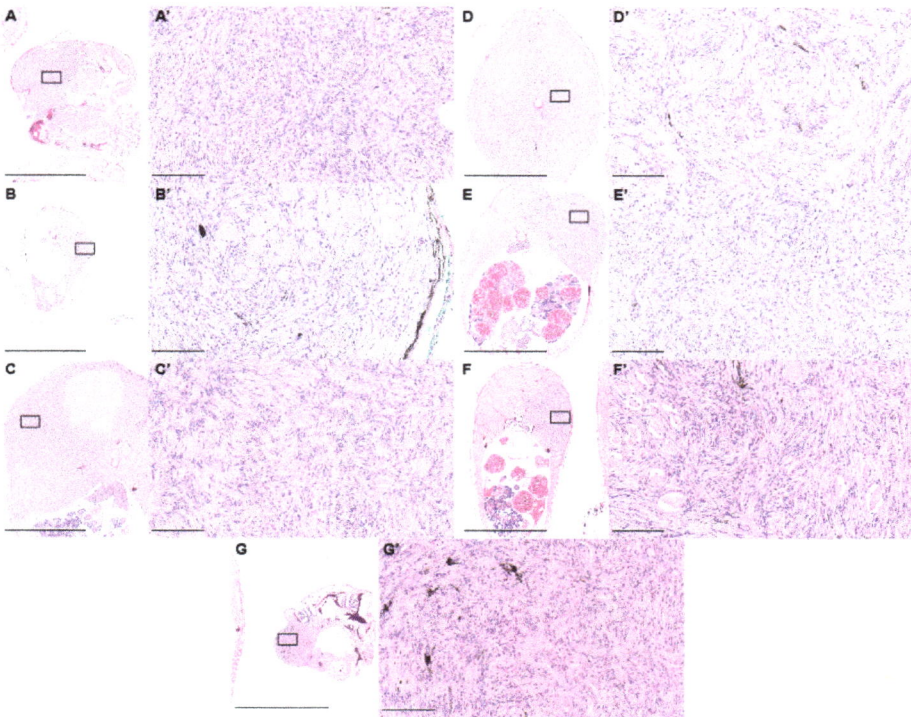

Figure 2. *lats2* CRISPR-injected fish develop tumors resembling PNSTs. (**A–G**) Low power image of tumors revealed anatomical position and size. Scale = 2.5 mm F and G represent two distinct tumors from the same fish. (**A'–F'**) Higher-power images of the sections in (**A–F**) revealing cellular morphology. Scale = 100 μm. Boxes in **A–G** denote the tumor area shown at higher magnification in **A'–G'**.

To confirm the diagnosis of PNST, we performed immunohistochemical (IHC) staining with two known markers of PNSTs, S100 and Sox10 [40,42,43]. We found that tumors were largely Sox10-positive, while diffusely and faintly positive for S100, supporting the diagnosis of PNST (Figure 3A–B'). IHC staining for proliferative cell nuclear antigen (PCNA) revealed that the tumors were highly proliferative, with large numbers of cells throughout the tumors staining positive (Figure 3C,C').

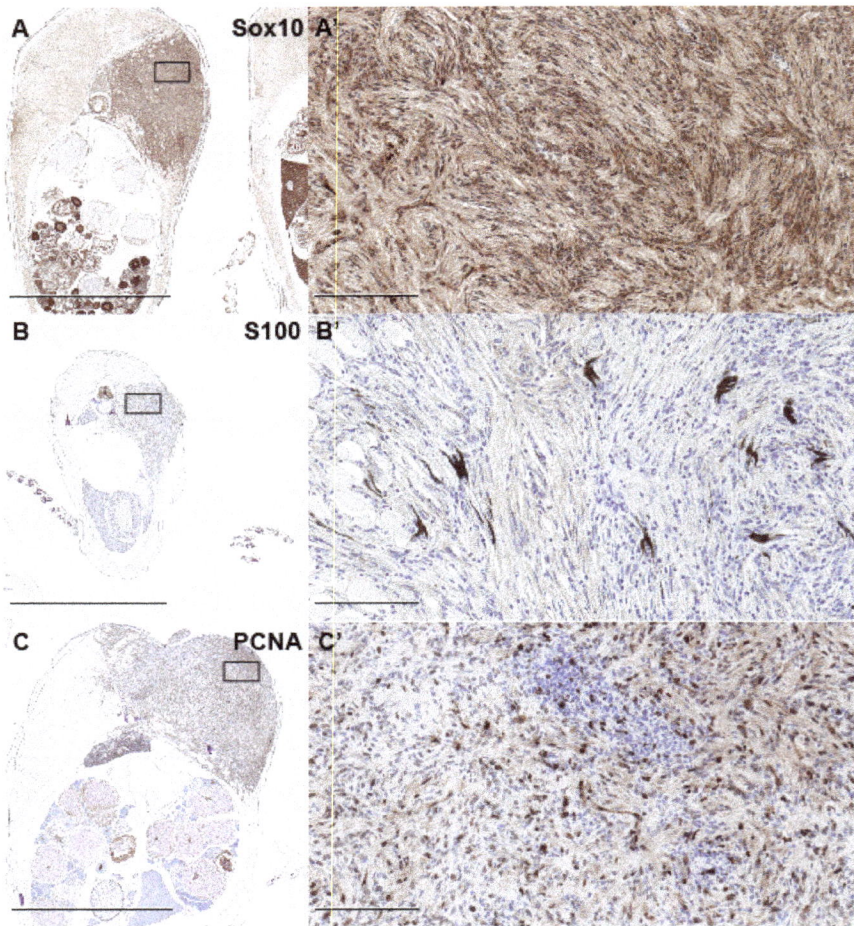

Figure 3. *lats2* CRISPR-injected fish tumors bear markers supporting PNST diagnosis. (**A,A′**) Representative images of Sox10 immunohistochemical staining in *lats2* CRISPR-injected fish. (**B,B′**) Representative images of S100 immunohistochemical staining in *lats2* CRISPR-injected fish. (**C,C′**) Representative images of PCNA immunohistochemical staining in *lats2* CRISPR-injected fish Scale = 2.5 mm for **A–C**, and 100 μm for **A′–C′**. Boxes in **A–C** denote the tumor area shown at higher magnification in **A′–C′**.

Yap and Taz are the main effectors of the Hippo pathway, and the targets of Lats1/2 phosphorylation. As phosphorylation of Yap and Taz by Lats1/2 leads to their cytoplasmic sequestration and eventual ubiquitin-mediated degradation, we also assessed Yap and Taz protein by IHC. To assay the Taz protein, we used a Yap/Taz antibody that we and others have previously shown to specifically recognize Taz in zebrafish [44–46]. Staining with this antibody revealed that all tumors analyzed were positive for Taz; however, we did see variability in the intensity of staining from sample to sample (Figure 4A–E). Closer inspection shows that, while most tumor cells show Taz staining, Taz was often primarily cytoplasmic and diffuse, with scattered cells displaying nuclear Taz. We also observed that areas of high cell density appeared to correlate to low or absent Taz staining (Figure 4A′–E′). Similar to Taz, staining for Yap revealed high levels of staining in all tumor samples (Figure 4F–J). In contrast to Taz staining, we found that Yap staining appeared more consistently nuclear. Most interestingly, the same areas of high cellularity that show low or absent Taz were consistently Yap-positive, and in some cases appeared to be enriched (Figure 4F′–J′). The variation in staining pattern between these two antibodies suggests that there may be differences between Yap and Taz and the forms in which they are found in PNST (i.e., nuclear vs. cytoplasmic, phosphorylated vs. unphosphorylated, etc.), and perhaps define different cell states within the tumor. To further assess these regions of high cell density, we assessed Sox10 and PCNA expression on adjacent sections. We found that these regions did not appear to be significantly different from the surrounding tissue in Sox10 or PCNA staining, suggesting that their identity and proliferative capacity are not greatly altered (Figure 5A–D). While Yap and Taz were found to have many redundant functions, compound genetic studies indicate that Taz may play a more prominent role in Schwann cell development and myelination [16–18]. This suggests that in PNST tumorigenesis, Yap and Taz could share many functions, but also display divergent roles. Our results showing the exclusion of Taz staining within regions of high cell density has yet to be described in existing models of PNST, which suggests that, within single PNSTs, separate populations of cells exist with distinct Hippo-Yap/Taz signaling states.

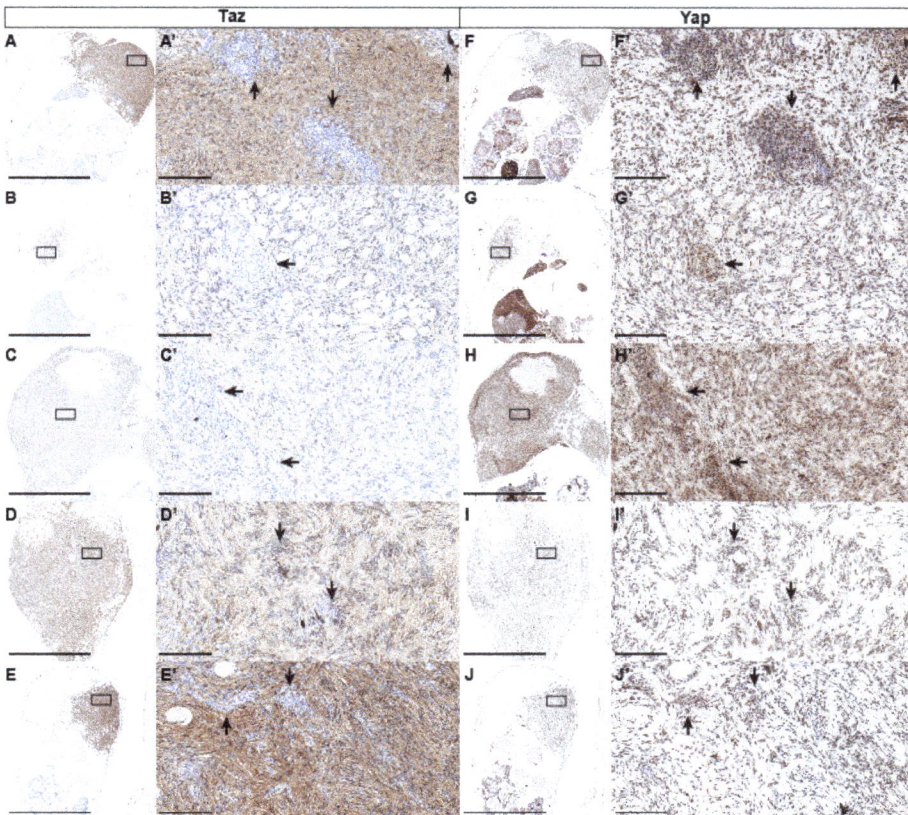

Figure 4. Yap/Taz protein expression in *lats2* CRISPR-injected fish tumors. (**A–E, A′–E′**) Representative images of Taz immunohistochemical staining in *lats2* CRISPR-injected fish. (**F–J, F′–J′**) Representative images of Yap immunohistochemical staining in *lats2* CRISPR-injected fish. Arrows denote regions of high cell density in **A′–J′**. Scale = 2.5 mm for **A–J**, and 100 μm for **A′–J′**. Boxes in **A–J** denote the tumor area shown at higher magnification in **A′–J′**.

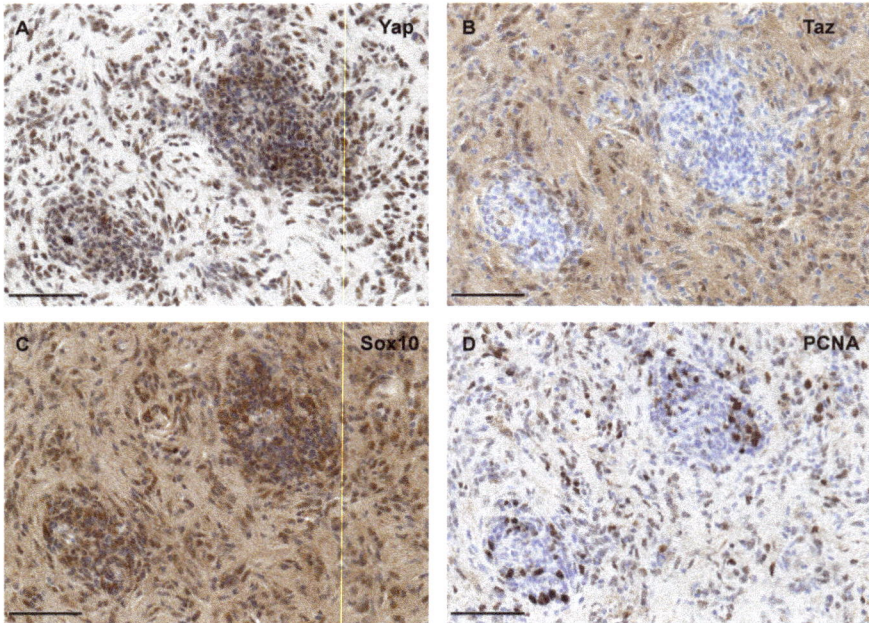

Figure 5. Yap/Taz, Sox10 and PCNA protein expression in areas of high cell density. (**A**) Representative image of Yap immunohistochemical staining in regions of high cell density with peripheral nerve sheath tumor. (**B**) Adjacent section to A with Taz immunohistochemical staining. (**C**) Adjacent section to A with Sox10 immunohistochemical staining. (**D**) Adjacent section to A with PCNA immunohistochemical staining. Scale = 50 µm.

Our study follows the publication of a similar report that found that conditional deletion of *Lats1/2* within the Schwann cell lineage led MPNST formation in mice [29]. The results presented here from zebrafish are largely consistent with the findings in mice, and therefore demonstrate a conserved role for Lats2 signaling in PNSTs across species. However, we note several differences between these two models. In the mouse model, knockout of *Lats1* or *Lats2* alone was insufficient for tumor formation. Zebrafish consistently developed tumors with somatic targeting of *lats2* alone. Importantly, our CRISPR gRNAs are targeted against regions specific to the *lats2* gene, and do not target *lats1* sequences. We also targeted *lats1* with a similar large deletion strategy using separate gRNAs and found that 0 of 33 F0 fish developed tumors (Figure S1). Interestingly, the earliest we identified tumorigenesis in the fish was 3 mpf, with some fish first showing visible tumors at one year of age, and still other fish never developing tumors. Conversely, Wu et al. found that inactivation of Lats1/2 in mice led to rapid tumor formation, with palpable tumors developing as early as three weeks of age and a maximum lifespan of 4–5 months [29]. This is particularly intriguing as genetic editing in our fish model likely occurs within the first 24 h post-fertilization. There are several potential explanations for the varying onset of tumorigenesis between these two models of PNST. The delayed tumorigenesis in the fish may be due to compensation by *lats1* in our model. Alternatively, the mosaic nature of our mutagenesis, including heterozygous and homozygous deletions, may result in fewer potential tumor of origin cells, while the Cre recombinase strategy used in the mouse can efficiently create homozygous mutations throughout the Schwann cell lineage. The temporal difference in tumorigenesis may also suggest that, in the fish model, additional mutations in other tumor suppressors or oncogenes may be necessary. We find this explanation most interesting. If it is true, further study of this model focused on identifying the co-occurring mutations necessary for tumorigenesis could prove beneficial.

The data presented here indicate that somatic deletion of *lats2* in zebrafish can serve as a complementary experimental system to the previously described mouse model for probing the mechanisms of Lats2-Yap/Taz PNST. One advantage of the zebrafish model, in comparison with the murine system, is the reduced amount of breeding required to generate large pedigrees and achieve high experimental sample sizes. Injections into large numbers of embryos allows for easy generation of a high number of tumor-bearing fish on any number of genetic or transgenic backgrounds. Another advantage is the increased number of transgenic fish available for characterization of cell signaling, stress responses, or metabolic state. For instance, injection into fluorescent reporter fish for Notch [47,48], BMP [49–51], Wnt [52,53], Hedgehog [54,55], Hippo [44,56,57], or other pathways would allow for sorting of tumor cells by reporter activity and subsequent analysis of the unique transcriptomic or proteomic signatures of those populations. Finally, as an aquatic species, drug screens and chemotherapeutic analysis is more easily accomplished.

In conclusion, we present here a novel model of PNSTs in zebrafish by somatic inactivation of the Hippo kinase Lats2. This methodology shows robust presentation of PNST formation in adult genetically mosaic fish. This work further solidifies the role of Hippo-Yap/Taz signaling, and specifically Lats2 in PNST biology, and offers a new model to study this form of cancer.

Supplementary Materials: The following are available online at http://www.mdpi.com/2073-4409/8/9/972/s1, Figure S1.

Author Contributions: Conceptualization, Z.J.B. and B.A.L.; methodology, Z.J.B. and B.A.L.; formal analysis, Z.J.B.; investigation, Z.J.B.; validation, Z.J.B.; resources, B.A.L.; data curation, Z.J.B.; writing—original draft preparation, Z.J.B.; writing—review and editing, Z.J.B., B.A.L. and P.N.N.; visualization, Z.J.B. and B.A.L.; supervision, B.A.L. and P.N.N.; project administration, B.A.L.; funding acquisition, B.A.L.

Funding: This work was supported in part through NIH grant no. R01 EY029267 and funds provided by the Cancer Center of the Medical College of Wisconsin.

Acknowledgments: Our appreciation extends to the Children's Research Institute Histology Core for their assistance with routine histology and immunohistochemistry.

Conflicts of Interest: The authors declare no conflict of interest.

References

1. Ostrom, Q.T.; Gittleman, H.; Truitt, G.; Boscia, A.; Kruchko, C.; Barnholtz-Sloan, J.S. CBTRUS Statistical Report: Primary Brain and Other Central Nervous System Tumors Diagnosed in the United States in 2011–2015. *Neuro-Oncology* **2018**, *20*, iv1–iv86. [CrossRef] [PubMed]

2. Louis, D.N.; Ohgaki, H.; Wiestler, O.D.; Cavenee, W.K.; Burger, P.C.; Jouvet, A.; Scheithauer, B.W.; Kleihues, P. The 2007 WHO Classification of Tumours of the Central Nervous System. *Acta Neuropathol.* **2007**, *114*, 97–109. [CrossRef] [PubMed]

3. Komori, T. The 2016 WHO Classification of Tumours of the Central Nervous System: The Major Points of Revision. *Neurol. Medico-Chirurgica* **2017**, *57*, 301–311. [CrossRef] [PubMed]

4. Sughrue, M.E.; Levine, J.; Barbaro, N.M. Pain as a symptom of peripheral nerve sheath tumors: Clinical significance and future therapeutic directions. *J. Brachial Plex. Peripher. Nerve Inj.* **2008**, *3*, 6. [CrossRef] [PubMed]

5. Bradford, D.; Kim, A. Current Treatment Options for Malignant Peripheral Nerve Sheath Tumors. *Curr. Treat. Options Oncol.* **2015**, *16*, 12. [CrossRef] [PubMed]

6. James, A.W.; Shurell, E.; Singh, A.; Dry, S.M.; Eilber, F.C. Malignant Peripheral Nerve Sheath Tumor. *Surg. Oncol. Clin.* **2016**, *25*, 789–802. [CrossRef] [PubMed]

7. Carroll, S.L. Molecular mechanisms promoting the pathogenesis of Schwann cell neoplasms. *Acta Neuropathol.* **2012**, *123*, 321–348. [CrossRef] [PubMed]

8. Korf, B.R. Pediatric Neurology Part I: Chapter 39—Neurofibromatosis. In *Handbook of Clinical Neurology*; Dulac, O., Lassonde, M., Sarnat, H.B., Eds.; Elsevier: Amsterdam, The Netherlands, 2013; Volume 111, pp. 333–340.

9. Messersmith, L.; Krauland, K. *Neurofibroma*; StatPearls Publishing: Treasure Island, FL, USA, 2019.

10. Kiuru, M.; Busam, K.J. The NF1 gene in tumor syndromes and melanoma. *Lab. Investig.* **2017**, *97*, 146–157. [CrossRef]

11. Rouleau, G.A.; Merel, P.; Lutchman, M.; Sanson, M.; Zucman, J.; Marineau, C.; Hoang-Xuan, K.; Demczuk, S.; Desmaze, C.; Plougastel, B.; et al. Alteration in a new gene encoding a putative membrane-organizing protein causes neuro-fibromatosis type 2. *Nature* **1993**, *363*, 515–521. [CrossRef]

12. Trofatter, J.A.; MacCollin, M.M.; Rutter, J.L.; Murrell, J.R.; Duyao, M.P.; Parry, D.M.; Eldridge, R.; Kley, N.; Menon, A.G.; Pulaski, K. A novel moesin-, ezrin-, radixin-like gene is a candidate for the neurofibromatosis 2 tumor suppressor. *Cell* **1993**, *72*, 791–800. [CrossRef]

13. Hamaratoglu, F.; Willecke, M.; Kango-Singh, M.; Nolo, R.; Hyun, E.; Tao, C.; Jafar-Nejad, H.; Halder, G. The tumour-suppressor genes NF2/Merlin and Expanded act through Hippo signalling to regulate cell proliferation and apoptosis. *Nat. Cell Biol.* **2006**, *8*, 27–36. [CrossRef] [PubMed]

14. Mindos, T.; Dun, X.; North, K.; Doddrell, R.D.S.; Schulz, A.; Edwards, P.; Russell, J.; Gray, B.; Roberts, S.L.; Shivane, A.; et al. Merlin controls the repair capacity of Schwann cells after injury by regulating Hippo/YAP activity. *J. Cell Biol.* **2017**, *216*, 495–510. [CrossRef] [PubMed]

15. Farid, M.; Demicco, E.G.; Garcia, R.; Ahn, L.; Merola, P.R.; Cioffi, A.; Maki, R.G. Malignant Peripheral Nerve Sheath Tumors. *The Oncologist* **2014**, *19*, 193–201. [CrossRef] [PubMed]

16. Deng, Y.; Wu, L.M.N.; Bai, S.; Zhao, C.; Wang, H.; Wang, J.; Xu, L.; Sakabe, M.; Zhou, W.; Xin, M.; et al. A reciprocal regulatory loop between TAZ/YAP and G-protein Gαs regulates Schwann cell proliferation and myelination. *Nat. Commun.* **2017**, *8*, 15161. [CrossRef] [PubMed]

17. Poitelon, Y.; Lopez-Anido, C.; Catignas, K.; Berti, C.; Palmisano, M.; Williamson, C.; Ameroso, D.; Abiko, K.; Hwang, Y.; Gregorieff, A.; et al. YAP and TAZ control peripheral myelination and the expression of laminin receptors in Schwann cells. *Nat. Neurosci.* **2016**, *19*, 879–887. [CrossRef] [PubMed]

18. Grove, M.; Kim, H.; Santerre, M.; Krupka, A.J.; Han, S.B.; Zhai, J.; Cho, J.Y.; Park, R.; Harris, M.; Kim, S.; et al. YAP/TAZ initiate and maintain Schwann cell myelination. *eLife* **2017**, *6*, e20982. [CrossRef]

19. Li, W.; Cooper, J.; Zhou, L.; Yang, C.; Erdjument-Bromage, H.; Zagzag, D.; Snuderl, M.; Ladanyi, M.; Hanemann, C.O.; Zhou, P.; et al. Merlin/NF2 Loss-Driven Tumorigenesis Linked to CRL4DCAF1-Mediated Inhibition of the Hippo Pathway Kinases Lats1 and 2 in the Nucleus. *Cancer Cell* **2014**, *26*, 48–60. [CrossRef]

20. Guerrant, W.; Kota, S.; Troutman, S.; Mandati, V.; Fallahi, M.; Stemmer-Rachamimov, A.; Kissil, J.L. YAP Mediates Tumorigenesis in Neurofibromatosis Type 2 by Promoting Cell Survival and Proliferation through a COX-2–EGFR Signaling Axis. *Cancer Res.* **2016**, *76*, 3507–3519. [CrossRef]

21. Boin, A.; Couvelard, A.; Couderc, C.; Brito, I.; Filipescu, D.; Kalamarides, M.; Bedossa, P.; De Koning, L.; Danelsky, C.; Dubois, T.; et al. Proteomic screening identifies a YAP-driven signaling network linked to tumor cell proliferation in human schwannomas. *Neuro-Oncology* **2014**, *16*, 1196–1209. [CrossRef]

22. Kim, Y.-H.; Ohta, T.; Oh, J.E.; Le Calvez-Kelm, F.; McKay, J.; Voegele, C.; Durand, G.; Mittelbronn, M.; Kleihues, P.; Paulus, W.; et al. TP53, MSH4, and LATS1 Germline Mutations in a Family with Clustering of Nervous System Tumors. *Am. J. Pathol.* **2014**, *184*, 2374–2381. [CrossRef]

23. Oh, J.-E.; Ohta, T.; Satomi, K.; Foll, M.; Durand, G.; McKay, J.; Calvez-Kelm, F.L.; Mittelbronn, M.; Brokinkel, B.; Paulus, W.; et al. Alterations in the NF2/LATS1/LATS2/YAP Pathway in Schwannomas. *J. Neuropathol. Exp. Neurol.* **2015**, *74*, 952–959. [CrossRef] [PubMed]

24. Zhao, F.; Yang, Z.; Chen, Y.; Zhou, Q.; Zhang, J.; Liu, J.; Wang, B.; He, Q.; Zhang, L.; Yu, Y.; et al. Deregulation of the Hippo Pathway Promotes Tumor Cell Proliferation Through YAP Activity in Human Sporadic Vestibular Schwannoma. *World Neurosurg.* **2018**, *117*, e269–e279. [CrossRef] [PubMed]

25. Faden, D.L.; Asthana, S.; Tihan, T.; DeRisi, J.; Kliot, M. Whole Exome Sequencing of Growing and Non-Growing Cutaneous Neurofibromas from a Single Patient with Neurofibromatosis Type 1. *PLoS ONE* **2017**, *12*, e0170348.

26. Tanahashi, K.; Natsume, A.; Ohka, F.; Motomura, K.; Alim, A.; Tanaka, I.; Senga, T.; Harada, I.; Fukuyama, R.; Sumiyoshi, N.; et al. Activation of Yes-Associated Protein in Low-Grade Meningiomas Is Regulated by Merlin, Cell Density, and Extracellular Matrix Stiffness. *J. Neuropathol. Exp. Neurol.* **2015**, *74*, 704–709. [CrossRef] [PubMed]

27. Tremblay, A.M.; Missiaglia, E.; Galli, G.G.; Hettmer, S.; Urcia, R.; Carrara, M.; Judson, R.N.; Thway, K.; Nadal, G.; Selfe, J.L.; et al. The Hippo Transducer YAP1 Transforms Activated Satellite Cells and Is a Potent Effector of Embryonal Rhabdomyosarcoma Formation. *Cancer Cell* **2014**, *26*, 273–287. [CrossRef] [PubMed]

28. Tranchant, R.; Quetel, L.; Tallet, A.; Meiller, C.; Renier, A.; de Koning, L.; de Reynies, A.; Pimpec-Barthes, F.L.; Zucman-Rossi, J.; Jaurand, M.-C.; et al. Co-occurring Mutations of Tumor Suppressor Genes, LATS2 and NF2, in Malignant Pleural Mesothelioma. *Clin. Cancer Res.* **2017**, *23*, 3191–3202. [CrossRef] [PubMed]

29. Wu, L.M.N.; Deng, Y.; Wang, J.; Zhao, C.; Wang, J.; Rao, R.; Xu, L.; Zhou, W.; Choi, K.; Rizvi, T.A.; et al. Programming of Schwann Cells by Lats1/2-TAZ/YAP Signaling Drives Malignant Peripheral Nerve Sheath Tumorigenesis. *Cancer Cell* **2018**, *33*, 292–308.e7. [CrossRef]

30. Bootorabi, F.; Manouchehri, H.; Changizi, R.; Barker, H.; Palazzo, E.; Saltari, A.; Parikka, M.; Pincelli, C.; Aspatwar, A. Zebrafish as a Model Organism for the Development of Drugs for Skin Cancer. *Int. J. Mol. Sci.* **2017**, *18*, 1550. [CrossRef]

31. Ablain, J.; Xu, M.; Rothschild, H.; Jordan, R.C.; Mito, J.K.; Daniels, B.H.; Bell, C.F.; Joseph, N.M.; Wu, H.; Bastian, B.C.; et al. Human tumor genomics and zebrafish modeling identify SPRED1 loss as a driver of mucosal melanoma. *Science* **2018**, *362*, 1055–1060. [CrossRef]

32. Durbin, A.D.; Ki, D.H.; He, S.; Look, A.T. Malignant Peripheral Nerve Sheath Tumors. In *Cancer and Zebrafish: Mechanisms, Techniques, and Models*; Langenau, D.M., Ed.; Springer International Publishing: New York, NY, USA, 2016; pp. 495–530, ISBN 978-3-319-30654-4.

33. Ki, D.H.; He, S.; Rodig, S.; Look, A.T. Overexpression of PDGFRA cooperates with loss of NF1 and p53 to accelerate the molecular pathogenesis of malignant peripheral nerve sheath tumors. *Oncogene* **2017**, *36*, 1058–1068. [CrossRef]

34. Solin, S.L.; Shive, H.R.; Woolard, K.D.; Essner, J.J.; McGrail, M. Rapid tumor induction in zebrafish by TALEN-mediated somatic inactivation of the retinoblastoma1 tumor suppressor rb1. *Sci. Rep.* **2015**, *5*, 13745. [CrossRef] [PubMed]

35. Shim, J.; Choi, J.-H.; Park, M.-H.; Kim, H.; Kim, J.H.; Kim, S.-Y.; Hong, D.; Kim, S.; Lee, J.E.; Kim, C.-H.; et al. Development of zebrafish medulloblastoma-like PNET model by TALEN-mediated somatic gene inactivation. *Oncotarget* **2017**, *8*, 55280–55297. [CrossRef] [PubMed]

36. Schultz, L.E.; Haltom, J.A.; Almeida, M.P.; Wierson, W.A.; Solin, S.L.; Weiss, T.J.; Helmer, J.A.; Sandquist, E.J.; Shive, H.R.; McGrail, M. Epigenetic regulators Rbbp4 and Hdac1 are overexpressed in a zebrafish model of RB1 embryonal brain tumor, and are required for neural progenitor survival and proliferation. *Dis. Model. Mech.* **2018**, *11*, dmm034124. [CrossRef] [PubMed]

37. Fukui, H.; Miyazaki, T.; Chow, R.W.-Y.; Ishikawa, H.; Nakajima, H.; Vermot, J.; Mochizuki, N. Hippo signaling determines the number of venous pole cells that originate from the anterior lateral plate mesoderm in zebrafish. *eLife* **2018**, *7*, e29106. [CrossRef] [PubMed]

38. El-Brolosy, M.A.; Kontarakis, Z.; Rossi, A.; Kuenne, C.; Günther, S.; Fukuda, N.; Kikhi, K.; Boezio, G.L.M.; Takacs, C.M.; Lai, S.-L.; et al. Genetic compensation triggered by mutant mRNA degradation. *Nature* **2019**, *568*, 193–197. [CrossRef]

39. Ma, Z.; Zhu, P.; Shi, H.; Guo, L.; Zhang, Q.; Chen, Y.; Chen, S.; Zhang, Z.; Peng, J.; Chen, J. PTC-bearing mRNA elicits a genetic compensation response via Upf3a and COMPASS components. *Nature* **2019**, *568*, 259–263. [CrossRef] [PubMed]

40. Pekmezci, M.; Reuss, D.E.; Hirbe, A.C.; Dahiya, S.; Gutmann, D.H.; von Deimling, A.; Horvai, A.E.; Perry, A. Morphologic and immunohistochemical features of malignant peripheral nerve sheath tumors and cellular schwannomas. *Mod. Pathol.* **2015**, *28*, 187–200. [CrossRef]

41. Rodriguez, F.J.; Folpe, A.L.; Giannini, C.; Perry, A. Pathology of peripheral nerve sheath tumors: Diagnostic overview and update on selected diagnostic problems. *Acta Neuropathol.* **2012**, *123*, 295–319. [CrossRef]

42. Karamchandani, J.; Nielsen, T.; van de Rijn, M.; West, R. Sox10 and S100 in the Diagnosis of Soft-tissue Neoplasms. *Appl. Immunohistochem. Mol. Morphol.* **2012**, *20*, 445–450. [CrossRef]

43. Miettinen, M.; McCue, P.A.; Sarlomo-Rikala, M.; Biernat, W.; Czapiewski, P.; Kopczynski, J.; Thompson, L.D.; Lasota, J.; Wang, Z.; Fetsch, J.F. Sox10—A marker for not only Schwannian and melanocytic neoplasms but also myoepithelial cell tumors of soft tissue. A systematic analysis of 5134 tumors. *Am. J. Surg. Pathol.* **2015**, *39*, 826–835. [CrossRef]

44. Miesfeld, J.B.; Gestri, G.; Clark, B.S.; Flinn, M.A.; Poole, R.J.; Bader, J.R.; Besharse, J.C.; Wilson, S.W.; Link, B.A. Yap and Taz regulate retinal pigment epithelial cell fate. *Development* **2015**, *142*, 3021–3032. [CrossRef] [PubMed]

45. Kimelman, D.; Smith, N.L.; Lai, J.K.H.; Stainier, D.Y. Regulation of posterior body and epidermal morphogenesis in zebrafish by localized Yap1 and Wwtr1. *eLife* **2017**, *6*, e31065. [CrossRef] [PubMed]

46. Lai, J.K.H.; Collins, M.M.; Uribe, V.; Jiménez-Amilburu, V.; Günther, S.; Maischein, H.-M.; Stainier, D.Y.R. The Hippo pathway effector Wwtr1 regulates cardiac wall maturation in zebrafish. *Development* **2018**, *145*, dev159210. [CrossRef] [PubMed]

47. Yeo, S.-Y.; Kim, M.; Kim, H.-S.; Huh, T.-L.; Chitnis, A.B. Fluorescent protein expression driven by her4 regulatory elements reveals the spatiotemporal pattern of Notch signaling in the nervous system of zebrafish embryos. *Dev. Biol.* **2007**, *301*, 555–567. [CrossRef] [PubMed]

48. Parsons, M.J.; Pisharath, H.; Yusuff, S.; Moore, J.C.; Siekmann, A.F.; Lawson, N.; Leach, S.D. Notch-responsive cells initiate the secondary transition in larval zebrafish pancreas. *Mech. Dev.* **2009**, *126*, 898–912. [CrossRef] [PubMed]

49. Collery, R.F.; Link, B.A. Dynamic smad-mediated BMP signaling revealed through transgenic zebrafish. *Dev. Dyn.* **2011**, *240*, 712–722. [CrossRef] [PubMed]

50. Laux, D.W.; Febbo, J.A.; Roman, B.L. Dynamic analysis of BMP-responsive smad activity in live zebrafish embryos. *Dev. Dyn.* **2011**, *240*, 682–694. [CrossRef] [PubMed]

51. Ramel, M.-C.; Hill, C.S. The ventral to dorsal BMP activity gradient in the early zebrafish embryo is determined by graded expression of BMP ligands. *Dev. Biol.* **2013**, *378*, 170–182. [CrossRef]

52. Dorsky, R.I.; Sheldahl, L.C.; Moon, R.T. A Transgenic Lef1/β-Catenin-Dependent Reporter Is Expressed in Spatially Restricted Domains throughout Zebrafish Development. *Dev. Biol.* **2002**, *241*, 229–237. [CrossRef]

53. Shimizu, N.; Kawakami, K.; Ishitani, T. Visualization and exploration of Tcf/Lef function using a highly responsive Wnt/β-catenin signaling-reporter transgenic zebrafish. *Dev. Biol.* **2012**, *370*, 71–85. [CrossRef]

54. Mich, J.K.; Payumo, A.Y.; Rack, P.G.; Chen, J.K. In Vivo Imaging of Hedgehog Pathway Activation with a Nuclear Fluorescent Reporter. *PLoS ONE* **2014**, *9*, e103661. [CrossRef] [PubMed]

55. Schwend, T.; Loucks, E.J.; Ahlgren, S.C. Visualization of Gli Activity in Craniofacial Tissues of Hedgehog-Pathway Reporter Transgenic Zebrafish. *PLoS ONE* **2010**, *5*, e14396. [CrossRef] [PubMed]

56. Astone, M.; Lai, J.K.H.; Dupont, S.; Stainier, D.Y.R.; Argenton, F.; Vettori, A. Zebrafish mutants and TEAD reporters reveal essential functions for Yap and Taz in posterior cardinal vein development. *Sci. Rep.* **2018**, *8*, 10189. [CrossRef] [PubMed]

57. Miesfeld, J.B.; Link, B.A. Establishment of transgenic lines to monitor and manipulate Yap/Taz-Tead activity in zebrafish reveals both evolutionarily conserved and divergent functions of the Hippo pathway. *Mech. Dev.* **2014**, *133*, 177–188. [CrossRef] [PubMed]

cells

MDPI

Review

Hippo Pathway in Mammalian Adaptive Immune System

Takayoshi Yamauchi [1] and Toshiro Moroishi [1,2,3,*]

[1] Department of Molecular Enzymology, Faculty of Life Sciences, Kumamoto University, Kumamoto 860-8556, Japan; yamauchi0@kumamoto-u.ac.jp
[2] Center for Metabolic Regulation of Healthy Aging, Faculty of Life Sciences, Kumamoto University, Kumamoto 860-8556, Japan
[3] Precursory Research for Embryonic Science and Technology (PRESTO), Japan Science and Technology Agency (JST), Kawaguchi 332-0012, Japan
* Correspondence: moroishi@kumamoto-u.ac.jp; Tel.: +81-96-344-2111

Received: 7 April 2019; Accepted: 28 April 2019; Published: 30 April 2019

Abstract: The Hippo pathway was originally identified as an evolutionarily-conserved signaling mechanism that contributes to the control of organ size. It was then rapidly expanded as a key pathway in the regulation of tissue development, regeneration, and cancer pathogenesis. The increasing amount of evidence in recent years has also connected this pathway to the regulation of innate and adaptive immune responses. Notably, the Hippo pathway has been revealed to play a pivotal role in adaptive immune cell lineages, as represented by the patients with T- and B-cell lymphopenia exhibiting defective expressions of the pathway component. The complex regulatory mechanisms of and by the Hippo pathway have also been evident as alternative signal transductions are employed in some immune cell types. In this review article, we summarize the current understanding of the emerging roles of the Hippo pathway in adaptive immune cell development and differentiation. We also highlight the recent findings concerning the dual functions of the Hippo pathway in autoimmunity and anti-cancer immune responses and discuss the key open questions in the interplay between the Hippo pathway and the mammalian immune system.

Keywords: Hippo pathway; innate immunity; adaptive immunity; cancer immunity; autoimmunity; YAP (yes-associated protein); TAZ (transcriptional co-activator with PDZ-binding motif); LATS (large tumor suppressor kinase); MST (mammalian STE20-like protein kinase)

1. Introduction

Since its initial discovery in *Drosophila*, the Hippo pathway has gained immense attention for being strongly involved in organ development [1–4], stem cell biology [5–7], regeneration [8–10], and cancer biology [11–14]. The Hippo pathway responds to a wide range of extracellular and intracellular physiological cues, sensing the entire cellular environment, orchestrating cellular responses, and thus, contributing to cell fate determination [15,16]. The Hippo pathway is now known to be composed of more than 30 components, including the core kinase module and the transcriptional module [17]. As shown in Figure 1, the kinase module includes 11 kinases, namely mammalian STE20-like protein kinase 1 (MST1, also known as STK4) and MST2 (also known as STK3), mitogen-activated protein kinase kinase kinase kinases (MAP4Ks, including seven kinases, namely, MAP4K1/2/3/4/5/6/7), large tumor suppressor kinase 1 (LATS1) and LATS2, and in addition, their activating adaptor proteins, salvador family WW domain-containing protein 1 (SAV1), Ras-related proteins RAP2s (including three GTPases RAP2A/B/C), and MOB kinase activator 1A (MOB1A) and MOB1B, respectively. The transcriptional module includes yes-associated protein (YAP) and transcriptional co-activator with PDZ-binding motif (TAZ, also known as WWTR1), in combination with their best-characterized target transcription factors, TEA domain family members (TEADs, including four transcription factors, namely, TEAD1/2/3/4). When the kinase module is "ON" (the Hippo pathway is "activated"), MST1/2 or MAP4Ks phosphorylate and activate the downstream kinases LATS1/2, which in turn promotes inhibitory phosphorylation of the transcriptional co-activators YAP/TAZ, resulting in the cytoplasmic sequestration or proteasomal degradation of YAP/TAZ [18]. In contrast, when the upstream kinase module is "OFF" (the Hippo pathway is "inactivated"), hypophosphorylated YAP/TAZ translocate into the nucleus wherein they bind to and thus activate TEADs transcription factors to promote the target gene transcription. YAP/TAZ-mediated transcription generally drives multiple aspects of cell behavior, including cell proliferation, survival, cell plasticity, and stemness, which is essential for tissue development and regeneration [19]. Thus, in short, the activation of LATS1/2 kinases and the inactivation of YAP/TAZ transcriptional co-activators represent the major molecular functions of the canonical Hippo pathway.

In addition to its roles in tissue development and tumorigenesis, numerous studies in recent years have revealed the extensive roles of the Hippo pathway in immune regulation, both in adaptive and innate immune systems (Figure 2). For example, *Drosophila* Hippo (Hpo) and its mammalian homologues MST1/2 have been revealed to mediate Toll-like receptor (TLR) signaling in both flies [20] and mammals [21,22]. YAP/TAZ bind and inhibit TBK1 (TANK binding kinase 1) or IRF3 (interferon regulatory factor 3) to antagonize the antiviral innate immune responses [23,24]. The critical functions of the Hippo pathway in innate immune responses have also been reviewed elsewhere [25–27]. In this review, we focus on the current knowledge about the functions of the core Hippo pathway components in adaptive immunity, particularly in lymphocyte homeostasis during their development and differentiation. Although the molecular functions of MST1/2 in the mammalian adaptive immune system have been extensively studied in previous works, characterization of the other components of the Hippo pathway is an emerging field of research. In contrast to their pivotal functions in adherent cell physiology, it appears that YAP/TAZ are dispensable for physiological and malignant hematopoiesis [28]. Recent studies have demonstrated that MST1/2 regulate the lymphocyte biology independently of the key Hippo pathway components YAP/TAZ and LATS1/2 [29,30]. Indeed, growing evidence suggests a crosstalk between the Hippo pathway and other pivotal signaling networks involved in immune regulation, such as MAPK (mitogen-activated protein kinase), p53, and the FOXO (forkhead box O) pathway [17,31,32]. We also discuss the complexity of the signal transduction mechanisms downstream of the Hippo pathway in immune cells, which appear to be distinct from those in adherent cells that have been widely used to draw the Hippo signaling network to date. Although the Hippo pathway takes its name from MST1/2—the mammalian homologs of the *Drosophila* Hippo (Hpo), MST1/2 are also known to regulate several proteins other than the key Hippo signaling components. Therefore, the functional outputs of MST1/2 are not limited to LATS1/2 or YAP/TAZ [18]. In this review, we define the signaling that specifically regulates LATS1/2 kinase activity and/or YAP/TAZ transcriptional activity as the "canonical" Hippo pathway. Other signaling cascades that involve the core Hippo pathway components (particularly MST1/2) but do not regulate LATS1/2 kinase or YAP/TAZ are defined as the "alternative" Hippo pathway (Figure 1).

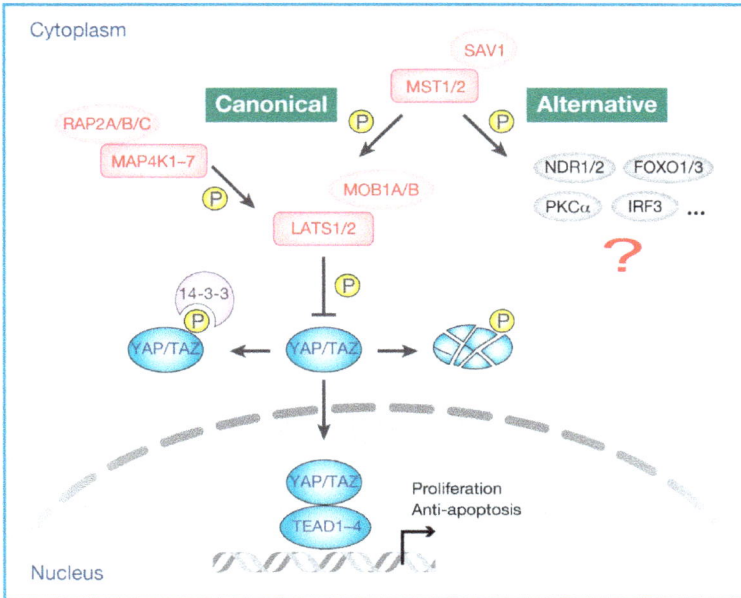

Figure 1. Canonical and alternative Hippo pathways. The heart of the Hippo pathway consists of the kinase module (indicated in red) and the transcriptional module (indicated in blue). The kinase module includes 11 kinases, namely mammalian STE20-like protein kinases (MST1/2), mitogen-activated protein kinase kinase kinase kinases (MAP4K1-7), and large tumor suppressor kinases (LATS1/2), as well as their activating adaptor proteins, salvador family WW domain-containing protein 1 (SAV1), Ras-related proteins (RAP2A/B/C), and MOB kinase activators (MOB1A/B). The transcriptional module includes the transcriptional co-activators, namely, yes-associated protein (YAP) and transcriptional co-activator with PDZ-binding motif (TAZ), and in addition, the transcription factors TEA domain family members (TEAD1–4). When the upstream signals are integrated to activate the Hippo pathway, LATS1/2 kinases phosphorylate and inhibit YAP/TAZ. Phosphorylation of YAP/TAZ promotes their proteasomal degradation or cytoplasmic retention via 14-3-3 binding. In contrast, when the kinase module is inactivated (the Hippo pathway is inactivated), hypophosphorylated YAP/TAZ translocate into the nucleus wherein they bind to TEAD1–4 and thus induce proliferative and anti-apoptotic gene transcription. In this review, we define the signaling that specifically regulates LATS1/2 kinase activity and/or YAP/TAZ transcriptional activity as the "canonical" Hippo pathway. The other signaling cascades that involve MST1/2 but do not regulate LATS1/2 kinases or YAP/TAZ are defined as the "alternative" Hippo pathway. MST1/2 have been shown to modulate a number of proteins, including nuclear Dbf2-related kinases (NDR1/2, also known as STK38/STK38L), forkhead box O (FOXO1/3), protein kinase C alpha (PKCα), and interferon regulatory factor 3 (IRF3) in the regulation of immune system.

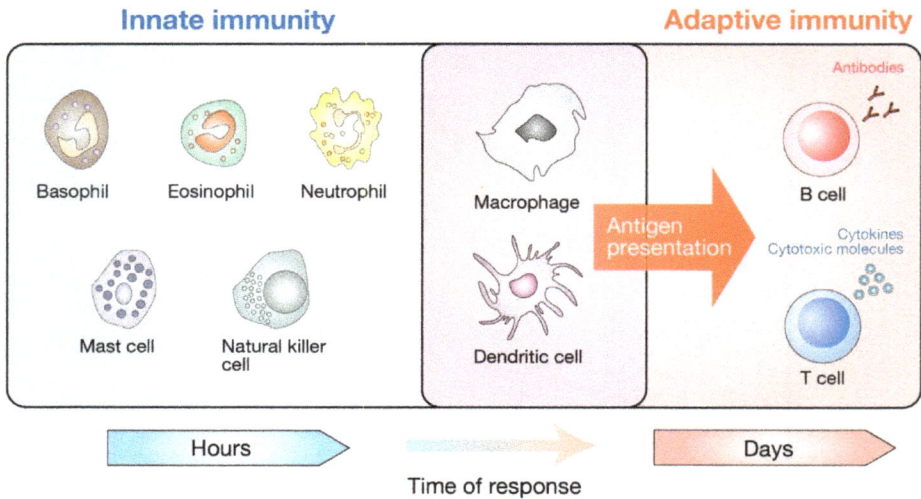

Figure 2. Cellular components of the mammalian immune system. The mammalian immune system consists of two distinct parts, innate and adaptive immunity. Basophils, eosinophils, neutrophils, mast cells, natural killer cells, macrophages, and dendritic cells mediate the innate immunity. They provide the first line of defense against bacteria, viruses, and cancer. The adaptive immune system refers to an antigen-specific defense mechanism that takes several days to develop but provides long-lasting protection. The adaptive immune system includes B cell-mediated humoral immunity and T cell-mediated cellular immunity, both of which are directed towards the specific antigens. Macrophages and dendritic cells are unique subsets that have both innate and adaptive immune cell traits. As professional antigen-presenting cells, macrophages and dendritic cells are critical in the induction of adaptive immunity by presenting the antigens to antigen-specific T and B lymphocytes.

2. Hippo Pathway in Adaptive Immune Cell Lineage and Functions

Adaptive immunity is defined by antigen-specific immune responses, consisting of cellular (cell-mediated) and humoral (antibody-mediated) responses. All T cells, B cells, and antigen-presenting cells cooperatively orchestrate this process. Recent studies have revealed pivotal functions of MST1/2 in T-cell development and differentiation, as well as in B cell homeostasis in the splenic marginal zone and the periphery. MST1/2 also play key roles in antigen-presenting cells, including macrophages and dendritic cells. It is likely that MST1/2 regulate the adhesion and trafficking of the immune cells and thus coordinate various immunological events. Mechanistically, MST1/2 appear to exert their biological functions via both canonical and alternative downstream effectors in the Hippo pathway, as suggested by a series of in vivo mice studies.

2.1. Clinical Significance of the Hippo Pathway in Adaptive Immunity

The importance of the Hippo signaling pathway in the adaptive immune system is supported by the clinical case reports of patients who have loss of heterozygosity mutations of the MST1 gene or hyper-methylation of the MST1 promoter region [33–38]. MST1-deficient patients show susceptibility to bacterial and viral infections, clinical signs of T and B cell lymphopenia, and in addition, although counterintuitive, autoimmune manifestations (this seemingly contradictory outcome of MST1 deficiency is discussed in Section 2.4.). MST1/2-deficient mice were found to recapitulate some of these symptoms in the patients, including lymphopenia and autoimmune symptoms such as autoantibody production [29,33,36,39]. These mouse models have greatly helped in our understanding of the biological functions and molecular mechanisms of immune regulation by MST1/2 kinases.

2.2. MST1/2 in T Cell Development

Hematopoietic stem cells (HSC) give rise to both common lymphoid progenitor cells (CLP) and common myeloid progenitor cells (CMP) inside the bone marrow. CLP produce T cell progenitors, known as early thymic progenitors (ETP), which develop and mature within the thymus (Figure 3).

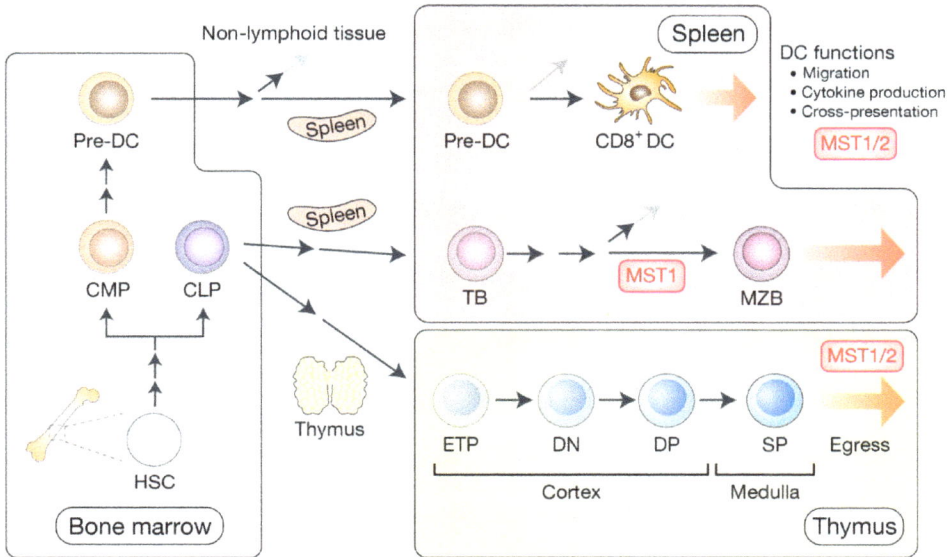

Figure 3. The Hippo pathway in the development of dendritic cells, B cells, and T cells. In bone marrow, hematopoietic stem cells (HSC) give rise to common myeloid progenitors (CMP) and common lymphoid progenitors (CLP). CMP generate pre-dendritic cells (pre-DC). Pre-DC are dendritic cell-restricted progenitors that routinely leave bone marrow for the lymphoid and non-lymphoid tissues to get differentiated into local dendritic cells. In the spleen, pre-DC further mature into CD8+ DC that possess a higher capacity of antigen cross-presentation to T cells. CLP give rise to transitional B cells (TB) that differentiate into marginal zone B cells (MZB) in the spleen. MZB provide the first line of defense against specific pathogens by mediating rapid antibody responses. CLP in the bone marrow migrate into the thymic cortex to generate early thymic progenitors (ETP), also known as double negative 1 (DN1) thymocytes. After the DN2–4 maturation stage, they subsequently become double positive (DP) thymocytes. DP thymocytes further mature into single positive (SP) T cells that migrate into the thymic medulla and egress to the periphery. MST1/2 are shown to mediate DC functions, B cell development, and the thymic egress of mature T cells via distinct mechanisms.

T lymphocytes originate in the thymus via three-steps, initiating as double-negative thymocytes (DN; CD4−CD8−), then changing to double-positive thymocytes (DP; CD4+CD8+), and finally, maturing into single-positive (SP; CD4+CD8− or CD4−CD8+) thymocytes. Single-positive mature thymocytes are then released from the thymus to the peripheral tissues to compose and start adequate adaptive immune responses [40,41]. MST1/2 are found to express at low levels in double-positive thymocytes, but their expression rises at the single-positive stage [30], suggesting the important functions of MST1/2 in later stage of T cell development.

Using MST1 single knockout mice (*Mst1*−/− mice), earlier studies have found the accumulation of *Mst1*−/− single-positive thymocytes in the perivascular space where T cells exit into the periphery, resulting in a decreased number of peripheral T cells [42,43]. In vitro lymphocyte adhesion cascade assays demonstrated that *Mst1*−/− T cells show reduced stopping times and disrupted integrin-mediated adhesion to endothelium, indicating that MST1 plays a major role in the efficient emigration of T cells from the thymus [42]. Mechanistically, the involvement of the alternative Hippo pathway has been proposed. MST1 in complex with RAP1 and RAPL (also known as RASSF5) phosphorylates NDR1 (nuclear Dbf2-related 1, also known as STK38) kinase, which recruits KINDLIN-3 (also known as FERMT3) to the immune synapse [44]. KINDLIN-3 was independently shown to mediate the high-affinity interaction between LFA-1 (also known as ITGB2; expressed on T cells) and ICAM1 (intercellular adhesion molecule 1; expressed on endothelial cells) to ensure efficient lymphocyte migration [45–47]. Another proposed mechanism suggests that MST1/2 directly phosphorylate and thus mediate MOB1A/B binding to DOCK8 (dedicator of cytokinesis 8) [30]. DOCK8 modulates the cytoskeletal remodeling as well as the cell migration in thymocytes by acting as a guanine nucleotide exchange factor (GEF) for RAC1 (Rac family small GTPase 1) [48]. It is to be noted that T cells from DOCK8-deficient mice shared some phenotypes of the MST1/2-deficient thymocytes, peculiarly a failure to polarize LFA-1 to the immune synapse [48], suggesting a key function of the Hippo pathway in small GTPase activation and cytoskeletal remodeling. Thus, collectively, MST1 functions in both cytoskeletal remodeling (through MOB1A/B phosphorylation) as well as cell adhesion (through NDR1 phosphorylation), ensuring the efficient emigration of T cells from the thymus.

2.3. Hippo Signaling in Effector T Cell Differentiation and Functions

Mature T lymphocytes that successfully pass the thymic selection incessantly migrate from the thymus towards the secondary peripheral lymphoid system to prepare for the antigen stimulation followed by the activation and differentiation phase (Figure 4). Naive T cells (Th0 cells) comprise CD4$^+$ and CD8$^+$ T cells that have not encountered its cognate antigen within the periphery and thus have not been differentiated. The expression levels of MST1 decrease as cell differentiation progresses from naive CD4$^+$ T cells to effector/memory T cells. In contrast, MST2 expression levels remain constant [29], indicative of a dominant role of MST1 in naive T cell functions. Although the precise mechanism remains to be elucidated, recent studies have suggested that MST1, via the alternative Hippo pathway, contributes to the maintenance of naive CD4$^+$ and CD8$^+$ T cells. MST1 promotes cell survival [34,49] and restricts antigen receptor-induced proliferation [29] of naive T cells in order to maintain their homeostatic state. In addition to its role in thymic egress, MST1 functions in naive T cell survival and maintenance, may also account for T cell lymphopenia observed in human patients with MST1 deficiency.

Upon stimulation by the T cell receptor signaling and specific cytokines in the surrounding environments, naive CD8$^+$ T cells become cytotoxic T lymphocytes (CTL), while naive CD4$^+$ T cells activate and differentiate into specialized subtypes, including type 1 helper (Th1) cells, Th2 cells, Th17 cells, or regulatory T cells (Treg) (Figure 4). We highlight the emerging roles of the Hippo pathway in cytotoxic T cells as well as Th17 and Treg cells below.

2.3.1. Hippo Pathway in Cytotoxic CD8$^+$ T Cell Differentiation and Functions

Although much is still unknown about the roles of the Hippo pathway in the regulation of cytotoxic T cell physiology, it is known that the T-cell-receptor and IL-2 (interleukin 2) signaling induce the expression and activation of the canonical Hippo pathway in CD8$^+$ T cells [50]. Activation of the T cell intrinsic Hippo signaling pathway suppresses YAP-mediated induction of the expression of transcriptional repressor BLIMP-1 (also known as PRDM1), thus resulting in terminal differentiation of the CD8$^+$ T cell [50]. However, another report has suggested that MST1 in CD8$^+$ T cells suppresses the cytotoxic function of CD8$^+$ T cells upon T cell receptor stimulation [51]. MST1 via FOXO suppresses the expression of the transcription factor T-bet (also known as TBX21). T-bet boosts cytotoxic T cell functions by inducing the effector cytokine IFNγ (interferon gamma) and the cytotoxic protease granzyme B expression [52]. Thus, MST1 deletion in CD8$^+$ T cells potentiates cytotoxic effector functions and prevents tumor growth in a syngeneic mouse model [51]. Together, these studies suggest seemingly paradoxical results that the Hippo pathway promotes terminal differentiation of CD8$^+$ T cell while suppressing their cytotoxic functions. Future studies validating the role of other Hippo pathway components in cytotoxic T cells will be required in order to clarify the complex regulatory mechanisms of cytotoxic T cell differentiation and functions by the Hippo pathway.

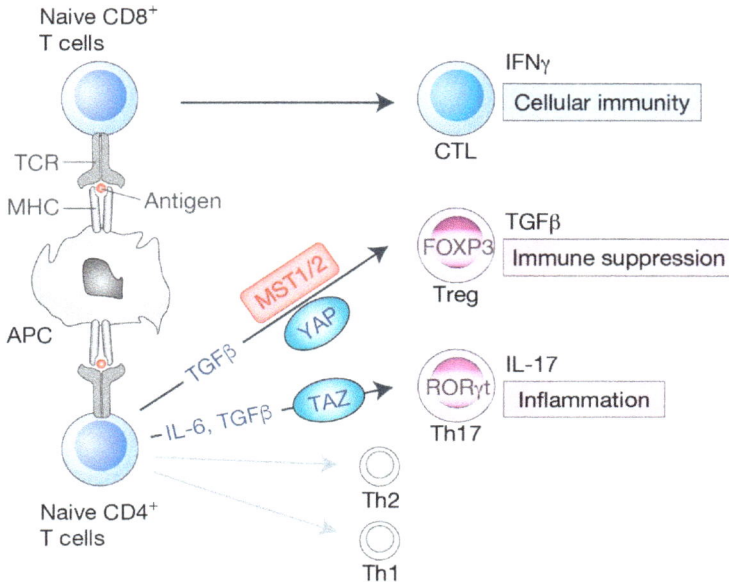

Figure 4. Roles of the Hippo pathway in effector T cell differentiation. The primary event of naive T cell activation is the interaction between T cell receptor (TCR) on T cells and an antigen loaded on antigen-presenting cells (APC). APC present antigenic peptides with major histocompatibility complex (MHC) class I (for CD8+ T cells) or class II (for CD4+ T cells) molecules. Upon activation, naive CD8+ T cells differentiate into cytotoxic T lymphocytes (CTL) that mediate cellular immunity and secrete series of cytokines such as interferon gamma (IFNγ). Naive CD4+ T cells differentiate into T helper (Th) cells, including effector T cells (Th1, Th2, Th17) and regulatory T cells (Treg). Transforming growth factor beta (TGFβ) signaling is required for Treg lineage, while interleukin 6 (IL-6) together with TGFβ cytokines are essential for Th17 differentiation. Key transcription factors have also been identified for each T cell lineage [forkhead box P3 (FOXP3) for Treg, RAR related orphan receptor gamma t (RORγt, also known as RORC) for Th17]. Treg cells produce TGFβ and suppress effector T cell activities, maintaining immune system homeostasis and self-tolerance. Th17 cells produce IL-17 cytokine and are involved in chronic and autoimmune inflammation. MST1/2 and YAP are shown to mediate Treg differentiation by diverse mechanisms. TAZ acts as a critical co-activator of RORγt for Th17 differentiation.

2.3.2. Hippo Pathway in Helper CD4+ Th17 and Treg Cells

Among CD4+ cell subsets, critical functions of the Hippo pathway in the regulation of Th17 and Treg cells have recently been revealed. IL-17-producing Th17 cells represent a pro-inflammatory subset, while FOXP3 (forkhead box P3)-expressing Treg cells are immunosuppressive [53,54]. The balance between Th17 and Treg cells is regulated by the inflammatory cytokines such as IL-6 and TGFβ (transforming growth factor beta), which has emerged as a striking factor in autoimmunity and cancer immune escape [53,54] (Figure 4).

Previous studies have demonstrated that MST1 promotes Treg differentiation and prevents autoimmunity and tissue damage [39,55]. MST1 appears to enhance the expression of FOXP3, the key transcriptional factor for Treg lineage, via the alternative Hippo pathway involving FOXO1/3 [39] or SIRT1 (sirtuin 1) [55]. In addition, another report suggested that MST1 also contributes to the contact-dependent suppression of Treg cells by mediating the immunological synapse formation between dendritic and Treg cells [56]. Collectively, these findings highlight the diverse functions of MST1 in Treg physiology.

A complementary recent study demonstrated that TAZ promotes Th17 lineage and inhibits Treg differentiation of naive CD4$^+$ T cells [57]. On one hand, TAZ binds to RORγt (RAR related orphan receptor gamma t) and enhances its transcriptional activity to promote Th17 differentiation. On the other hand, TAZ competes with FOXP3 for its binding to the histone acetyltransferase TIP60 (also known as KAT5), thus destabilizing FOXP3 and inhibiting Treg differentiation [57]. Interestingly, the expression of TAZ is increased during Th17 differentiation in vitro [57]. This is in contrast with YAP, which is not expressed in Th17 cells, but is highly expressed in Treg cells [58]. Another study found that YAP in Treg cells induces the genes involved in the TGFβ superfamily member activin pathway and thus reinforces TGFβ–SMAD signaling, upregulating FOXP3 expression and Treg functions. Genetic inhibition of YAP dramatically dampened the immune suppressive function of Treg cells, and therefore boosted the anti-cancer immune responses [58]. Thus, these studies suggest opposite roles of YAP and TAZ in Th17/Treg differentiation. It is clear that YAP and TAZ share similar molecular activities [59] and show functional redundancy during embryonic development [3] and regeneration [60] as the simultaneous deletion of YAP and TAZ generally results in a more severe phenotype than their single depletion. However, several lines of evidence in recent years indicate their distinct functions [61–63]. This divergence adds an extra level of complexity to the functions of the Hippo pathway in different cell-types in a context-dependent manner. The precise molecular mechanisms underlying these superficially contradictory regulations remain to be elucidated.

2.4. Critical Roles of MST1 in B cell Development and Functions

B cell progenitors arise from common lymphoid progenitors (CLP) in the bone marrow, and further develop into mature B cells in the secondary lymphoid tissue, such as the spleen [64] (Figure 3). MST1 is expressed in B-cell lineages and its deficiency causes B cell lymphopenia as well as autoantibody production in humans and in mouse models [29,33,36,39]. MST1-deficient mice show reduced CD19 expression, and disrupted B cell receptor clustering/downstream signaling, which results in a dramatic reduction of B cell viability and a developmental defect of marginal zone B cells (MZB) [65].

As described above, MST1 plays pivotal roles in both T cell and B cell development. Therefore, MST1 deficiency causes T and B cell lymphopenia in human patients, resulting in combined immunodeficiency. However, MST1-deficient patients also exhibit autoimmune-like symptoms, including autoantibody production. These symptoms are likely due to the impaired, MST1-lacking T cell functions, instead of the intrinsic signals in B cells. While T cell-specific MST1-deficient (*Lck*-Cre/*Mst1*$^{F/F}$) mice demonstrated autoantibody production, B cell-specific MST1 deficiency (*Mb1*-Cre/*Mst1*$^{F/F}$) did not result in autoantibody production for up to 18–20 months [66]. As MST1 is critical for the proper activities of FOXO, defective MST1–FOXO signaling impairs the differentiation and function of Treg cells, collapsing immune tolerance and thereby provoking autoimmunity [39,55]. Another study demonstrated that IL-4-rich environments created by MST1-deficient CD4$^+$ T cells also contribute towards the uncontrolled B cell responses [67]. Together, these results not only highlight the critical functions of MST1 in B cell development, but also demonstrate its role in maintaining immune tolerance to prevent B cell overactivation through the regulation of CD4$^+$ T cells. Despite the critical functions of MST1 in B cell regulation, other components of the Hippo pathway have been poorly explored in B cell lineage. It would be of interest to elucidate the involvement of other Hippo pathway components in B cell development and functions in future studies.

2.5. MST1/2 in Dendritic Cell Functions

Dendritic cells arise from bone marrow-resided pre-dendritic cells (pre-DC) that migrate to non-lymphoid (such as payer's patches and dermis) or lymphoid (mainly spleen) tissues where they further differentiate into lymphoid dendritic cells [68,69] (Figure 3). Differentiated mature dendritic cells in peripheral tissues capture exogenous antigens and migrate to the draining lymph node through CCR7 (C-C motif chemokine receptor 7)-dependent chemotaxis [70]. Among lymphoid dendritic cells, CD8$^+$ dendritic cell subset represents the population with a higher capacity of antigen cross-presentation to T cells [71].

An early study found that MST1-deficient mice show impaired trafficking of dendritic cells from the skin to the draining lymph node [42]. MST1-deficient dendritic cells demonstrated reduced attachments to the extracellular matrix in vitro [42]. Another study has suggested that MST1 mediates CCR7-dependent chemotaxis of human mature dendritic cells through the regulation of actin cytoskeleton [72]. Thus, MST1 regulates adhesion and cell motility, orchestrating the efficient migration of dendritic cells. MST1 also plays an important role in directing the T cell lineage by regulating the production of dendritic-cell-derived cytokines. MST1-deficient dendritic cells (*Cd11c*-Cre/*Mst1*$^{F/F}$) produce more IL-6 due to the activation of the p38 MAPK pathway, which in turn stimulates IL-6–STAT3 (signal transducer and activator of transcription 3) signaling in CD4$^+$ T cells to facilitate Th17 differentiation [73]. More recently, another study identified MST1/2 as crucial regulators of CD8$^+$ dendritic cells that reside in lymphoid tissue and elicit efficient antigen cross-presentation [74]. MST1/2 promote oxidative metabolism and contribute to the maintenance of bioenergetic activities in CD8$^+$ dendritic cells. Mechanistically, although not fully elucidated, MST1/2 appear to exert their functions via the alternative Hippo pathway as the dendritic cell-specific ablation of LATS1/2 or YAP/TAZ failed to phenocopy MST1/2 deletion. MST1/2 orchestrate selective expression of the T-cell-activating cytokine IL-12 via crosstalk with the non-canonical NF-κB (nuclear factor-kappa B) signaling pathway in CD8$^+$ dendritic cells [74]. Together, these results highlight the important roles of MST1/2 in dendritic cellular functions. Further in vivo analyses involving other Hippo pathway components in each dendritic cell subset will bring advances in our understanding of lineage commitment, differentiation, and the function of dendritic cells.

2.6. Hippo Pathway in Macrophages

Macrophages are one of the fastest immune cells that are capable of interacting with antigen at sites of infection or tumor initiation. Macrophage provides both MHC (major histocompatibility complex) class II and peptide complex, together with co-stimulatory signals, to T cells. The production of inflammatory cytokines in response to antigenic stimulation to initiate and maintain the inflammatory response is also a key function of the macrophages [75].

Critical roles of the Hippo pathway in the biological functions of macrophages have been highlighted in recent studies. MST1/2 boost the phagocytic induction of reactive oxygen species and the anti-bacteria response in macrophages. MST1/2 promote the recruitment of mitochondria to phagosome by phosphorylating PKCα (protein kinase C alpha), and thus regulating the mitochondrial trafficking and mitochondrion-phagosome juxtaposition, which is required for effective reactive oxygen species generation to kill bacteria [21]. Intriguingly, several recent studies have revealed the reciprocal interaction between the Hippo pathway and endocytic trafficking [76–80]. It is therefore possible that the Hippo pathway may modulate intracellular trafficking and, in turn, regulate the functions and integrity of organelles in macrophages as well as other immune cell-types.

YAP is also implicated in the regulation of antiviral responses in macrophages. In antiviral immune responses, viral DNA or RNA in cytosol activates TBK1 that phosphorylates and activates the key antiviral transcription factor IRF3 to induce the type I interferon response [81]. YAP binds to IRF3 and thus blocks its dimerization and translocation to the nucleus after viral infection [23]. However, since YAP/TAZ are rarely expressed in a variety of immune cells [24], their regulation in macrophages requires further validation. Another study consistently demonstrated that YAP/TAZ, through their direct binding, impair ubiquitylation-dependent TBK1 activation and antiviral responses in YAP/TAZ-abundant adherent cells, including HEK293, mouse embryonic fibroblasts (MEFs), and NMuMG epithelial cells [24]. In contrast, however, another study suggested that MST1 directly phosphorylates IRF3 to inhibit its dimerization and activation, impeding cytosolic antiviral defense in the aforementioned cell lines [82]. The precise role of the Hippo pathway in regulating antiviral immune response is thus not fully elucidated, but the interplay between canonical (through YAP/TAZ) and the alternative (through IRF3) Hippo pathway downstream of MST1/2 may contribute to these counterintuitive results. Future studies clarifying the role of other Hippo pathway components in antibacterial and antiviral responses will provide a clear mechanistic insight into the functions of the Hippo pathway in innate immune responses.

3. Dual Functions of the Intracellular Hippo Signaling in Regulating Immune Responses

As we have highlighted above, the Hippo pathway plays crucial roles in the adaptive immune cell development and functions. Recent studies have revealed that the Hippo pathway in non-immune cells also contributes to the induction and direction of the immune responses in both immunostimulatory and immunosuppressive ways. These involve cytokine production, immune checkpoint molecule expression, and extracellular vesicle release from the non-immune cells, allowing for intercellular communication and orchestrating the immune responses.

3.1. Immunostimulatory Role of the Hippo Pathway in Non-Immune Cells

Apart from its physiological roles, recent studies have also suggested the involvement of the Hippo pathway in several aspects of the tumor cell-intrinsic mechanisms of immune suppression in tumor microenvironments (Figure 5). YAP-mediated transcription is shown to promote cytokine/chemokine production in many types of tumors. For example, active YAP in murine prostate adenocarcinoma promotes CXCL5 (C-X-C motif chemokine ligand 5) secretion, which attracts CXCR2 (C-X-C motif chemokine receptor 2)-expressing myeloid derived suppressor cells (MDSC) to suppress the immune responses [83]. Pharmacological inhibition of the CXCL5–CXCR2 axis as well as MDSC depletion by neutralizing antibodies therefore impeded tumor progression [83]. Another study using the KRAS/p53 mutant pancreatic ductal adenocarcinoma model found that active YAP contributes to the differentiation and accumulation of MDSC in tumor microenvironments by promoting the expression and secretion of multiple cytokines and chemokines, including CXCL1/2 and C-C motif chemokine 2 (CCL2) [84]. YAP has further been found to function downstream of the ovarian cancer-specific oncogene PRKCI (protein kinase C iota) to up-regulate TNF (tumor necrosis factor) expression, recruiting MDSC to inhibit cytotoxic T cell functions. [85]. An additional mechanism for YAP-mediated immune-suppression is suggested to be through the recruitment of type II (M2) macrophages that suppress the immune clearance of cancer cells [86]. Tumor-initiating cells with YAP hyperactivation recruit M2 macrophages at the early phases of cancer development, mainly through direct YAP–TEAD targeting of CCL2 and CSF1 (colony stimulating factor 1) [87]. Interestingly, a knockdown of CCL2/CSF1 in tumor-initiating cells blocked M2 macrophage recruitment and abolished tumorigenesis in an immune system-dependent manner, suggesting that YAP-activated tumor-initiating cells are eliminated without the protection of M2 macrophages. Similarly, another in vitro study suggested that YAP in DLD-1 colon cancer cells contributes to the M2 macrophage differentiation of co-cultured THP-1 monocytes [88]. Collectively, these studies demonstrate that the activation of YAP in cancer cells promotes cytokine/chemokine production and thus contributes to the recruitment of immunosuppressive cells in tumor microenvironments.

In addition to the alteration of the cellular composition in tumor microenvironments, a context and cell-type dependent involvement of an immune checkpoint molecule has been implicated in YAP-mediated immune-suppression. The immune checkpoint receptor, programmed cell death 1 (PD-1, also known as PDCD1), and its ligand PD-L1 (programmed death-ligand-1, also known as CD274) provide a negative regulatory pathway that prevents self-antigen recognition by T cells [89]. Cancer cells hijack this built-in regulatory pathway to evade the host immunity by upregulating PD-L1, which results in the apoptosis or anergy of T cells by stimulating suppressive PD-1 signaling in T cells [89]. Several studies suggest that YAP/TAZ suppress T-cell-mediated killing of cancer cells by directly transcribing PD-L1 in human melanoma, lung cancer, and breast cancer cells [90–93]. YAP–TEAD transcription complex binds to the PD-L1 promoter or enhancer, directly inducing PD-L1 expression. In contrast, another study demonstrated that YAP inhibits IFNγ-inducible PD-L1 expression in murine syngeneic cancer models [94]. YAP–TEAD transcriptional complex induces the expression of its target gene miR-130a, which in turn suppresses the expression of IRF1, the major transcriptional factor for PD-L1 expression, and thus inhibits tumor growth in mice [94]. Immunosuppressive functions of the Hippo pathway will be discussed in the following section.

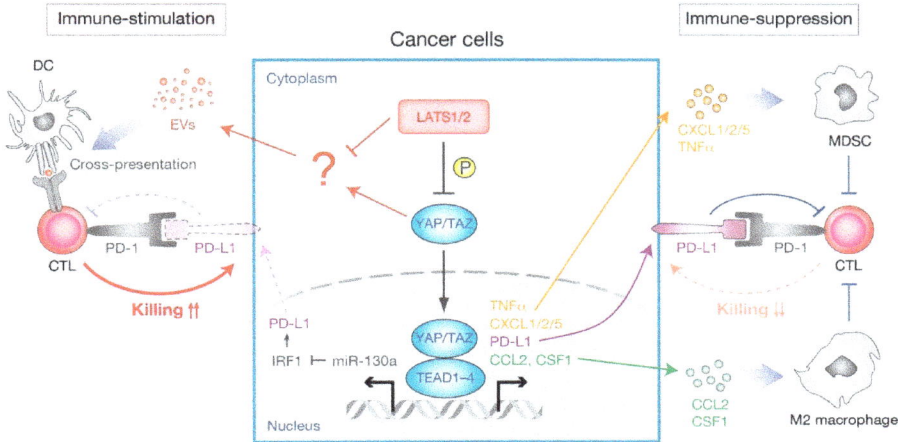

Figure 5. Dual functions of the cancer cell-intrinsic Hippo pathway in immune modulation. (Left) Immunnostimulatory effects mediated by YAP activation or LATS1/2 inactivation. In cancer cells, YAP inhibits IFNγ-inducible PD-L1 (programmed death-ligand-1, also known as CD274) expression partially through miR-130a-mediated suppression of IRF1. IRF1 up-regulates the expression of PD-L1 that binds to PD-1 (programmed cell death 1, also known as PDCD1) on T cells and provides inhibitory signals to cytotoxic T cells (CTL). Upon deletion of LATS1/2, extracellular vesicles (EVs) secreted from cancer cells trigger an anti-cancer immune response by stimulating the host nucleic-acid-sensing pathway and enhancing antigen cross-presentation. (Right) Immunosuppressive effects mediated by YAP activation. Active YAP induces the expression of PD-L1, as well as cytokines [TNFα (tumor necrosis factor alpha), CSF1 (colony stimulating factor 1)] and chemokine ligands [CXCL1 (C-X-C motif chemokine ligand 1), CXCL2, CXCL5, and CCL2 (C-C motif chemokine 2)]. Those cytokines or chemokines recruit immunosuppressive cells, such as myeloid derived suppressor cells (MDSC) and M2-type macrophages, to inhibit CTL functions.

3.2. Immunosuppressive Role of the Hippo Pathway in Non-Immune Cells

It is largely accepted that the Hippo pathway acts as a tumor suppressor in adherent cells that inhibits cell proliferation and survival, preventing tumorigenesis [11–14]. Inactivation of the Hippo pathway drives a cell-intrinsic program to promote cell proliferation and survival, allowing cellular expansion and migration. Hyperactivation of YAP/TAZ is wide-spread in human neoplasia [8], and numerous reports indicate gene amplification and epigenetic modulation of the YAP/TAZ loci in cancer [12], implying that YAP/TAZ-mediated transcription drives the development and sustainability of human cancer. However, this assumption is in contrast with the fact that germline or somatic mutations in key components of the canonical Hippo pathway are relatively rare in human cancers [11–14]. So far mutations in NF2 or LATS2 loci have been highlighted in specific types of human cancers, such as mesothelioma, schwannoma, and meningioma. However, another conundrum is that these mutations only occur in specific cancer histotypes and are not broadly distributed [13]. Therefore, it is not surprising that the Hippo pathway has divergent functions in cancer growth. For instance, YAP has been shown to have a tumor suppressive function in colorectal cancer [95], multiple myeloma [96], breast cancer [97,98], and lung squamous cell carcinoma [99]. Moreover, the cell type-dependent functions of LATS1/2 in promoting or suppressing cancer cell growth became apparent [100]. These studies suggest that the precise role of the Hippo pathway in human cancer is context dependent, and that the Hippo pathway has dual functions in both cancer progression and suppression.

Another, perhaps more feasible, possibility for the low mutation rate of the Hippo pathway components in human neoplasia is that this pathway may possess built-in feedback mechanisms that prevent the overgrowth of undesirable cells in the organism. These include cell-intrinsic feedback mechanisms that prevent the Hippo pathway dysregulation [101–103], the selective elimination of YAP-activated cells from the neighboring cells [104], and immunomodulation by the Hippo pathway (Figure 5) [94,105]. As mentioned above, YAP inhibits IFNγ-inducible PD-L1 expression and thus inhibits tumor growth in syngeneic mouse models [94]. Another study revealed that LATS1/2-deficient cancer cells induce a type I interferon response via the host TLR signaling, enhancing cross-presentation of tumor antigens to boost the anti-cancer immune response [105]. While LATS1/2 deficiency showed a significant increase in anchorage-independent cancer cell growth in vitro, their growth in vivo in immune-competent mice is severely compromised due to the induction of strong anti-cancer immune responses. Though the precise mechanism remains to be elucidated, it is likely that nucleic-acid-rich extracellular vesicles (EVs), released from LATS1/2-deficient cancer cells, stimulate the host nucleic-acid-sensing TLR signaling to induce a type I interferon response [105]. Therefore, enhanced immunogenicity unmasked by the LATS1/2 deletion in cancer cells induces strong immune responses and overwhelms any growth advantage that might be gained due to LATS1/2 deletion, leading to a strong inhibition of tumor growth in the immune-competent host. Thus, collectively, immunosuppressive functions of the Hippo pathway provide a built-in homeostatic control mechanism that prevents tissue overgrowth and tumorigenesis. Interestingly, while genetic ablation of MST1/2 in liver [106–108] and intestine [109] promotes tumorigenesis, deletion of LATS1/2 in liver [110,111] or kidney [112] does not result in the development of cancer. These observations imply that the molecular functions of MST1/2 and LATS1/2 are not necessarily the same, and canonical and alternative Hippo pathways in non-immune cells may exert different functions on immunomodulation in a context-dependent manner. This complexity adds to the already divergent functions of the Hippo pathway in the regulation of the mammalian immune system.

4. Conclusions

In recent years, much attention has been drawn toward the elucidation of the diverse roles of Hippo signaling in adaptive immunity. Especially, MST1/2 have been revealed to be broadly involved in maintaining proper adaptive immune responses by regulating cell survival, differentiation, migration, and function in diverse immune cell types. It is interesting to note that other key Hippo pathway components, such as LATS1/2 and YAP/TAZ, seem to be less important and only partially participate in these processes, as is evident by their negative expression or unaffected phosphorylation status upon immune cell activation [29,57]. Moreover, studies utilizing conditional deletion of YAP/TAZ or LATS1/2 in mice revealed that these genes are dispensable for some immune cell functions [28,74]. Although it requires further validation, previous studies thus imply that the alternative Hippo signal transduction may play key roles along with MST1/2 in regulating immune cell functions in several contexts. This includes the characterization of NDR1/2 in T cell biological functions [44,113]. NDR1/2 (also known as STK38/STK38L) belong to the NDR/LATS subfamily of AGC serine/threonine kinases. NDR1/2 and LATS1/2 share many overlapping molecular functions and regulatory mechanisms, such as phosphorylation by MST1/2 kinases, binding to MOB1A/B, and kinase activity towards YAP (reviewed in [114]). In addition, unbiased proteomic studies of the Hippo pathway interactome have consistently placed NDR1/2 in the Hippo pathway network [76,77,79]. Therefore, in future studies, it is worth validating whether NDR1/2 act as the main mediators of the alternative Hippo pathway in vivo and investigating how these diverse regulations of the Hippo pathway result in different biological outputs downstream of NDR1/2- or LATS1/2-dependent signaling branches in the adaptive immune system.

Accumulating evidence, using semisynthetic substrates and model systems, suggests that the Hippo pathway involves important mechano-regulated factors that integrate physical cues to gene expression and cellular responses. Mechanical signals such as cell shape, extracellular matrix stiffness, and shear flow, modulate activities of RHO GTPases, which in turn lead to actin cytoskeleton remodeling (reviewed in [115]). The actin cytoskeleton network can control nuclear–cytoplasmic shuttling and the transcriptional activities of YAP/TAZ [116–118]. In particular, focal adhesion regulates the Hippo signaling through RAP2 GTPases [119]. The integrin signaling also transduces mechano-signals to the Hippo pathway [120–124]. Given that the immune cells are continuously exposed to stresses that came from extracellular matrix and liquid shear-flow, it is of particular interest to investigate the involvement of the Hippo pathway in sensing mechanical forces and processing physical cues in lymphocytes. While we have made substantial advances in our understanding of the Hippo signaling in the adaptive immune system, the research must continue in order to broaden our knowledge on the interplay between the Hippo signaling pathway and the immune system. Delineating these interactions will have important clinical implications in autoimmune diseases and cancer.

Author Contributions: T.Y. and T.M. wrote and edited the manuscript.

Funding: Work in the Moroishi laboratory is supported by JST PRESTO [JPMJPR17HA], JSPS KAKENHI [JP18K19433 and JP18H02438], Mitsubishi Foundation, Naito Foundation, Nakajima Foundation, Daiichi Sankyo Foundation of Life Science, Mochida Memorial Foundation for Medical and Pharmaceutical Research, MSD Life Science Foundation Public Interest Incorporated Foundation, Astellas Foundation for Research on Metabolic Disorders, Bristol-Myers Squibb Foundation, Yasuda Medical Foundation, Shinnihon Foundation of Advanced Medical Treatment Research, Uehara Memorial Foundation, Kao Foundation for Arts and Sciences, Taiju Life Social Welfare Foundation, and Japan Foundation for Applied Enzymology (all to T.M.). T.Y. is supported by JSPS KAKENHI [JP19K15810], Astellas Foundation for Research on Metabolic Disorders, and The Nakatomi Foundation.

Acknowledgments: The authors sincerely apologize to colleagues whose work could not be cited owing to space limitations. The authors thank all the members of the Moroishi laboratory for insightful discussions and critical comments pertaining to this review article.

Conflicts of Interest: The authors declare no conflict of interest.

References

1. Irvine, K.D.; Harvey, K.F. Control of organ growth by patterning and hippo signaling in Drosophila. *Cold Spring Harb. Perspect. Biol.* **2015**, *7*. [CrossRef]
2. Pan, D. The hippo signaling pathway in development and cancer. *Dev. Cell* **2010**, *19*, 491–505. [CrossRef]
3. Varelas, X. The Hippo pathway effectors TAZ and YAP in development, homeostasis and disease. *Development* **2014**, *141*, 1614–1626. [CrossRef] [PubMed]
4. Sasaki, H. Roles and regulations of Hippo signaling during preimplantation mouse development. *Dev. Growth Differ.* **2017**, *59*, 12–20. [CrossRef] [PubMed]
5. Barry, E.R.; Camargo, F.D. The Hippo superhighway: Signaling crossroads converging on the Hippo/Yap pathway in stem cells and development. *Curr. Opin. Cell Biol.* **2013**, *25*, 247–253. [CrossRef]
6. Mo, J.S.; Park, H.W.; Guan, K.L. The Hippo signaling pathway in stem cell biology and cancer. *EMBO Rep.* **2014**, *15*, 642–656. [CrossRef]
7. Fu, V.; Plouffe, S.W.; Guan, K.L. The Hippo pathway in organ development, homeostasis, and regeneration. *Curr. Opin. Cell Biol.* **2017**, *49*, 99–107. [CrossRef] [PubMed]
8. Johnson, R.; Halder, G. The two faces of Hippo: Targeting the Hippo pathway for regenerative medicine and cancer treatment. *Nat. Rev. Drug Discov.* **2014**, *13*, 63–79. [CrossRef]
9. Wang, Y.; Yu, A.; Yu, F.X. The Hippo pathway in tissue homeostasis and regeneration. *Protein Cell* **2017**, *8*, 349–359. [CrossRef]
10. Moya, I.M.; Halder, G. Hippo-YAP/TAZ signalling in organ regeneration and regenerative medicine. *Nat. Rev. Mol. Cell Biol.* **2019**, *20*, 211–226. [CrossRef] [PubMed]
11. Moroishi, T.; Hansen, C.G.; Guan, K.L. The emerging roles of YAP and TAZ in cancer. *Nat. Rev. Cancer* **2015**, *15*, 73–79. [CrossRef] [PubMed]
12. Harvey, K.F.; Zhang, X.; Thomas, D.M. The Hippo pathway and human cancer. *Nat. Rev. Cancer* **2013**, *13*, 246–257. [CrossRef] [PubMed]

13. Zanconato, F.; Cordenonsi, M.; Piccolo, S. YAP/TAZ at the Roots of Cancer. *Cancer Cell* **2016**, *29*, 783–803. [CrossRef]

14. Moon, S.; Yeon Park, S.; Woo Park, H. Regulation of the Hippo pathway in cancer biology. *Cell Mol. Life Sci.* **2018**, *75*, 2303–2319. [CrossRef] [PubMed]

15. Yu, F.X.; Zhao, B.; Guan, K.L. Hippo Pathway in Organ Size Control, Tissue Homeostasis, and Cancer. *Cell* **2015**, *163*, 811–828. [CrossRef]

16. Mo, J.S. The role of extracellular biophysical cues in modulating the Hippo-YAP pathway. *BMB Rep.* **2017**, *50*, 71–78. [CrossRef]

17. Hansen, C.G.; Moroishi, T.; Guan, K.L. YAP and TAZ: A nexus for Hippo signaling and beyond. *Trends Cell Biol.* **2015**, *25*, 499–513. [CrossRef]

18. Meng, Z.; Moroishi, T.; Guan, K.L. Mechanisms of Hippo pathway regulation. *Genes Dev.* **2016**, *30*, 1–17. [CrossRef]

19. Totaro, A.; Panciera, T.; Piccolo, S. YAP/TAZ upstream signals and downstream responses. *Nat. Cell Biol.* **2018**, *20*, 888–899. [CrossRef]

20. Liu, B.; Zheng, Y.; Yin, F.; Yu, J.; Silverman, N.; Pan, D. Toll Receptor-Mediated Hippo Signaling Controls Innate Immunity in Drosophila. *Cell* **2016**, *164*, 406–419. [CrossRef]

21. Geng, J.; Sun, X.; Wang, P.; Zhang, S.; Wang, X.; Wu, H.; Hong, L.; Xie, C.; Li, X.; Zhao, H.; et al. Kinases Mst1 and Mst2 positively regulate phagocytic induction of reactive oxygen species and bactericidal activity. *Nat. Immunol.* **2015**, *16*, 1142–1152. [CrossRef]

22. Boro, M.; Singh, V.; Balaji, K.N. Mycobacterium tuberculosis-triggered Hippo pathway orchestrates CXCL1/2 expression to modulate host immune responses. *Sci. Rep.* **2016**, *6*, 37695. [CrossRef] [PubMed]

23. Wang, S.; Xie, F.; Chu, F.; Zhang, Z.; Yang, B.; Dai, T.; Gao, L.; Wang, L.; Ling, L.; Jia, J.; et al. YAP antagonizes innate antiviral immunity and is targeted for lysosomal degradation through IKKε-mediated phosphorylation. *Nat. Immunol.* **2017**, *18*, 733–743. [CrossRef] [PubMed]

24. Zhang, Q.; Meng, F.; Chen, S.; Plouffe, S.W.; Wu, S.; Liu, S.; Li, X.; Zhou, R.; Wang, J.; Zhao, B.; et al. Hippo signalling governs cytosolic nucleic acid sensing through YAP/TAZ-mediated TBK1 blockade. *Nat. Cell Biol.* **2017**, *19*, 362–374. [CrossRef] [PubMed]

25. Taha, Z.; Janse van Rensburg, H.J.; Yang, X. The Hippo Pathway: Immunity and Cancer. *Cancers (Basel)* **2018**, *10*, 94. [CrossRef]

26. Zhang, Y.; Zhang, H.; Zhao, B. Hippo Signaling in the Immune System. *Trends Biochem. Sci.* **2018**, *43*, 77–80. [CrossRef]

27. Hong, L.; Li, X.; Zhou, D.; Geng, J.; Chen, L. Role of Hippo signaling in regulating immunity. *Cell. Mol. Immunol.* **2018**, *15*, 1003–1009. [CrossRef]

28. Donato, E.; Biagioni, F.; Bisso, A.; Caganova, M.; Amati, B.; Campaner, S. YAP and TAZ are dispensable for physiological and malignant haematopoiesis. *Leukemia* **2018**, *32*, 2037–2040. [CrossRef]

29. Zhou, D.; Medoff, B.D.; Chen, L.; Li, L.; Zhang, X.F.; Praskova, M.; Liu, M.; Landry, A.; Blumberg, R.S.; Boussiotis, V.A.; et al. The Nore1B/Mst1 complex restrains antigen receptor-induced proliferation of naive T cells. *Proc. Natl. Acad. Sci. USA* **2008**, *105*, 20321–20326. [CrossRef]

30. Mou, F.; Praskova, M.; Xia, F.; Van Buren, D.; Hock, H.; Avruch, J.; Zhou, D. The Mst1 and Mst2 kinases control activation of rho family GTPases and thymic egress of mature thymocytes. *J. Exp. Med.* **2012**, *209*, 741–759. [CrossRef]

31. Fallahi, E.; O'Driscoll, N.A.; Matallanas, D. The MST/Hippo Pathway and Cell Death: A Non-Canonical Affair. *Genes (Basel)* **2016**, *7*, 28. [CrossRef]

32. Furth, N.; Aylon, Y.; Oren, M. p53 shades of Hippo. *Cell Death Differ.* **2018**, *25*, 81–92. [CrossRef] [PubMed]

33. Abdollahpour, H.; Appaswamy, G.; Kotlarz, D.; Diestelhorst, J.; Beier, R.; Schaffer, A.A.; Gertz, E.M.; Schambach, A.; Kreipe, H.H.; Pfeifer, D.; et al. The phenotype of human STK4 deficiency. *Blood* **2012**, *119*, 3450–3457. [CrossRef] [PubMed]

34. Nehme, N.T.; Schmid, J.P.; Debeurme, F.; Andre-Schmutz, I.; Lim, A.; Nitschke, P.; Rieux-Laucat, F.; Lutz, P.; Picard, C.; Mahlaoui, N.; et al. MST1 mutations in autosomal recessive primary immunodeficiency characterized by defective naive T-cell survival. *Blood* **2012**, *119*, 3458–3468. [CrossRef] [PubMed]

35. Dang, T.S.; Willet, J.D.; Griffin, H.R.; Morgan, N.V.; O'Boyle, G.; Arkwright, P.D.; Hughes, S.M.; Abinun, M.; Tee, L.J.; Barge, D.; et al. Defective Leukocyte Adhesion and Chemotaxis Contributes to Combined Immunodeficiency in Humans with Autosomal Recessive MST1 Deficiency. *J. Clin. Immunol.* **2016**, *36*, 117–122. [CrossRef]

36. Crequer, A.; Picard, C.; Patin, E.; D'Amico, A.; Abhyankar, A.; Munzer, M.; Debre, M.; Zhang, S.Y.; de Saint-Basile, G.; Fischer, A.; et al. Inherited MST1 deficiency underlies susceptibility to EV-HPV infections. *PLoS ONE* **2012**, *7*, e44010. [CrossRef] [PubMed]

37. Fukuhara, T.; Tomiyama, T.; Yasuda, K.; Ueda, Y.; Ozaki, Y.; Son, Y.; Nomura, S.; Uchida, K.; Okazaki, K.; Kinashi, T. Hypermethylation of MST1 in IgG4-related autoimmune pancreatitis and rheumatoid arthritis. *Biochem. Biophys. Res. Commun.* **2015**, *463*, 968–974. [CrossRef]

38. Halacli, S.O.; Ayvaz, D.C.; Sun-Tan, C.; Erman, B.; Uz, E.; Yilmaz, D.Y.; Ozgul, K.; Tezcan, I.; Sanal, O. STK4 (MST1) deficiency in two siblings with autoimmune cytopenias: A novel mutation. *Clin. Immunol.* **2015**, *161*, 316–323. [CrossRef]

39. Du, X.; Shi, H.; Li, J.; Dong, Y.; Liang, J.; Ye, J.; Kong, S.; Zhang, S.; Zhong, T.; Yuan, Z.; et al. Mst1/Mst2 regulate development and function of regulatory T cells through modulation of Foxo1/Foxo3 stability in autoimmune disease. *J. Immunol.* **2014**, *192*, 1525–1535. [CrossRef]

40. Takahama, Y. Journey through the thymus: Stromal guides for T-cell development and selection. *Nat. Rev. Immunol.* **2006**, *6*, 127–135. [CrossRef]

41. Shah, D.K.; Zuniga-Pflucker, J.C. An overview of the intrathymic intricacies of T cell development. *J. Immunol.* **2014**, *192*, 4017–4023. [CrossRef]

42. Katagiri, K.; Katakai, T.; Ebisuno, Y.; Ueda, Y.; Okada, T.; Kinashi, T. Mst1 controls lymphocyte trafficking and interstitial motility within lymph nodes. *EMBO J.* **2009**, *28*, 1319–1331. [CrossRef] [PubMed]

43. Dong, Y.; Du, X.; Ye, J.; Han, M.; Xu, T.; Zhuang, Y.; Tao, W. A cell-intrinsic role for Mst1 in regulating thymocyte egress. *J. Immunol.* **2009**, *183*, 3865–3872. [CrossRef] [PubMed]

44. Kondo, N.; Ueda, Y.; Kita, T.; Ozawa, M.; Tomiyama, T.; Yasuda, K.; Lim, D.S.; Kinashi, T. NDR1-Dependent Regulation of Kindlin-3 Controls High-Affinity LFA-1 Binding and Immune Synapse Organization. *Mol. Cell. Biol.* **2017**, *37*, 8. [CrossRef]

45. Malinin, N.L.; Zhang, L.; Choi, J.; Ciocea, A.; Razorenova, O.; Ma, Y.Q.; Podrez, E.A.; Tosi, M.; Lennon, D.P.; Caplan, A.I.; et al. A point mutation in KINDLIN3 ablates activation of three integrin subfamilies in humans. *Nat. Med.* **2009**, *15*, 313–318. [CrossRef] [PubMed]

46. Svensson, L.; Howarth, K.; McDowall, A.; Patzak, I.; Evans, R.; Ussar, S.; Moser, M.; Metin, A.; Fried, M.; Tomlinson, I.; et al. Leukocyte adhesion deficiency-III is caused by mutations in KINDLIN3 affecting integrin activation. *Nat. Med.* **2009**, *15*, 306–312. [CrossRef]

47. Moser, M.; Bauer, M.; Schmid, S.; Ruppert, R.; Schmidt, S.; Sixt, M.; Wang, H.V.; Sperandio, M.; Fassler, R. Kindlin-3 is required for beta2 integrin-mediated leukocyte adhesion to endothelial cells. *Nat. Med.* **2009**, *15*, 300–305. [CrossRef]

48. Randall, K.L.; Chan, S.S.; Ma, C.S.; Fung, I.; Mei, Y.; Yabas, M.; Tan, A.; Arkwright, P.D.; Al Suwairi, W.; Lugo Reyes, S.O.; et al. DOCK8 deficiency impairs CD8 T cell survival and function in humans and mice. *J. Exp. Med.* **2011**, *208*, 2305–2320. [CrossRef]

49. Choi, J.; Oh, S.; Lee, D.; Oh, H.J.; Park, J.Y.; Lee, S.B.; Lim, D.S. Mst1-FoxO signaling protects Naive T lymphocytes from cellular oxidative stress in mice. *PLoS ONE* **2009**, *4*, e8011. [CrossRef]

50. Thaventhiran, J.E.; Hoffmann, A.; Magiera, L.; de la Roche, M.; Lingel, H.; Brunner-Weinzierl, M.; Fearon, D.T. Activation of the Hippo pathway by CTLA-4 regulates the expression of Blimp-1 in the CD8+ T cell. *Proc. Natl. Acad. Sci. USA* **2012**, *109*, E2223–E2229. [CrossRef]

51. Yasuda, K.; Ueda, Y.; Ozawa, M.; Matsuda, T.; Kinashi, T. Enhanced cytotoxic T-cell function and inhibition of tumor progression by Mst1 deficiency. *FEBS Lett.* **2016**, *590*, 68–75. [CrossRef]

52. Lazarevic, V.; Glimcher, L.H.; Lord, G.M. T-bet: A bridge between innate and adaptive immunity. *Nat. Rev. Immunol.* **2013**, *13*, 777–789. [CrossRef]

53. Knochelmann, H.M.; Dwyer, C.J.; Bailey, S.R.; Amaya, S.M.; Elston, D.M.; Mazza-McCrann, J.M.; Paulos, C.M. When worlds collide: Th17 and Treg cells in cancer and autoimmunity. *Cell Mol. Immunol.* **2018**, *15*, 458–469. [CrossRef]

54. Nishikawa, H.; Sakaguchi, S. Regulatory T cells in tumor immunity. *Int. J. Cancer* **2010**, *127*, 759–767. [CrossRef]

55. Li, J.; Du, X.; Shi, H.; Deng, K.; Chi, H.; Tao, W. Mammalian Sterile 20-like Kinase 1 (Mst1) Enhances the Stability of Forkhead Box P3 (Foxp3) and the Function of Regulatory T Cells by Modulating Foxp3 Acetylation. *J. Biol. Chem.* **2015**, *290*, 30762–30770. [CrossRef]

56. Tomiyama, T.; Ueda, Y.; Katakai, T.; Kondo, N.; Okazaki, K.; Kinashi, T. Antigen-specific suppression and immunological synapse formation by regulatory T cells require the Mst1 kinase. *PLoS ONE* **2013**, *8*, e73874. [CrossRef]

57. Geng, J.; Yu, S.; Zhao, H.; Sun, X.; Li, X.; Wang, P.; Xiong, X.; Hong, L.; Xie, C.; Gao, J.; et al. The transcriptional coactivator TAZ regulates reciprocal differentiation of TH17 cells and Treg cells. *Nat. Immunol.* **2017**, *18*, 800–812. [CrossRef]

58. Ni, X.; Tao, J.; Barbi, J.; Chen, Q.; Park, B.V.; Li, Z.; Zhang, N.; Lebid, A.; Ramaswamy, A.; Wei, P.; et al. YAP Is Essential for Treg-Mediated Suppression of Antitumor Immunity. *Cancer Discov.* **2018**, *8*, 1026–1043. [CrossRef]

59. Plouffe, S.W.; Lin, K.C.; Moore, J.L., 3rd; Tan, F.E.; Ma, S.; Ye, Z.; Qiu, Y.; Ren, B.; Guan, K.L. The Hippo pathway effector proteins YAP and TAZ have both distinct and overlapping functions in the cell. *J. Biol. Chem.* **2018**, *293*, 11230–11240. [CrossRef]

60. Xin, M.; Kim, Y.; Sutherland, L.B.; Murakami, M.; Qi, X.; McAnally, J.; Porrello, E.R.; Mahmoud, A.I.; Tan, W.; Shelton, J.M.; et al. Hippo pathway effector Yap promotes cardiac regeneration. *Proc. Natl. Acad. Sci. USA* **2013**, *110*, 13839–13844. [CrossRef]

61. Sun, C.; De Mello, V.; Mohamed, A.; Ortuste Quiroga, H.P.; Garcia-Munoz, A.; Al Bloshi, A.; Tremblay, A.M.; von Kriegsheim, A.; Collie-Duguid, E.; Vargesson, N.; et al. Common and Distinctive Functions of the Hippo Effectors Taz and Yap in Skeletal Muscle Stem Cell Function. *Stem Cells* **2017**, *35*, 1958–1972. [CrossRef]

62. Mohamed, A.; Sun, C.; De Mello, V.; Selfe, J.; Missiaglia, E.; Shipley, J.; Murray, G.I.; Zammit, P.S.; Wackerhage, H. The Hippo effector TAZ (WWTR1) transforms myoblasts and TAZ abundance is associated with reduced survival in embryonal rhabdomyosarcoma. *J. Pathol.* **2016**, *240*, 3–14. [CrossRef]

63. Reginensi, A.; Scott, R.P.; Gregorieff, A.; Bagherie-Lachidan, M.; Chung, C.; Lim, D.S.; Pawson, T.; Wrana, J.; McNeill, H. Yap- and Cdc42-dependent nephrogenesis and morphogenesis during mouse kidney development. *PLoS Genet.* **2013**, *9*, e1003380. [CrossRef]

64. Montecino-Rodriguez, E.; Dorshkind, K. B-1 B cell development in the fetus and adult. *Immunity* **2012**, *36*, 13–21. [CrossRef]

65. Bai, X.; Huang, L.; Niu, L.; Zhang, Y.; Wang, J.; Sun, X.; Jiang, H.; Zhang, Z.; Miller, H.; Tao, W.; et al. Mst1 positively regulates B-cell receptor signaling via CD19 transcriptional levels. *Blood Adv.* **2016**, *1*, 219–230. [CrossRef]

66. Ueda, Y.; Katagiri, K.; Tomiyama, T.; Yasuda, K.; Habiro, K.; Katakai, T.; Ikehara, S.; Matsumoto, M.; Kinashi, T. Mst1 regulates integrin-dependent thymocyte trafficking and antigen recognition in the thymus. *Nat. Commun.* **2012**, *3*, 1098. [CrossRef]

67. Park, E.; Kim, M.S.; Song, J.H.; Roh, K.H.; Lee, R.; Kim, T.S. MST1 deficiency promotes B cell responses by CD4(+) T cell-derived IL-4, resulting in hypergammaglobulinemia. *Biochem. Biophys. Res. Commun.* **2017**, *489*, 56–62. [CrossRef]

68. Murphy, T.L.; Grajales-Reyes, G.E.; Wu, X.; Tussiwand, R.; Briseno, C.G.; Iwata, A.; Kretzer, N.M.; Durai, V.; Murphy, K.M. Transcriptional Control of Dendritic Cell Development. *Annu. Rev. Immunol.* **2016**, *34*, 93–119. [CrossRef]

69. Merad, M.; Sathe, P.; Helft, J.; Miller, J.; Mortha, A. The dendritic cell lineage: Ontogeny and function of dendritic cells and their subsets in the steady state and the inflamed setting. *Annu. Rev. Immunol.* **2013**, *31*, 563–604. [CrossRef]

70. Ohl, L.; Mohaupt, M.; Czeloth, N.; Hintzen, G.; Kiafard, Z.; Zwirner, J.; Blankenstein, T.; Henning, G.; Forster, R. CCR7 governs skin dendritic cell migration under inflammatory and steady-state conditions. *Immunity* **2004**, *21*, 279–288. [CrossRef]

71. Alloatti, A.; Kotsias, F.; Magalhaes, J.G.; Amigorena, S. Dendritic cell maturation and cross-presentation: Timing matters! *Immunol. Rev.* **2016**, *272*, 97–108. [CrossRef]

72. Torres-Bacete, J.; Delgado-Martin, C.; Gomez-Moreira, C.; Simizu, S.; Rodriguez-Fernandez, J.L. The Mammalian Sterile 20-like 1 Kinase Controls Selective CCR7-Dependent Functions in Human Dendritic Cells. *J. Immunol.* **2015**, *195*, 973–981. [CrossRef]

73. Li, C.; Bi, Y.; Li, Y.; Yang, H.; Yu, Q.; Wang, J.; Wang, Y.; Su, H.; Jia, A.; Hu, Y.; et al. Dendritic cell MST1 inhibits Th17 differentiation. *Nat. Commun.* **2017**, *8*, 14275. [CrossRef] [PubMed]

74. Du, X.; Wen, J.; Wang, Y.; Karmaus, P.W.F.; Khatamian, A.; Tan, H.; Li, Y.; Guy, C.; Nguyen, T.M.; Dhungana, Y.; et al. Hippo/Mst signalling couples metabolic state and immune function of CD8alpha(+) dendritic cells. *Nature* **2018**, *558*, 141–145. [CrossRef]

75. Murray, P.J.; Wynn, T.A. Protective and pathogenic functions of macrophage subsets. *Nat. Rev. Immunol.* **2011**, *11*, 723–737. [CrossRef] [PubMed]

76. Kwon, Y.; Vinayagam, A.; Sun, X.; Dephoure, N.; Gygi, S.P.; Hong, P.; Perrimon, N. The Hippo signaling pathway interactome. *Science* **2013**, *342*, 737–740. [CrossRef] [PubMed]

77. Wang, W.; Li, X.; Huang, J.; Feng, L.; Dolinta, K.G.; Chen, J. Defining the protein-protein interaction network of the human hippo pathway. *Mol. Cell. Proteomics* **2014**, *13*, 119–131. [CrossRef]

78. Gilbert, M.M.; Tipping, M.; Veraksa, A.; Moberg, K.H. A screen for conditional growth suppressor genes identifies the Drosophila homolog of HD-PTP as a regulator of the oncoprotein Yorkie. *Dev. Cell* **2011**, *20*, 700–712. [CrossRef]

79. Couzens, A.L.; Knight, J.D.; Kean, M.J.; Teo, G.; Weiss, A.; Dunham, W.H.; Lin, Z.Y.; Bagshaw, R.D.; Sicheri, F.; Pawson, T.; et al. Protein interaction network of the mammalian Hippo pathway reveals mechanisms of kinase-phosphatase interactions. *Sci. Signal.* **2013**, *6*, rs15. [CrossRef]

80. Rausch, V.; Bostrom, J.R.; Park, J.; Bravo, I.R.; Feng, Y.; Hay, D.C.; Link, B.A.; Hansen, C.G. The Hippo Pathway Regulates Caveolae Expression and Mediates Flow Response via Caveolae. *Curr. Biol.* **2019**, *29*, 242–255 e246. [CrossRef]

81. McNab, F.; Mayer-Barber, K.; Sher, A.; Wack, A.; O'Garra, A. Type I interferons in infectious disease. *Nat. Rev. Immunol.* **2015**, *15*, 87–103. [CrossRef] [PubMed]

82. Meng, F.; Zhou, R.; Wu, S.; Zhang, Q.; Jin, Q.; Zhou, Y.; Plouffe, S.W.; Liu, S.; Song, H.; Xia, Z.; et al. Mst1 shuts off cytosolic antiviral defense through IRF3 phosphorylation. *Genes Dev.* **2016**, *30*, 1086–1100. [CrossRef]

83. Wang, G.; Lu, X.; Dey, P.; Deng, P.; Wu, C.C.; Jiang, S.; Fang, Z.; Zhao, K.; Konaparthi, R.; Hua, S.; et al. Targeting YAP-Dependent MDSC Infiltration Impairs Tumor Progression. *Cancer Discov.* **2016**, *6*, 80–95. [CrossRef]

84. Murakami, S.; Shahbazian, D.; Surana, R.; Zhang, W.; Chen, H.; Graham, G.T.; White, S.M.; Weiner, L.M.; Yi, C. Yes-associated protein mediates immune reprogramming in pancreatic ductal adenocarcinoma. *Oncogene* **2017**, *36*, 1232–1244. [CrossRef] [PubMed]

85. Sarkar, S.; Bristow, C.A.; Dey, P.; Rai, K.; Perets, R.; Ramirez-Cardenas, A.; Malasi, S.; Huang-Hobbs, E.; Haemmerle, M.; Wu, S.Y.; et al. PRKCI promotes immune suppression in ovarian cancer. *Genes Dev.* **2017**, *31*, 1109–1121. [CrossRef]

86. Noy, R.; Pollard, J.W. Tumor-associated macrophages: From mechanisms to therapy. *Immunity* **2014**, *41*, 49–61. [CrossRef]

87. Guo, X.; Zhao, Y.; Yan, H.; Yang, Y.; Shen, S.; Dai, X.; Ji, X.; Ji, F.; Gong, X.G.; Li, L.; et al. Single tumor-initiating cells evade immune clearance by recruiting type II macrophages. *Genes Dev.* **2017**, *31*, 247–259. [CrossRef]

88. Huang, Y.J.; Yang, C.K.; Wei, P.L.; Huynh, T.T.; Whang-Peng, J.; Meng, T.C.; Hsiao, M.; Tzeng, Y.M.; Wu, A.T.; Yen, Y. Ovatodiolide suppresses colon tumorigenesis and prevents polarization of M2 tumor-associated macrophages through YAP oncogenic pathways. *J. Hematol. Oncol.* **2017**, *10*, 60. [CrossRef] [PubMed]

89. Sharma, P.; Hu-Lieskovan, S.; Wargo, J.A.; Ribas, A. Primary, Adaptive, and Acquired Resistance to Cancer Immunotherapy. *Cell* **2017**, *168*, 707–723. [CrossRef]

90. Kim, M.H.; Kim, C.G.; Kim, S.K.; Shin, S.J.; Choe, E.A.; Park, S.H.; Shin, E.C.; Kim, J. YAP-Induced PD-L1 Expression Drives Immune Evasion in BRAFi-Resistant Melanoma. *Cancer Immunol. Res.* **2018**. [CrossRef]

91. Lee, B.S.; Park, D.I.; Lee, D.H.; Lee, J.E.; Yeo, M.K.; Park, Y.H.; Lim, D.S.; Choi, W.; Lee, D.H.; Yoo, G.; et al. Hippo effector YAP directly regulates the expression of PD-L1 transcripts in EGFR-TKI-resistant lung adenocarcinoma. *Biochem. Biophys. Res. Commun.* **2017**, *491*, 493–499. [CrossRef]

92. Janse van Rensburg, H.J.; Azad, T.; Ling, M.; Hao, Y.; Snetsinger, B.; Khanal, P.; Minassian, L.M.; Graham, C.H.; Rauh, M.J.; Yang, X. The Hippo Pathway Component TAZ Promotes Immune Evasion in Human Cancer through PD-L1. *Cancer Res.* **2018**, *78*, 1457–1470. [CrossRef]

93. Miao, J.; Hsu, P.C.; Yang, Y.L.; Xu, Z.; Dai, Y.; Wang, Y.; Chan, G.; Huang, Z.; Hu, B.; Li, H.; et al. YAP regulates PD-L1 expression in human NSCLC cells. *Oncotarget* **2017**, *8*, 114576–114587. [CrossRef]

94. Wu, A.; Wu, Q.; Deng, Y.; Liu, Y.; Lu, J.; Liu, L.; Li, X.; Liao, C.; Zhao, B.; Song, H. Loss of VGLL4 suppresses tumor PD-L1 expression and immune evasion. *EMBO J.* **2019**, *38*. [CrossRef]

95. Barry, E.R.; Morikawa, T.; Butler, B.L.; Shrestha, K.; de la Rosa, R.; Yan, K.S.; Fuchs, C.S.; Magness, S.T.; Smits, R.; Ogino, S.; et al. Restriction of intestinal stem cell expansion and the regenerative response by YAP. *Nature* **2013**, *493*, 106–110. [CrossRef]

96. Cottini, F.; Hideshima, T.; Xu, C.; Sattler, M.; Dori, M.; Agnelli, L.; ten Hacken, E.; Bertilaccio, M.T.; Antonini, E.; Neri, A.; et al. Rescue of Hippo coactivator YAP1 triggers DNA damage-induced apoptosis in hematological cancers. *Nat. Med.* **2014**, *20*, 599–606. [CrossRef]

97. Matallanas, D.; Romano, D.; Yee, K.; Meissl, K.; Kucerova, L.; Piazzolla, D.; Baccarini, M.; Vass, J.K.; Kolch, W.; O'Neill, E. RASSF1A elicits apoptosis through an MST2 pathway directing proapoptotic transcription by the p73 tumor suppressor protein. *Mol. Cell* **2007**, *27*, 962–975. [CrossRef]

98. Yuan, M.; Tomlinson, V.; Lara, R.; Holliday, D.; Chelala, C.; Harada, T.; Gangeswaran, R.; Manson-Bishop, C.; Smith, P.; Danovi, S.A.; et al. Yes-associated protein (YAP) functions as a tumor suppressor in breast. *Cell Death Differ.* **2008**, *15*, 1752–1759. [CrossRef]

99. Huang, H.; Zhang, W.; Pan, Y.; Gao, Y.; Deng, L.; Li, F.; Li, F.; Ma, X.; Hou, S.; Xu, J.; et al. YAP Suppresses Lung Squamous Cell Carcinoma Progression via Deregulation of the DNp63-GPX2 Axis and ROS Accumulation. *Cancer Res.* **2017**, *77*, 5769–5781. [CrossRef]

100. Pan, W.W.; Moroishi, T.; Koo, J.H.; Guan, K.L. Cell type-dependent function of LATS1/2 in cancer cell growth. *Oncogene* **2018**. [CrossRef]

101. Moroishi, T.; Park, H.W.; Qin, B.; Chen, Q.; Meng, Z.; Plouffe, S.W.; Taniguchi, K.; Yu, F.X.; Karin, M.; Pan, D.; et al. A YAP/TAZ-induced feedback mechanism regulates Hippo pathway homeostasis. *Genes Dev.* **2015**, *29*, 1271–1284. [CrossRef]

102. Chen, Q.; Zhang, N.; Xie, R.; Wang, W.; Cai, J.; Choi, K.S.; David, K.K.; Huang, B.; Yabuta, N.; Nojima, H.; et al. Homeostatic control of Hippo signaling activity revealed by an endogenous activating mutation in YAP. *Genes Dev.* **2015**, *29*, 1285–1297. [CrossRef]

103. Dai, X.; Liu, H.; Shen, S.; Guo, X.; Yan, H.; Ji, X.; Li, L.; Huang, J.; Feng, X.H.; Zhao, B. YAP activates the Hippo pathway in a negative feedback loop. *Cell Res.* **2015**, *25*, 1175–1178. [CrossRef]

104. Miyamura, N.; Hata, S.; Itoh, T.; Tanaka, M.; Nishio, M.; Itoh, M.; Ogawa, Y.; Terai, S.; Sakaida, I.; Suzuki, A.; et al. YAP determines the cell fate of injured mouse hepatocytes in vivo. *Nat. Commun.* **2017**, *8*, 16017. [CrossRef]

105. Moroishi, T.; Hayashi, T.; Pan, W.W.; Fujita, Y.; Holt, M.V.; Qin, J.; Carson, D.A.; Guan, K.L. The Hippo Pathway Kinases LATS1/2 Suppress Cancer Immunity. *Cell* **2016**, *167*, 1525–1539 e1517. [CrossRef]

106. Lu, L.; Li, Y.; Kim, S.M.; Bossuyt, W.; Liu, P.; Qiu, Q.; Wang, Y.; Halder, G.; Finegold, M.J.; Lee, J.S.; et al. Hippo signaling is a potent in vivo growth and tumor suppressor pathway in the mammalian liver. *Proc. Natl. Acad. Sci. USA* **2010**, *107*, 1437–1442. [CrossRef]

107. Zhou, D.; Conrad, C.; Xia, F.; Park, J.S.; Payer, B.; Yin, Y.; Lauwers, G.Y.; Thasler, W.; Lee, J.T.; Avruch, J.; et al. Mst1 and Mst2 maintain hepatocyte quiescence and suppress hepatocellular carcinoma development through inactivation of the Yap1 oncogene. *Cancer Cell* **2009**, *16*, 425–438. [CrossRef]

108. Song, H.; Mak, K.K.; Topol, L.; Yun, K.; Hu, J.; Garrett, L.; Chen, Y.; Park, O.; Chang, J.; Simpson, R.M.; et al. Mammalian Mst1 and Mst2 kinases play essential roles in organ size control and tumor suppression. *Proc. Natl. Acad. Sci. USA* **2010**, *107*, 1431–1436. [CrossRef]

109. Zhou, D.; Zhang, Y.; Wu, H.; Barry, E.; Yin, Y.; Lawrence, E.; Dawson, D.; Willis, J.E.; Markowitz, S.D.; Camargo, F.D.; et al. Mst1 and Mst2 protein kinases restrain intestinal stem cell proliferation and colonic tumorigenesis by inhibition of Yes-associated protein (Yap) overabundance. *Proc. Natl. Acad. Sci. USA* **2011**, *108*, E1312–1320. [CrossRef]

110. Lee, D.H.; Park, J.O.; Kim, T.S.; Kim, S.K.; Kim, T.H.; Kim, M.C.; Park, G.S.; Kim, J.H.; Kuninaka, S.; Olson, E.N.; et al. LATS-YAP/TAZ controls lineage specification by regulating TGFbeta signaling and Hnf4alpha expression during liver development. *Nat. Commun.* **2016**, *7*, 11961. [CrossRef] [PubMed]

111. Yi, J.; Lu, L.; Yanger, K.; Wang, W.; Sohn, B.H.; Stanger, B.Z.; Zhang, M.; Martin, J.F.; Ajani, J.A.; Chen, J.; et al. Large tumor suppressor homologs 1 and 2 regulate mouse liver progenitor cell proliferation and maturation through antagonism of the coactivators YAP and TAZ. *Hepatology* **2016**, *64*, 1757–1772. [CrossRef] [PubMed]

112. Reginensi, A.; Enderle, L.; Gregorieff, A.; Johnson, R.L.; Wrana, J.L.; McNeill, H. A critical role for NF2 and the Hippo pathway in branching morphogenesis. *Nat. Commun.* **2016**, *7*, 12309. [CrossRef]

113. Tang, F.; Gill, J.; Ficht, X.; Barthlott, T.; Cornils, H.; Schmitz-Rohmer, D.; Hynx, D.; Zhou, D.; Zhang, L.; Xue, G.; et al. The kinases NDR1/2 act downstream of the Hippo homolog MST1 to mediate both egress of thymocytes from the thymus and lymphocyte motility. *Sci. Signal.* **2015**, *8*, ra100. [CrossRef]

114. Hergovich, A. The Roles of NDR Protein Kinases in Hippo Signalling. *Genes (Basel)* **2016**, *7*, 21. [CrossRef] [PubMed]

115. Panciera, T.; Azzolin, L.; Cordenonsi, M.; Piccolo, S. Mechanobiology of YAP and TAZ in physiology and disease. *Nat. Rev. Mol. Cell Biol.* **2017**, *18*, 758–770. [CrossRef] [PubMed]

116. Dupont, S.; Morsut, L.; Aragona, M.; Enzo, E.; Giulitti, S.; Cordenonsi, M.; Zanconato, F.; Le Digabel, J.; Forcato, M.; Bicciato, S.; et al. Role of YAP/TAZ in mechanotransduction. *Nature* **2011**, *474*, 179–183. [CrossRef]

117. Yu, F.X.; Zhao, B.; Panupinthu, N.; Jewell, J.L.; Lian, I.; Wang, L.H.; Zhao, J.; Yuan, H.; Tumaneng, K.; Li, H.; et al. Regulation of the Hippo-YAP pathway by G-protein-coupled receptor signaling. *Cell* **2012**, *150*, 780–791. [CrossRef] [PubMed]

118. Aragona, M.; Panciera, T.; Manfrin, A.; Giulitti, S.; Michielin, F.; Elvassore, N.; Dupont, S.; Piccolo, S. A mechanical checkpoint controls multicellular growth through YAP/TAZ regulation by actin-processing factors. *Cell* **2013**, *154*, 1047–1059. [CrossRef]

119. Meng, Z.; Qiu, Y.; Lin, K.C.; Kumar, A.; Placone, J.K.; Fang, C.; Wang, K.C.; Lu, S.; Pan, M.; Hong, A.W.; et al. RAP2 mediates mechanoresponses of the Hippo pathway. *Nature* **2018**, *560*, 655–660. [CrossRef] [PubMed]

120. Tang, Y.; Rowe, R.G.; Botvinick, E.L.; Kurup, A.; Putnam, A.J.; Seiki, M.; Weaver, V.M.; Keller, E.T.; Goldstein, S.; Dai, J.; et al. MT1-MMP-dependent control of skeletal stem cell commitment via a beta1-integrin/YAP/TAZ signaling axis. *Dev. Cell* **2013**, *25*, 402–416. [CrossRef]

121. Serrano, I.; McDonald, P.C.; Lock, F.; Muller, W.J.; Dedhar, S. Inactivation of the Hippo tumour suppressor pathway by integrin-linked kinase. *Nat. Commun.* **2013**, *4*, 2976. [CrossRef] [PubMed]

122. Kim, N.G.; Gumbiner, B.M. Adhesion to fibronectin regulates Hippo signaling via the FAK-Src-PI3K pathway. *J. Cell Biol.* **2015**, *210*, 503–515. [CrossRef]

123. Wang, L.; Luo, J.Y.; Li, B.; Tian, X.Y.; Chen, L.J.; Huang, Y.; Liu, J.; Deng, D.; Lau, C.W.; Wan, S.; et al. Integrin-YAP/TAZ-JNK cascade mediates atheroprotective effect of unidirectional shear flow. *Nature* **2016**, *540*, 579–582. [CrossRef] [PubMed]

124. Hu, J.K.; Du, W.; Shelton, S.J.; Oldham, M.C.; DiPersio, C.M.; Klein, O.D. An FAK-YAP-mTOR Signaling Axis Regulates Stem Cell-Based Tissue Renewal in Mice. *Cell Stem Cell* **2017**, *21*, 91–106 e106. [CrossRef]

MDPI

St. Alban-Anlage 66

4052 Basel

Switzerland

Tel. +41 61 683 77 34

Fax +41 61 302 89 18

www.mdpi.com

Cells Editorial Office

E-mail: cells@mdpi.com

www.mdpi.com/journal/cells